The Scope and History of Commutative and Noncommutative Harmonic Analysis

Titles in This Series

The Scope and History of Commutative and Noncommutative Harmonic Analysis

George W. Mackey

HISTORY OF MATHEMATICS

Volume 5

AMERICAN MATHEMATICAL SOCIETY
LONDON MATHEMATICAL SOCIETY

2000 *Mathematics Subject Classification.* Primary 00B60; Secondary 22D30, 01–02, 11–02, 81Q99.

Library of Congress Cataloging-in-Publication Data

Mackey, George Whitelaw, 1916–
  The scope and history of commutative and noncommutative harmonic analysis / George W. Mackey.
    p. cm. — (History of mathematics, ISSN 0899-2428 ; v. 5)
  Consists of six reprinted articles originally written as expanded versions of talks given at various conferences and published between 1978 and 1990.
  Includes bibliographical references.
  ISBN 0-8218-9903-1 (acid-free paper)
  ISBN 0-8218-3790-7 (soft cover)
  1. Harmonic analysis—History. I. Title. II. Series.
QA403.M28 1992   92-12857
515′.2433—dc20   CIP

Harmonic Analysis as the Exploitation of Symmetry—A Historical Survey, Rice University Studies, volume 64, numbers 2 and 3, pages 73–228. Copyright © 1978. Reprinted by permission of Rice University Press.

Herman Weyl and the Application of Group Theory to Quantum Mechanics, Exact Sciences and their Philosophical Foundations/Exakte Wissenschaften und ihre philosophische Grundlegung, Vorträge des Internationalen Hermann-Weyl-Kongresses, Kiel, 1985, edited by Wolfgang Deppert, Kurt Hübner, Arnold Oberschelp, and Volker Weidemann, pages 131–159. Copyright © 1985. Reprinted by permission of Verlag Peter Lang GmbH.

The Significance of Invariant Measures for Harmonic Analysis, Colloquia Mathematica Societatis János Bolyai, volume 49, pages 551–609. Copyright © 1985. Reprinted by permission of János Bolyai Mathematical Society.

Weyl's Program and Modern Physics, Differential Geometrical Methods in Theoretical Physics, edited by K. Bleuleer and M. Werner, pages 11–36. Copyright © 1988. Reprinted by permission of Kluwer Academic Publishers.

Induced Representations and the Applications of Harmonic Analysis, Springer Lecture Notes in Mathematics, volume 1359, edited by P. Eymand and J. P. Pier, pages 16–51. Copyright © 1988. Reprinted by permission of Springer Verlag.

Von Neumann and the Early Days of Ergodic Theory, Proceedings of Symposia in Pure Mathematics, volume 50, edited by James Glimm, John Impagliazzo, Isadore Singer, pages 25–38. Copyright © 1990 American Mathematical Society.

Contents

alibris

BRANCAMP BOOKS LLC
8 WERNER WAY
BATESVILLE, IN 47006
UNITED STATES

**To: ALIBRIS APEX DC 76524100-60
APEX
800 AVONDALE AVE.
GRANDVIEW HEIGHTS, OH 43212-3473**

Shipping Instructions for TIMBRANC

Print this Packing Slip and enclose inside the front cover of the book.

Please ship this item no later than Thu Jul 16, 2026.

Ship to:

ALIBRIS APEX DC 76524100-60
APEX
800 AVONDALE AVE.
GRANDVIEW HEIGHTS, OH 43212-3473
UNITED STATES

PN #	Item ID	Alibris ID	Media Type	Title / Author	Seller List Price	Order Date
76524100-60	260429032	B083672517	BOOK	The Scope and History of Commutative and Noncommutative Harm G. W. Mackey; Mackey, George W.	$40.00	Jul, 6 2026

Good Paperback Size: 7x1x10; Text is mostly Very Clean. No writing or tears to text. Has a small stain to top edge of back cover and to the last few pages.

76524100-60

Introduction

This book consists mainly of reprints of six articles of mine, all originally written as expanded versions of talks given at various conferences held between 1977 and 1988 and first published in the relevant conference proceedings. All of them include a fair amount of historical material and are primarily, but not exclusively, expository in character. They are intimately related to one another and may be thought of as overlapping presentations of various aspects of a single theme. In addition, it contains a lengthy section entitled "Final Remarks" whose nature is explained in its own introduction.

The first and earliest is by far the longest and most comprehensive. To some extent, the others may be regarded as fuller treatments or updatings of parts of the first with the special purposes of the corresponding conference in mind. For instance, three of the conferences were organized to do honor to the memories of Hermann Weyl, Alfred Haar, and John von Neumann, respectively, and my talks naturally put emphasis on their contributions to my central theme. However, in each case I had one or more new things to say as a result of insights and points of view gained since the first was written.

Before discussing the six articles individually, it will be useful to comment on the background of the first. In the summer of 1965, Jack Feldman invited me to spend six weeks at the University of California in Berkeley and to give some lectures on my work. For some time I had become increasingly impressed by the extent and interrelatedness of the applications of the theory of unitary group representations to physics, probability, and number theory and I seized the opportunity to organize my thoughts on these matters. I gave twelve 90 minute lectures and wrote extensive lecture notes which were typed and mimeographed. The first paragraph of the introduction to these notes reads as follows:

> In these lectures I am going to attempt something a little unusual. Instead of giving a detailed account with proofs of a relatively small body of mathematical theory, I propose to give a series of interrelated survey lectures. These will be designed to show the extent to which the theory of infinite dimensional group representations is a universal tool with significant applications in subjects as

diverse as number theory, ergodic theory, quantum physics, probability theory, and the theory of automorphic functions. In fact, to speak of significant applications is to understate the case. Large sections of some of these subjects may be looked upon as nearly identical with certain branches of the theory of group representations. Moreover, one obtains a clearer view of the many known relationships between the subjects in question by looking at them in this way.

While I was in Berkeley, I received a formal invitation (with Michael Atiyah as sponsor) to go to Oxford for the academic year 1966–1967 and give a course of lectures as George Eastman Visiting Professor. By this time, my enthusiasm for my new program had increased, and I lost no time in deciding to treat the same theme in my Oxford lectures, taking advantage of a several-fold increase in lecturing time, to do a more complete and thorough job. Once again mimeographed lecture notes were produced and distributed on a small scale. More than a decade later (in 1978), these lecture notes, slightly revised and supplemented by some 9000 words of "Notes and References," were published in book form [5].

The first paper in this collection may be regarded as being a presentation of the point of view of the Berkeley and Oxford lectures, but with less detailed development, much more emphasis on the period before 1940, especially before 1896, and organized as a history of the applications of harmonic analysis. Recall that group representations were not invented until 1896 and that it was only in 1927 that Peter and Weyl pointed out and emphasized the (still insufficiently appreciated) fact that classical Fourier analysis can be illuminatingly regarded as a chapter in the representation theory of compact commutative Lie groups.

When I was invited to speak at the conference on the history of analysis given at Rice University on March 12–13, 1977, I decided that it might be interesting to review the history of mathematics and physics in the last three hundred years or so, with heavy emphasis on those parts in which harmonic analysis had played a decisive or, at least, a major role. I was pleased and somewhat astonished to find how much of both subjects could be included under this rubric. To put it slightly differently, I decided to sketch the history of harmonic analysis—not as an interesting self-contained branch of mathematics, but as a widely applicable method with considerable unifying power. My subject turned out to be much too vast to be treated adequately in two lectures. Thus, the talks actually given were short on concrete details and long on vague "handwaving." In writing my talks up for publication, I decided to fill in some of these gaps, and found myself drawn into a much more extensive project than I had anticipated. It took several months of full-time work but turned out to be an extremely rewarding experience. The picture that gradually emerged as the various details fell into place was one that I found very beautiful, and the process of seeing it do so left me in an almost constant state of euphoria. I would like to believe that others can be led to see this picture by reading my paper; to facilitate this, I have included a large number of short expositions of topics which are not widely understood by non-specialists. However, I fear that there is little hope of achieving my goal for those not willing to take the time to go through the paper rather slowly and read each exposition with care. The paper is written for the mathematical public at large and, moreover, can be read selectively. However, its full message is only available to those who are willing to read the whole paper.

The papers given at the Kiel conference honoring the one hundredth birthday of Hermann Weyl and at the Como conference on Differential Geometrical Methods in Theoretical Physics deal exclusively with the applications of unitary group representations to quantum mechanics. These began in 1927 with independent papers by Weyl and Wigner published in volumes 46 and 43, respectively, of *Zeitschrift für Physik*. While Wigner was concerned with the use of group representations in the solution of concrete physical problems, Weyl was interested in using the same subject to clarify the foundations. In these two papers we discuss only Weyl's program.

The first 40% of the Weyl centenary paper is devoted to a detailed account of the historical background for Weyl's fundamental work of 1924–1925 on group representation theory, its application to quantum mechanics, and its unification with Fourier analysis. The middle 25% is a review of the contents of his 1927 paper in the *Zeitschrift für Physik* and the now classic book [13] which appeared a year later. The final 35% first explains how this paper stimulated a fundamental paper by M. H. Stone, which in turn (almost two decades later) stimulated the present author to prove a general theorem about group representations known as the imprimitivity theorem. It then goes on to explain how the imprimitivity theorem makes it possible to give a much more satisfactory answer to the fundamental foundational question posed by Weyl in the *Zeitschrift* paper of 1927 than Weyl could give with the tools available at the time.

The paper prepared for the Como conference overlaps the Weyl centenary paper in that the first half of the former is a review of parts of the latter. However, the author believes that the exposition of section IV of the Como paper is a substantial improvement on the corresponding exposition in the centenary paper. In the second half of the Como paper, it is explained how one can go on from the account of free particle quantum statics, contained in the first part, to include dynamics, particle interactions, isotopic spin, etc., and to make contact with such relatively recent developments in elementary particle quantum mechanics as gauge fields and supersymmetry.

The Haar centenary paper is, in a sense, a shortened version of the Rice article. However it is organized differently. The role of Haar measure is underlined and certain topics not treated at all in the Rice article are included.

GEORGE W. MACKEY   ix

Among these are (a) a detailed argument supporting the author's philosophy that harmonic analysis "is" the decomposition of induced representations—provided that one includes the representations induced by representations of the "virtual subgroups" defined by properly ergodic actions, and (b) a detailed account of the nature of Wigner's use of group representations to solve specific quantum mechanical problems. Furthermore, the applications to number theory are presented from a somewhat more general and systematic point of view, and the account of Artin's work using group representations to factor the zeta functions of algebraic number fields into L functions is improved.

The Luxembourg article is heavily weighted on the side of the applications of harmonic analysis to number theory and to some extent redresses the imbalance produced by the emphasis on physics in the Weyl centenary and Como papers. The physics is confined to the five pages of section 3 while the twenty-five pages of sections 4 and 5 are entirely devoted to number theory. Section 1 is introductory and only one page long, and the five pages of section 2 are essentially a condensation of the twenty pages of the Haar centenary article presenting the author's philosophy of the central role of induced representations in harmonic analysis. Most of section 3 is a slight variant of section IV of the Como paper.

Sections 4 and 5 are of quite a different character in that a large part of what they contain was not known to the author until the late 1980s, and much of this was worked out in early 1986 after the Weyl and Haar centenary articles had already been written. Section 4 is an exposition of the author's rather idiosyncratic point of view toward the nature of number theory. It is based on his observation that certain statements of Fermat about the Diophantine equations $x^2 + ay^2 = n$ where $a = 1, 2$, or 3 can be reformulated in such a way that a very large part of eighteenth, nineteenth, and early twentieth century number theory can be looked upon as providing steps toward natural generalizations of these early theorems of Fermat. It is carried out in the form of a brief history of number theory from Fermat to the mid-twentieth century. Section 5 is a progress report on the author's attempts to understand the power of harmonic analysis in number theory in a systematic way. Both sections are addressed to nonexperts.

The von Neumann article is relevant to our main theme via its connections with applications of harmonic analysis to probability theory. See for example section 18 of the Rice article. After reviewing von Neumann's key role in the development of ergodic theory and some historical background, this article concludes with the author's views on the significance of certain later developments.

In concluding this introduction, the author would like to mention a paper of K. Gross [2] which came to his attention in preprint form just as the Rice article was being completed. Although much shorter and consequently less comprehensive, Gross's article has the same theme and might well be read in advance of the latter in order to get a brief preliminary overview.

BIBLIOGRAPHY *

1. C. Chevalley, *Généralisation de la théorie du corps de classes pour les extensions infinies*, J. Math. Pures Appl. (9) **15** (1936), 359–371.

2. K. I. Gross, *On the evolution of noncommutative harmonic analysis*, Amer. Math. Monthly 85 (1978), 525–548.

3. R. P. Langlands, *Problems in the theory of automorphic forms*, in Lectures in Modern Analysis and Applications, III, Lecture Notes in Math. No. 170 (Springer-Verlag, Berlin-Heidelberg-New York, 1970), pp. 18–61.

4. ____, *Euler products*, Yale Mathematical Monographs No. 1 (Yale University Press, New Haven, C.T., 1971), pp.1–53.

5. G. W. Mackey, *Unitary group representations in physics, probability, and number theory* (Benjamin/Cummings, Reading, M.A., 1978; Second Ed., Addison-Wesley, Reading, M.A., 1989).

6. ____, *Ergodic theory and its significance for statistical mechanics and probability theory*, Advances in Math. **12** (1974), 178–268.

7. ____, *Quantum mechanics from the point of view of the theory of representations*, in Applications of Group Theory in Physics and Mathematical Physics (Chicago, 1982), Lectures in Appl. Math., vol. 21 (Amer. Math. Soc., Providence, R.I., 1985), pp. 219–253.

8. L. Pontrjagin, *Topological groups* (Princeton Univ. Press, Princeton, N.J., 1939).

9. A. Selberg, *Harmonic analysis and discontinuous groups in weakly symmetric Riemannian spaces with applications to Dirichlet series*, J. Indian Math. Soc. **20** (1956), 47–87.

10. M. H. Stone, *Linear transformations in Hilbert space III: Operational methods and group theory*, Proc. Nat. Acad. Sci. U.S.A. 16 (1930), 172–175.

11. J. Tate, *Fourier analysis in number fields and Hecke's zeta functions* [reproduction of doctoral thesis, Princeton Univ., 1950], in Algebraic Number Theory (J.W.S. Cassels and A. Frohlich, editors; Thompson Book Co., Washington, D.C., 1967), Chapter 15.

12. A. Weil, *L'intégration dans les groupes topologiques et ses applications*, Actualités Sci. Indust. No. 869 (Hermann et Cie, Paris, 1940), pp. 1–58; Second Ed., ibid. No. 1145 (Paris, 1951), pp. 1–162.

13. H. Weyl, *The theory of groups and quantum mechanics* (E. P. Dutton, New York, 1931) [original German edition: S. Hirzel, Leipzig, 1928].

14. E. P. Wigner, *Über nicht kombinierende Terme in der neueren Quanten theorie*, II, Zeit. für Physik **40** (1927), 883–892.

15. ____, *Einige folgerungen aus der Schrödingerschen theorie für die termstrukturen*, Zeit. für Physik **43** (1927), 624–652; ibid. **45** (1927), 601–602.

16. ____, *Group theory and its application to the quantum mechanics of atomic spectra*, (Academic Press, New York, 1959) [original German edition: F. Vieweg & Sohn, Braunschweig, 1931].

---

*This bibliography applies only to the six reprinted articles. The bibliography for the final remarks is internal.

BULLETIN (New Series) OF THE
AMERICAN MATHEMATICAL SOCIETY
Volume 3, Number 1, July 1980

HARMONIC ANALYSIS AS THE EXPLOITATION OF SYMMETRY–A HISTORICAL SURVEY

BY GEORGE W. MACKEY

CONTENTS

Reprinted from *Rice University Studies* (Volume 64, Numbers 2 and 3, Spring–Summer 1978, pages 73 to 228), with the permission of the publisher.

1980 *Mathematics Subject Classification.* Primary 01, 10, 12, 20, 22, 26, 28, 30, 35, 40, 42, 43, 45, 46, 47, 60, 62, 70, 76, 78, 80, 81, 82.

GEORGE W. MACKEY   1

PREFACE

This paper is an expansion (by a factor of twelve) of two talks that I gave in the spring of 1977 at the Rice University Conference on the history of analysis. I am not a historian in the usual professional sense of the word and I did not pretend to be reporting on the results of a careful scholarly investigation. My talks consisted rather of an informal account of the knowledge and impressions I have gained over the years as I have tried to satisfy my curiosity about the origins and interrelationships of the parts of mathematics and science which most interest me. Moreover, in contradistinction to most (if not all) professional historians of science and mathematics, I was much more interested in getting an overall approximate idea of how our present understanding unfolded than in studying the fine structure of particular discoveries. I totally ignored such questions as false starts, the thought processes of individual scientists, and the intellectual climate of the times. The question constantly in my mind was this: How much of what we understand now had they grasped by then?

When I began in early August 1977 to concentrate on the job of writing up my talks for publication, I found that I could not say what I wanted to say without making a number of assertions about matters of fact that were not always easy to verify but whose exactitude was at most marginally relevant to the story I was trying to tell. I made some attempt to make only correct assertions (sometimes by being deliberately vague) but my time and patience were limited and I am sure I did not always succeed.

The length of the finished product is not all due to striving for local historical accuracy (this would require at least fifty books), but rather to my desire to make clear the relationship of a large part of mathematics to my main theme. To this end I composed a large number of brief introductory expositions, which I hope the average mathematician will find intelligible. It is these expositions and their connectedness in time and intellectual content which is the real point of the paper.[1]

1. INTRODUCTION

In this article I shall sketch the history, applications, and ramifications of a certain method. For want of a better word I shall call it the method of harmonic analysis, although I am aware that many people use these words to denote a different class of generalizations of the classical harmonic analysis of Fourier. In crude terms the method may be described as follows: Let S be a "space" or "set" and let G be a group of one-to-one transformations of S onto itself. Let $[s]x$ denote the transform of s in S by x in G. Ordinarily S will have further structure which will be preserved by the transformations of G so that the transformations $s \to [s]x$ are symmetries of S. It will be convenient to allow members of G other than the identity e to define the identity map so that some quotient group G/N is the actual transformation group. Now let $\mathfrak{F}$ be some vector space of complex valued functions on S which is G invariant in the sense that $s \to f([s]x)$, the translate of f by x, is in $\mathfrak{F}$ whenever f is in $\mathfrak{F}$. Then for each x in G, the mapping $f \to g$ where $g(s) = f([s]x)$ is a linear transformation V_x of $\mathfrak{F}$ onto $\mathfrak{F}$ and $V_{xy} = V_x V_y$ for all x and y in G. The mapping $x \to V_x$ is thus an example of what is called a (linear) representation of the group G. More generally, a (linear) representation of a group G is by definition any homomorphism $x \to W_x$ of G into the group of all bijective linear transformations of some vector space $\mathfrak{H}$ (W). The method I propose to discuss in this article consists (in its simplest form) in attempting to find subspaces M_λ of the space $\mathfrak{F}$ such that

(1) $V_x(M_\lambda) = M_\lambda$ for all x and λ.
(2) Every element f in $\mathfrak{F}$ is uniquely a finite or infinite sum $f = \Sigma f_\lambda$ where each $f_\lambda \epsilon M_\lambda$.
(3) The subspaces M_λ are either not susceptible of further decomposition or are somehow much simpler in structure than $\mathfrak{F}$. Of course, one must have a topology in $\mathfrak{F}$ in order to make sense of infinite sums. More generally one also considers "continuous direct sums" or direct integrals and vector-valued as well as complex-valued functions. Of course, each M_λ is the space of a new representation V^λ, which is a so-called subrepresentation, and one speaks of the direct sum or direct integral decomposition of V. It turns out that the decomposition of functions in $\mathfrak{F}$ into sums and integrals of functions associated with the components of V is a decomposition that greatly simplifies many problems.

GEORGE W. MACKEY   3

2. THE CHARACTERS OF FINITE COMMUTATIVE GROUPS
AND THE CONNECTION WITH FOURIER ANALYSIS

Let W be a (linear) representation of the group G. If $\mathfrak{H}(W)$ is one-dimensional then each W_x is some complex number $\chi(x)$ times the identity, and $x \to \chi(x)I$ is a representation of G if and only if $\chi(xy) = \chi(x)\chi(y)$ for all x and y in G. When G is a finite commutative group, such functions χ are called characters. It is easy to see that the representations $x \to \chi(x)$ are the only representations of G that are *irreducible* in the sense that no proper subrepresentations exist. Let $\mathfrak{F}(G)$ be the vector space of all complex- valued functions on G and let $V_x f(y) = f(yx)$. Then V is a representation of G such that $\mathfrak{H}(V) = \mathfrak{F}(G)$ and the one-dimensional subspace generated by each character is clearly an invariant subspace. Indeed every one-dimensional invariant subspace is of this form, and one shows easily that V is a direct sum of one-dimensional representations each associated with a distinct character χ. Thus every member f of $\mathfrak{F}$ is uniquely of the form

$$\sum_{\chi \in \hat{G}} c_\chi \chi(x)$$

where $\hat{G}$ denotes the set of all characters on G. Evidently the product of two characters is again such, and the set $\hat{G}$ of all characters is itself a finite commutative group. Since $\chi(x^n) = \chi(x)^n = 1$ when n is the order of x, it follows that $|\chi(x)| \equiv 1$ for all χ and all x, so $\chi^{-1} = \overline{\chi}$ for all χ. Consider $\sum_{x \in G} \chi(x)$. This sum is clearly equal to $\sum_{x \in G} \chi(xy) = \chi(y) \sum_{x \in G} \chi(x)$. Thus whenever $\chi(y) \neq 1$, we have $\sum_{x \in G} \chi(x) = 0$. It follows that $\sum_{x \in G} \chi_1(x)\overline{\chi_2}(x) = 0$ whenever $\chi_1 \neq \chi_2$, so that the characters of G are *orthogonal* with respect to the inner product $f \cdot g = \sum_{x \in G} f(x) \overline{g(x)}$.

Now consider the expansion formula

$$(2.1) \qquad\qquad f(x) = \sum_{\chi \in \hat{G}} c_\chi \chi(x)$$

Multiplying each side by $\overline{\chi'(x)}$ summing over G and using the orthogonality of the characters leads at once to a formula for $c_{\chi'}$.

$$(2.2) \qquad\qquad c_{\chi'} = \frac{1}{o(G)} \sum_{x \in G} f(x)\overline{\chi}(x)$$

where $o(G)$ is the number of elements in G. The formulae (2.1) and (2.2) are strikingly similar to the formulae (2.3) and (2.4) below, which occur in the theory of Fourier series when sines and cosines are replaced by complex exponentials.

$$(2.3) \qquad f(x) = \sum_{n=-\infty}^{\infty} c_n e^{inx}$$

$$(2.4) \qquad c_n = \frac{1}{2\pi} \int_0^{2\pi} f(x)e^{-inx}\,dx$$

and the analogy becomes closer when one notices 1) that functions on the real line with period 2π may be identified with functions on the compact topological group T which the additive real line becomes when the subgroup of all integer multiples of 2π is factored out; 2) that the functions $x \rightarrow e^{inx}$ are precisely the continuous characters on T; and 3) that $c_x = \frac{1}{o(G)}\sum_{x\in G} f(x)\overline{\chi}(x)$ may be written as $c_x = \frac{1}{\mu(G)} \int f(x)\overline{\chi}(x)d\mu(x)$, where μ is the measure on G such that the measure of every subset is the number of elements it contains.

What we have called the method of harmonic analysis thus includes classical harmonic analysis (in the sense of expansion in Fourier series) as a very special case.

The formula (2.2) may be looked upon as defining a linear transformation of the vector space of all complex-valued functions on G onto the vector space of all complex-valued functions on $\hat{G}$. Similarly, formula (2.1) may be looked upon as defining the inverse of this transformation. A key property of the resulting one-to-one correspondence between functions on G and functions on $\hat{G}$ is that the operation of translation of functions on G is carried over into the operation of multiplication by a fixed function for the functions on $\hat{G}$. Similarly, the formulae (2.3) and (2.4) define a one-to-one linear correspondence between certain (not all) periodic functions on the line and certain functions on the additive group of all integers; and this correspondence also carries translation into multiplication. More significantly and for the same formal reasons it converts differentiation into multiplication by a function and so converts differential equations into algebraic equations.

A considerable part of the utility of classical harmonic analysis may be traced to this simple fact.

Looking at the functions e^{inx} as group characters and Fourier analysis as a special case of the decomposition of group representations are of course twentieth-century viewpoints. Indeed the very concept of a group representation was not formulated until the closing years of the nineteenth century. On the other hand, characters of finite commutative groups go back to the work of Gauss, and the analogue of Fourier expansions for functions on such groups has played a key role in number theory since the beginning of the nineteenth century. Moreover, the use of functional transforms involving characters to convert translation into multiplication may be traced back to early eighteenth-century work in probability. Thus the method of har-

GEORGE W. MACKEY   5

monic analysis has at least three independent origins; in probability, in number theory, and in mathematical physics. In the immediately following sections I shall sketch the history of these subjects with emphasis on the rise of the method of harmonic analysis.

3. PROBABILITY THEORY BEFORE THE TWENTIETH CENTURY

The basic facts of elementary probability theory were clarified, systematized, and to some extent discovered in a celebrated correspondence between Fermat (1601-1665) and Pascal (1623-1662), which began in 1654. The first book on the subject appeared in 1657 and was a short pamphlet by Huygens (1629-1695) entitled "De ratiocinio in ludo aleve." It applied the principles discovered by Fermat and Pascal to various gambling problems. The first major treatise on the subject was J. Bernoulli's (1654-1705) *Ars Conjectandi,* published posthumously in 1713. In addition to a commentary on Huygens's pamphlet (which was reprinted in full), it contained a statement and proof of the weak law of large numbers. This work was soon followed by another: the publication in 1718 of *Doctrine of Chances* by A. de Moivre (1667-1754). De Moivre's book is noteworthy for three things: In the form of an approximation to a formula of Bernoulli it contains an early intimation of the central limit theorem; it introduced the technique of solving problems in probability by reducing them to difference equations; and (at least implicitly) it introduced the technique of using "generating functions" to solve difference equations. No other major book on probability appeared until 1812, when Laplace (1749-1827) published his great treatise *Théorie Analytique des probabilités.* Based on a series of nine memoirs published between 1771 and 1786, Laplace's treatise developed and synthesized the work of his predecessors and put probability into a form which was to be more or less unchanged until the twentieth century. Two important new ideas were introduced between the appearance of the book by de Moivre and the first of Laplace's memoirs. In 1756 Thomas Simpson (1710-1761) began the application of probability theory to the study of errors of measurement, and in 1763 Bayes introduced the concept of inverse probability or probability of causes. Laplace's treatise included a development of both these ideas as well as a very extensive development of the ideas of de Moivre concerning difference equations, generating functions, and the central limit theorem. Indeed among Laplace's chief original contributions are 1) a formulation and heuristic proof of the central limit theorem; 2) an extension of the theory of difference equations to equations in several variables; and 3) a systematic use of generating functions in dealing with difference equations in one and several variables.

The nineteenth century was far richer in new applications of probability theory than in the development of new methods and principles. It was the century in which Quetelet (1796-1874), Galton (1822-1911), and Pearson (1857-1926) began the probabilistic study of human variation, in which Mendel (1822-1884) applied probability to genetics, and in which Maxwell (1831-1879), Boltzmann (1844-1906), and Gibbs (1839-1903) developed a statistical theory of heat and thermodynamics. As far as purer aspects are concerned, there were two chief contributors after the early and independent work of Gauss (1777-1855) and Legendre (1752-1833) on the theory of errors and the method of least squares. These were Poisson (1781-1840) and Tchebycheff (1821-1894). Poisson recognized the importance of the distribution which bears his name, generalized Bernoulli's work to the case of probabilities that vary from trial to trial, and published a book on probability in 1837. Tchebycheff was the first to think systematically in terms of "random variables" and their "expectations" and "moments." Using these concepts he discovered a simple inequality in 1867 that led to a remarkably simple proof of Bernoulli's law of large numbers. Moreover, he inaugurated a program for using the moment concept to give a rigorous proof of the central limit theorem. This program was completed by his student Markov (1856-1922), who became one of the leading probabilists of the early twentieth century. Shortly thereafter Liapunov (1858-1918), another pupil of Tchebycheff, found a simpler and better proof of the central limit theorem using "characteristic functions" instead of moment sequences.

4. THE METHOD OF GENERATING FUNCTIONS IN PROBABILITY THEORY

During the first third of the twentieth century, it became clear that both the central limit theorem and the law of large numbers are essentially corollaries of theorems in harmonic analysis. In particular, the so-called "characteristic function" of a probability distribution is just its Fourier transform, and Liapunov's proof of the central limit theorem essentially exhibits the theorem as a corollary of the (nonobvious) fact that the Fourier transform is a homeomorphism between appropriately topologized function spaces. Moreover (as I shall explain in some detail in a later section), the pioneering work of Norbert Wiener in the 1920s led over the next few decades to a rather profound development and intermingling of concepts from probability theory with those of harmonic analysis on the line. Thus it is interesting that harmonic analysis as a method seems to have first been used to deal with problems in probability theory.

Let r be a positive integer and for each $n = 1,2,3,\cdots$, let p_n^r denote the probability that at least one "run" of r heads will occur during n coin tosses. We assume that the tosses are independent and that the probability of getting heads on any given toss is q where $0 < q < 1$. The problem is to compute p_n^r for each n. This problem was first posed and solved by de Moivre. His method was to consider p_n^r as a function of n and show that this function satisfies a certain "difference equation." Suppressing the r and writing $p_n^r = P(n)$, application of elementary principles of probability theory leads to the conclusion that

$$P(n + 1) = P(n) + (1 - P(n - r))\, q^r(1 - q)$$

for all $n \geq r$. Setting $\tilde{P}(n) = 1 - P(n)$ thus yields the homogeneous difference equation

$$\tilde{P}(n + 1) = \tilde{P}(n) - q^r(1 - q)\,\tilde{P}(n - r)\,.$$

Our unknown function $\tilde{P}$ is clearly the unique solution of this equation which satisfies the "initial conditions" $\tilde{P}(1) = \tilde{P}(2) = \cdots = \tilde{P}(r - 1) = 1$, $\tilde{P}(r) = 1 - q^r$.

Many problems in probability theory may be thus reduced to the solution of difference equations. De Moivre's book *Doctrine of Chances* is rich in examples and in methods for solving such equations. One important method—the method of generating functions—occurs at least implicitly in de Moivre's work, but was first developed and used systematically by Laplace in a paper published in 1782. It was Laplace who coined the term "generating function." The first chapter of his treatise on probability is largely a reprint of the contents of the 1782 paper.

The idea of the method is very simple. Let $f(n)$ be an unknown function of the non-negative integer variable n, and suppose that $f(n + 2) = af(n + 1) + bf(n)$ for all $n \geq 0$ where a and b are known constants. Laplace calls the power series

$$\tilde{f}(t) = f(0) + tf(1) + t^2 f(2) + \cdots$$

the generating function of f and shows that determining f can be reduced to solving an algebraic equation. The point is that if $g(n) = f(n + 1)$ then $\tilde{g}(t) = f(1) + tf(2) + t^2 f(3) + \cdots = \dfrac{\tilde{f}(t) - f(0)}{t}$, and if $h(n) = f(n + 2)$ then $\tilde{h}(t) = f(2) + tf(3) + \cdots = \dfrac{\tilde{f}(t) - f(0) - tf(1)}{t^2}$. Thus f satisfies the given difference equation if and only if

$$\frac{\tilde{f}(t) - f(0) - tf(1)}{t^2} = a\left(\frac{\tilde{f}(t) - f(0)}{t}\right) + b\tilde{f}(t),$$ and solving a simple algebraic equation gives us a formula for $\tilde{f}$ as a rational function of t whose coefficients depend on $f(0)$ and $f(1)$. To find f for other values of n, we need only expand this rational function in a power series.

The method can obviously be applied to any kth order difference equation with one variable and constant coefficients, but, as Laplace emphasized, it can also be used for the difference equations in several variables to which one is led by more complicated problems in probability.

The relationship to what we have called the method of harmonic analysis is almost evident. Since $t^{n+m} = t^n t^m$, the functions $n \to t^n$ are characters on the group Z of all integers, and the generating function $\bar{f}(t) = f(0) + tf(1) + \cdots$ bears just the same relationship to the function f (extended to be defined on all of G by letting it be zero for negative n) that f and c bear to one another when G is a finite commutative group. The difference equation becomes an algebraic equation because translation becomes multiplication by a function.

5. NUMBER THEORY BEFORE 1801

The theory of numbers is one of the very oldest branches of mathematics, going back at least to work of Euclid around 300 B.C. Euclid already knew a proof that there are an infinite number of primes, and by A.D. 300 Diophantus had written a treatise on methods for finding the integral solutions of indeterminate equations. The six surviving "books" of this treatise were translated into Latin in 1621, and modern number theory is considered to have begun when the same Fermat who helped found probability theory read this translation and began to study the subject for himself. He announced his results in letters to others and made marginal notes in his copy of the works of Diophantus, all too often neglecting to explain how he had arrived at them or how they might be proved. A century later Leonard Euler (1707-1783) became interested in the challenge presented by the unproved assertions of Fermat and produced the first published proofs of a number of them—often with great difficulty. Consider for example the problem of finding the integer solutions of $x^2 + y^2 = n$ where n is a given positive integer. It is not difficult to reduce the problem to the case in which n is a prime p, and it is also quite easy to show that there can be no solution when p is of the form $4l + 3$. Every other prime is either 2 (where there clearly is a solution) or of the form $4l + 1$. Fermat asserted that every prime of the form $4l + 1$ is a sum of two squares, and Euler managed to demonstrate this in 1754, but reportedly only after many years of effort. Later he was able to prove a number of analogous assertions of Fermat, such as the solvability of $x^2 + 3y^2 = p$ where p is a prime of the form $6n + 1$ and of $x^2 + 2y^2 = p$ when p is a prime of the form $8n + 1$. Rather earlier Euler made a very original and important contribution to number theory by noticing that an infinite series of the form $\sum_{n=1}^{\infty} \frac{1}{n^s}$ (with s a real number greater than one) can be written as an infinite product $\prod_p \left(\dfrac{1}{1 - \dfrac{1}{p^s}} \right)$ where p runs over all the primes. Using this and the fact that $\sum \dfrac{1}{n}$ diverges, he was able to show that $\sum \dfrac{1}{p}$ diverges, thus strengthening and re-proving Euclid's discovery that there is an infinite number of primes. Similar arguments enabled him to show that there is an infinite number of primes of the form $4n + 1$ as well as of the form $4n + 3$.

The next major advances were made by J.L. Lagrange (1736-1813), the first mathematician after Euler to reach comparable stature. If one starts to study Diophantine equations systematically, one finds that the theory of linear equations can be completely worked out rather easily. The simplest case presenting a genuine challenge is that of quadratic equations in two unknowns—the most general being the equation

$$Ax^2 + Bxy + Cy^2 + Dx + Ey + F = 0$$

where A, B, C, D, E, and F are given integers. The linear terms can be eliminated by simple transformations and one is confronted with equations of the form $Ax^2 + Bxy + Cy^2 = -F$, that is with the problem of deciding whether and in how many ways a given integer $-F$ may be represented by the "binary quadratic form" $Ax^2 + Bxy + Cy^2$. Special cases of the problem were studied and solved or partially solved by Fermat and Euler as described above. Lagrange's contribution was to attack the general case and to discover a number of important theorems about it. His main results appear in two long memoirs in publications dated 1773 and 1775 respectively.

Of key importance in Lagrange's work is a natural notion of equivalence between quadratic forms. If $\left(\begin{smallmatrix}a&b\\c&d\end{smallmatrix}\right)$ is a matrix of integers such that $ad - bc = \pm 1$, then substituting $x = ax' + by'$, $y = cx' + dy'$ converts $Ax^2 + Bxy + Cy^2$ into another form $A'x'^2 + B'x'y' + C'y'^2$, which evidently represents precisely the same integers that $Ax^2 + Bxy + Cy^2$ does. Calling two forms equivalent when they may be obtained from one another in this way, one sees that in the representation problem one need only consider a single form in each class. The integer $D = B^2 - 4AC$ is called the discriminant of the form and is easily verified to depend only on the class of the form. It is possible for inequivalent forms to have the same discriminant, however, and two of Lagrange's more important discoveries may be formulated as follows: 1) For a given value of D there can be only a finite number of distinct classes of forms having D as a discriminant; 2) when the number of classes of forms with discriminant D is greater than one they must all be considered together if one wants to reduce the problem of solving $Ax^2 + Bxy + Cy^2 = n$ to the case in which n is a prime. For example if $Ax^2 + Bxy + Cy^2 = p_1 p_2$ where p_1 and p_2 are distinct primes and x and y are integers, one cannot conclude in general that either of the equations $Ax^2 + Bxy + Cy^2 = p_1$ or $Ax^2 + Bxy + Cy^2 = p_2$ has a solution in integers. One can only conclude that there exist forms $A'x^2 + B'xy + C'y^2$ and $A''x^2 + B''xy + C''y^2$ where $(B'')^2 - 4A''C'' = (B')^2 - 4A'C' = B^2 - 4AC$ such that

$$A'x^2 + B'xy + C'y^2 = p_1 \text{ and } A''x^2 + B''xy + C''y^2 = p_2$$

both have integer solutions. To get an elegant theory one has to be less ambitious and seek to evaluate the sum $\phi_{Q_1}(n) + \phi_{Q_2}(n) + \cdots + \phi_{Q_h}(n)$ where $\phi_Q(n)$ is the (suitably normalized) number of solutions of $Q(x,y) = n$ in integers and $Q_1, \cdots, Q_h$ constitutes a complete set of mutually inequivalent forms having the common discriminant D. (With suitable normalization even indefinite forms ($D > 0$) lead to equations with a finite number of solutions.) One can reduce to the case in which D is square free and then the key facts may be formulated as follows (Lagrange presented them differently):

(1) $\phi_D = \phi_{Q_1} + \phi_{Q_2} + \cdots + \phi_{Q_h}$ is "multiplicative" in the sense that $\phi_D(n_1 n_2) = \phi_D(n_1)\,\phi_D(n_2)$ whenever n_1 and n_2 are relatively prime.

(2) If p is a prime and $k = 1,2\cdots$ then $\phi_D(p^k) = 1 + \chi_D(p) + \cdots + (\chi_D(p))^k$ where $\chi_D(p) = \phi_D(p) - 1$. It follows from (1) and (2) that $\phi_D(n)$ can be computed from the factorization of n when $\phi_D(p)$ is known for all primes p.

(3) $\phi_D(p)$ is 0,2, or 1, and which it is depends only on the solvability of the equations $Q_j(x,y) = 0 \bmod p$ and hence the value of $D \bmod p$.

To get an overview of the values of $\phi_D(p)$ for fixed D, one needs to apply the celebrated quadratic reciprocity law, which together with its two supplements allows one to compute the behavior of $D \bmod p$ from that of $p \bmod D$. The probable truth of this law was known to both Euler and Lagrange, but it was Legendre (1752-1833) who first clearly stated it in a paper published in 1785. He also offered a proof, but it relied on a lemma that was first properly proved by Dirichlet over half a century later. Let p be an odd prime and for each nonzero integer n let us define the "Legendre symbol $\left(\dfrac{n}{p}\right)$ to be 1 or -1 according as the nonzero integer n is or is not a square mod p. It is trivial that $\left(\dfrac{nm}{p}\right) = \left(\dfrac{n}{p}\right)\left(\dfrac{m}{p}\right)$ so that it suffices to know $\left(\dfrac{n}{p}\right)$ when n is 2, -1, or an odd prime to know it for all n. The quadratic reciprocity law asserts that $\left(\dfrac{q}{p}\right)\left(\dfrac{p}{q}\right) = (-1)^{(p-1)(q-1)/4}$ whenever q is an odd prime. Its two supplements state that $\left(\dfrac{-1}{p}\right) = (-1)^{(p-1)/2}$ and $\left(\dfrac{2}{p}\right) = (-1)^{(p^2-1)/8}$ and are easier to prove.

Legendre published an important book on number theory in 1798. He presented Lagrange's theory with various improvements, including the quadratic reciprocity law, and also made a start on a theory of ternary quadratic forms in a celebrated treatise published in 1798.

6. THE WORK OF GAUSS AND DIRICHLET AND THE INTRODUCTION OF CHARACTERS AND HARMONIC ANALYSIS INTO NUMBER THEORY

The first person to carry on the work begun by Lagrange and Legendre on the general theory of binary quadratic forms was C.F. Gauss (1777-1855). By his own account Gauss did not become aware of the work of his predecessors until he entered the university at Göttingen in the fall of 1795. Earlier the same year he accidentally discovered that if p is an odd prime, then -1 is a square mod p if and only if p is of the form $4n + 1$. This excited him tremendously and, determined to get to the bottom of such phenomena, he had found and proved the quadratic reciprocity law by the end of March 1795. One thing led to another, and before arriving in Göttingen and beginning to study the works of Euler, Lagrange, and Legendre, Gauss had rediscovered many of their results. All this is explained in the introduction to his classic treatise *Disquisitiones Arithmeticae* published in 1801, which set the course for the future development of number theory. He claims that most of the material in the first four of the seven sections of his treatise was known to him before he arrived in Göttingen and that a large part of the book had been set up in type before Legendre's book of 1798 appeared. While he acknowledges that he was inspired to study quadratic forms by the work of Lagrange and Legendre, he reworked the whole subject in his own way, giving new proofs and introducing important new ideas and concepts. The seventh section contains Gauss's famous proof that for any prime p of the form $2^n + 1$ (e.g., 17) one can give a ruler and compass construction of a regular polygon with p sides. He is reported to have definitely made up his mind to be a mathematician when he found this beautiful result in the spring of 1796. At this point the celebrated Galois theory of equations was thirty-five years in the future, but Gauss's proof is imbedded in what amounts to a complete development of that theory for the equation $x^p - 1 = 0$. ($x^p - 1$ factors into $x - 1$ and an irreducible polynomial of degree $p - 1$. If $p - 1 = 2^k$, it follows from Galois theory that the equation can be solved by rational operations and the taking of square roots.)

In his work on the theory of binary quadratic forms, Gauss confined himself to the case in which the middle coefficient is even, writing $Ax^2 + 2Bxy + Cy^2$ and defining the discriminant to be $B^2 - AC$ instead of $B^2 - 4AC$. Also he pointed out that simplifications ensue if one makes a distinction between "proper" and "improper" equivalence of forms; two forms being said to be properly equivalent only when the transformation matrix $\left(\begin{smallmatrix} a & b \\ c & d \end{smallmatrix}\right)$ has determinant 1 rather than -1. By far his most important contribution, however, was his observation that there is a natural (but not obvious) composition law for proper equivalence classes of forms having a fixed discriminant and that this composition law converts the finite set of classes with square free discriminant D into a finite commutative group. Of course, group theory did not then exist (cf. section 11) and Gauss did not use this terminology, but this is in effect what he proved about his composition law. Legendre had already divided the classes with a given square free discriminant into subsets that Gauss called genera and related in an interesting way to his composition law.

In modern terminology, the genera are just the cosets of the subgroup of squares in the group of all classes with discriminant D. As defined by Gauss, Q_1 and Q_2 are in the same genus if they have the same "total character" where the "total character" is a system of ones and minus ones that is canonically associated to each form. In effect Gauss defined a finite set χ_1, . . ., χ_l of functions from classes of forms to ± 1 and called them characters. He then put two forms Q_1 and Q_2 in the same genus when $\chi_j(Q_1) = \chi_j(Q_2)$ for all j. It turns out that Gauss's characters are characters in the modern sense for the finite commutative group in question. The fact that the genera are cosets of some subgroup which contains all squares follows at once from simple group theoretical considerations. To say that it contains only squares is equivalent to saying that every character of order two is a product of Gauss characters. Gauss proved this by a difficult argument using ternary forms. As he expressed it, every form in the principal genus can be obtained by composing some form with itself. This is known as Gauss's theorem on duplication. That all genera have the same number of elements and that the genera inherit a composition law are evident from the group theoretical interpretation.

To understand the significance of the division of classes into genera from the point of view of solving $Q(x,y) = n$ in integers and at the same time to appreciate the usefulness of finite Fourier analysis in number theory, it will be convenient to anticipate the future and introduce the characters of the group of classes that take on values other than ± 1. Let $Q_1, Q_2, \cdots, Q_h$ be a complete set of inequivalent binary forms of square free discriminant D and let $\phi_j(n)$ denote the (suitably normalized) number of representations of n by Q_j. We have already remarked in section 5 that if $\phi = \phi_1 + \phi_2 + \cdots + \phi_h$, then ϕ is multiplicative and there is a simple formula expressing $\phi(p^k)$ in terms of $\phi(p)$. More generally, one can show that for each character χ of the group C_D defined by composition of classes, the function $n \rightarrow \phi_\chi(n) = \Sigma \chi(Q_j)\, \phi_j(n)$ has these same properties. That is, $\phi_\chi(n_1\, n_2) = \phi_\chi(n_1)\, \phi_\chi(n_2)$ when n_1 and n_2 are relatively prime, and $\phi_\chi(p^k) = a(p)^k + b(p)a(p)^{k-1} + b(p)^2\, a(p)^{k-2} + \cdots + b(p)^k$ where $a(p) + b(p) = \phi_\chi(p)$ and $a(p)b(p) = \phi(p) - 1$. Thus if $\phi_j(p)$ is known for all primes p and all $j = 1,2, \cdots, h$, then $\phi_\chi(p)$ can be computed for all χ and p and hence $\phi_\chi(n)$ for all χ and n. But by finite Fourier analysis $\phi_j(n) = h\, \Sigma\, \overline{\chi(Q_j)}\, \phi_\chi(n)$. Thus one has an explicit formula for each $\phi_j(n)$ as a finite linear combination of the multiplicative functions $\phi_\chi(n)$, each of which can be computed once one knows the ϕ_j at the primes.

Unfortunately it seems to be quite difficult to find an analogue for general χ of the fact that $p \rightarrow \phi(p)$ is the restriction to the primes of a periodic function on the integers and correspondingly difficult to get an explicit expression for $\phi_j(n)$ in the general case. But this difficulty disappears when χ is of order two. Thus in the special case in which all characters are of order two (i.e., when there is just one class in each genus), one can get a simple explicit formula for each ϕ_j. In general one needs to sum the ϕ_j only over a genus rather than a class in order to obtain such formulae.

The word *character* as used today stems directly from Gauss's use of the term in his theory of binary quadratic forms. As a homomorphism of an *abstract* finite commutative group into the group of roots of unity it was first defined in 1882 by Weber, who refers to the version of Dedekind's ideal theory for algebraic number fields published in 1879. In that reference Dedekind makes the same definition for the special case of the ideal class group of an algebraic number field. As Dedekind himself had pointed out earlier, the ideal class group is a generalization of Gauss's group of equivalence classes of binary quadratic forms. Dedekind's mode of expression is such as to make it clear that he regards his definition as a generalization of that of Gauss.

Characters and Fourier analysis on finite commutative groups occur implicitly in other parts of Gauss's work. For example, let R_N denote the ring of integers mod N for $N = 2,3,4 \ldots$ and let $\phi_N(k)$ be the number of members l of R_N such that $l^2 = k$ in R_N. The characters on the additive group of R_N are the functions $k \rightarrow \chi_q(k) = e^{2\pi i k q/N}$ where $q = 0,1, \cdots, N-1$. Thus the Fourier transform $\hat{\phi}_N$ of ϕ_N is the function $\chi_q \rightarrow \sum_{k=0}^{N-1} \chi_q(k)\, \phi_N(k) = \sum_{s=0}^{N-1} e^{2\pi i q s^2/N}$. When N is an odd prime or a product of two such and q does not divide N, it is easy to see that $\hat{\phi}_N(\chi_1) = \left(\dfrac{q}{N}\right) \hat{\phi}_N(\chi_1)$ where $\left(\dfrac{q}{N}\right)$ is the Legendre symbol, so that it suffices to know $\hat{\phi}_N(\chi_1)$ to know $\hat{\phi}_N$. It is also easy to see that $\hat{\phi}_N(\chi_1)^2 = \pm N$. But Gauss found it quite difficult to determine the unknown signs and show that $\hat{\phi}_N(\chi_1) = \sqrt{N}$ or $i\sqrt{N}$ depending on whether N is of the form $4n + 1$ or $4n + 3$. Indeed, Gauss's fourth proof of the quadratic reciprocity law consists in showing it to be an easy corollary of the precise determination of the argument of the "Gauss sum" $\sum_{s=0}^{N-1} e^{2\pi i s^2/N} = \hat{\phi}_N(\chi_1)$. According to Davenport ([4], p. 14), the most satisfactory evaluation of the Gauss sums is one given by Dirichlet in 1835 and based on a straightforward application of the so-called Poisson summation formula in the theory of Fourier transforms on the line. Thus the quadratic reciprocity law itself may be regarded as resulting from an application of harmonic analysis to number theory.

Dirichlet (1805-1859) was an assiduous student of Gauss's *Disquisitiones* and apparently the first to understand some of its more obscure parts. He simplified and amplified many arguments, but above all managed to complete and extend Gauss's work in two important respects. First of all he gave the first valid proof of the fact that whenever a and m are relatively prime positive integers, then the arithmetic progression $a, a + m, a + 2m \cdots$ contains infinitely many primes. Second, he found an explicit formula for computing the number $h(D)$ of equivalence classes of binary quadratic forms of given discriminant D. The two results are closely related and depend on using limits and other concepts from analysis—in particular infinite series of the form $\sum_{n=1}^{\infty} \frac{a_n}{n^s}$ where s is > 1. Such series are now known as Dirichlet series and Dirichlet is often credited with being the founder of analytic number theory. His results were published in several installments between 1837 and 1840.

Dirichlet's proof of the existence of infinitely many primes in an arithmetic progression was inspired by Euler's proof of the existence of infinitely many primes and is at the same time a beautiful example of an application of Fourier analysis on finite commutative groups. Recall that Euler based his proof on the factorization $\sum_{n=1}^{\infty} \frac{1}{n^s} = \prod_{p} \frac{1}{1 - \frac{1}{p^s}}$ for $s > 1$. It is natural to try to copy Euler by replacing $\sum_{n=1}^{\infty} \frac{1}{n^s}$ by $\sum_{n=1}^{\infty} \frac{\theta(n)}{n^s}$ where $\theta(n)$ is 1 when n is in the given progression and zero otherwise. But $\sum_{n=1}^{\infty} \frac{\theta(n)}{n^s}$ does not factor. On the other hand it is easy to see that $\sum_{n=1}^{\infty} \frac{\chi(n)}{n^s} = \prod_{p} \frac{1}{1 - \frac{\chi(p)}{p^s}}$ whenever χ is a complex valued function on the positive integers which is strongly multiplicative in the sense that $\chi(n_1 n_2) = \chi(n_1)\chi(n_2)$ and is also such that $|\chi(n)| \leq 1$. Consider then the ring R_m of integers mod m. The elements of R_m which have multiplicative inverses form a finite commutative group under multiplication. If χ is any character on this group, we may extend it to be defined on all of R_m by making it zero where it is not already defined, and then regard it as a periodic function on the integers. Such functions are strongly multiplicative and are called *Dirichlet characters* mod m. It follows at once from finite Fourier analysis (and the fact that $\theta(n)$ is zero when n fails to have an inverse in R_m) that θ is a unique linear combination of Dirichlet characters mod m and hence that $\sum_{n=1}^{\infty} \frac{\theta(n)}{n^s}$ is a linear combination of Dirichlet series $\sum_{n=1}^{\infty} \frac{\chi(n)}{n^s}$ which factor into so-called "Euler products"

$\prod_p \dfrac{1}{1 - \dfrac{\chi(p)}{p^s}}$. Exploiting this fact, Dirichlet was able to adapt Euler's argument.

This adaptation, however, was not entirely straightforward. It depended on being able to prove that $L(s,\chi) = \sum\limits_{n=1}^{\infty} \dfrac{\chi(n)}{n^s}$ tends to a finite nonzero value as $s \to 1$ whenever $\chi \neq 1$. This can be done without great difficulty except when χ is a character of order 2. To take care of this case Dirichlet used a very ingenious argument. Let $Q_1, \cdots, Q_h$ be a complete set of inequivalent binary quadratic forms with a fixed square free discriminant D, and for each $n = 1, 2, \cdots$ let $\phi_j(n)$ denote the (suitably normalized) number of representations of n by Q_j. Then as indicated above, the function $n \to \phi(n) = \phi_1(n) + \cdots + \phi_h(n)$ is multiplicative and $\phi(p^k) = 1 + \chi(p) + \chi(p)^2 + \cdots + \chi(p)^k$ where $\chi(p) = \phi(p) - 1$. Consider now the Dirichlet series $\sum\limits_{n=1}^{\infty} \dfrac{\phi(n)}{n^s}$. The multiplicativity of ϕ shows that it factors as $\prod\limits_p \left(\sum\limits_{k=0}^{\infty} \dfrac{\phi(p^k)}{p^{ks}} \right)$

and the formula for $\phi(p^k)$ lets one replace this by

$$\prod_p \frac{1}{\left(1 - \dfrac{1}{p^s}\right)\left(1 - \dfrac{\chi(p)}{p^s}\right)} .$$

Hence

$$\sum_{n=1}^{\infty} \frac{\phi(n)}{n^s} = \left(\sum_{n=1}^{\infty} \frac{1}{n^s} \right) \left(\sum_{n=1}^{\infty} \frac{\phi_\chi(n)}{n^s} \right)$$

where $n \to \phi_\chi(n)$ is the unique strongly multiplicative function that coincides on the primes with $\chi(p) = \phi(p) - 1$. But the quadratic reciprocity law tells us that each ϕ_χ is a Dirichlet character mod m for some m, and it is easy to verify that every Dirichlet character of order 2 occurs. In other words, the function $L(s,\chi)$ whose behavior at $s = 1$ is to be investigated may be written as a quotient $\sum\limits_{n=1}^{\infty} \dfrac{\phi(n)}{n^s} \Big/ \sum\limits_{n=1}^{\infty} \dfrac{1}{n^s}$ where ϕ is as indicated above for some square free discriminant D. The behavior of $\sum\limits_{n=1}^{\infty} \dfrac{a_n}{n^s}$ at $s = 1$ is determined by the asymptotic behavior of $a_1 + a_2 + \cdots + a_n$ as $n \to \infty$. Using simple geometric arguments, Dirichlet showed that

$$\frac{\phi_j(1) + \phi_j(2) + \cdots + \phi_j(n)}{n}$$

has a limit τ as $n \to \infty$, which is the same for all j and can be computed when D is known. Thus $\dfrac{\phi(1) + \cdots + \phi(n)}{n}$ has $h\tau$ as a limit when $n \to \infty$.

Putting all of this together, he was able to deduce that $\sum\limits_{n=1}^{\infty} \dfrac{\phi(n)}{n^s} \Big/ \sum\limits_{n=1}^{\infty} \dfrac{1}{n^s}$ has a finite nonzero limit as $s \to 1$. His formula for h in terms of D was a byproduct of these constructions.

In connection with these applications of analysis to number theory, it is interesting to recall that Dirichlet was also the first to prove (1829) that a Fourier series of a function actually converges when the function satisfies suitable weak conditions. It should also be noted that passing from $n \to \phi(n)$ to the Dirichlet series $\Sigma \dfrac{\phi(n)}{n^s}$ is a form of harmonic analysis in itself, since the functions $n \to n^s$ are the restrictions to the integers of characters on the multiplicative group of all rationals. In particular the explicit formula for $\phi(n)$ that follows from its multiplicativity and the fact that $\phi(p^k) = 1 + \chi(p) + \cdots + \chi(p)^k$ is equivalent to the rather simple statement that

$$\Sigma \frac{\phi(n)}{n^s} = \prod_p \left(\frac{1}{1 - \dfrac{1}{p^s}} \right) \prod_p \left(\frac{1}{1 - \dfrac{\chi(p)}{p^s}} \right).$$

7. MATHEMATICAL PHYSICS BEFORE 1807

Mathematical physics in simple form goes back at least to Archimedes (287–212 B.C.), who formulated the laws governing the magnification of forces by levers and pulleys and the magnitudes of the forces that fluids exert on bodies immersed in them. However, except for the ideas of Copernicus (1473–1503) concerning the central position of the sun among the planets, little further progress seems to have been made until near the end of the sixteenth century when Tycho Brahe (1546–1601) made extremely accurate naked-eye observations of planetary motions, and Stevinus (1548–1626) and Galileo (1564–1642) began to study swinging pendulums, falling bodies, etc. The observations of Tycho Brahe led to Kepler's (1571–1630) empirical laws of planetary motion in 1609, and one may think of modern mathematical physics as being formally inaugurated in 1637 and 1638 with the respective publications of Descartes's (1596–1650) "Discourse on Method" and Galileo's "Two new Sciences." The first introduced analytic geometry (independently invented by Fermat) to the world, and the second clarified the foundations of mechanics. A good idea of the state of knowledge at the time can be had from the following words of Galileo: "Some superficial observations have been made, as, for instance, that the free motion of a heavy falling body is continuously accelerated but to just what extent this acceleration occurs has not yet been announced"; and "It has been observed that missiles and projectiles describe a curved path of some sort; however no one has pointed out the fact that this path is a parabola."

The work begun by Galileo was enormously advanced in the 1660s by the work of Isaac Newton (1642–1727), who showed that Kepler's laws of planetary motion and Galileo's laws of falling bodies are both consequences of a set of simple laws concerning a) motion in general, and b) the magnitude of the force attracting any two masses in the universe toward one another. To deal with the variable accelerations predicted by these laws, Newton invented the differential and integral calculus. This investion was also made (apparently independently) by Leibniz (1646–1716), and Newton acknowledges getting the idea from a method of Fermat for finding tangents to curves. Newton's *Principia*, containing a systematic account of the inventions and discoveries just alluded to, was published in 1687, half a century after the books of Descartes and Galileo. Newton also concerned himself with the theory of light and published a book on the subject in 1704, based on the hypothesis that light consists of rapidly moving particles. The opposing view, that light is a form of wave motion, was defended and developed in a book published in 1690 by Newton's slightly older contemporary Huygens (1629–1695). Newton's view was shown to be untenable in the early nineteenth century, but until then it was the view accepted by a majority of scientists.

Working out the full consequences of Newton's laws for planetary motion presents enormous mathematical difficulties, and celestial mechanics has been a source of profound and challenging mathematical problems since the appearance of the *Principia*. It will probably continue to be so for the foreseeable future. However, there is also the problem of applying similar ideas to other kinds of motion—in particular the relative motion of the parts of elastic bodies and fluids. The main steps in laying the foundations for such a continuum mechanics were taken in the mid-eighteenth century by Daniel Bernoulli (1700-1782), Euler (1707-1783), and D'Alembert (1717-1783). Bernoulli is usually considered to be the founder of fluid mechanics. His book *Hydrodynamica* appeared in 1738, originating the term *hydrodynamics*. On the other hand it is D'Alembert who is considered to be the originator of the idea of reducing problems in continuum mechanics to the study of partial differential equations, and Euler who first wrote down the system of partial differential equations governing the flow of a nonviscous (but possibly compressible) fluid. D'Alembert's study of the partial differential equation governing the motion of a vibrating string $\frac{\partial^2 f}{\partial t^2} = \mu^2 \frac{\partial^2 f}{\partial x^2}$ was published in 1747 and Euler's equations for nonviscous fluid flow came out in 1755.

If one formulates Newton's laws for gravitating "particles" in a suitable manner, they have a straightforward generalization that encompasses continuum mechanics as well. Let $x_1, y_1, z_1, \cdots, x_n, y_n, z_n$ be the coordinates in some rectangular coordinate system of n "particles." Newton's laws then assert the existence of $n + 1$ positive constants $m_1, \cdots, m_n, G$ such that when the particles move under their mutual attractions the coordinates as functions of the time satisfy the differential equations

$$m_j \, \frac{d^2 x_j}{dt^2} \;=\; -\, \frac{\partial V}{\partial x_j}$$

$$m_j \, \frac{d^2 y_j}{dt^2} \;=\; -\, \frac{\partial V}{\partial y_j}$$

$$m_j \, \frac{d^2 z_j}{dt^2} \;=\; -\, \frac{\partial V}{\partial z_j}$$

where V is the function of $x_1, \cdots, z_n$ given by the formula

$$V(x_1, \cdots, z_n) \;=\; \sum_{\substack{i,j = 1 \cdots n \\ i \neq j}} \frac{G\, m_i m_j}{\sqrt{(x_i - x_j)^2 + (y_i - y_j)^2 + (z_i - z_j)^2}} \; .$$

It is clear that replacing $m_1, \cdots, m_n, G$ *by* $\lambda m_1, \cdots, \lambda m_n, G/\lambda$ does not change the allowed trajectories, and that on the other hand the trajectories uniquely determine the products $m_1 G, m_2 G, \cdots, m_n G$. Thus the ratios of the m_j are uniquely determined and are by definition the *relative masses* of the particles. By assigning an arbitrary number as the mass of an arbitrarily chosen particle, all other particles acquire a well-defined positive mass. One says that one has chosen a unit of mass, and once this is chosen, G has a uniquely determined value called the *gravitational constant*. An important and easy consequence of the above differential equations is that the function

$$\sum_{j=1}^{n} \frac{m_j}{2}\left[\left(\frac{dx_j}{dt}\right)^2 + \left(\frac{dy_j}{dt}\right)^2 + \left(\frac{dz_j}{dt}\right)^2\right] + V(x_1, y_1, z_1, \cdots, x_n, y_n, z_n)$$

remains constant throughout time and is accordingly what is called an integral of the motion. Its value is called the *energy* of the motion on that particular trajectory. The term

$$\frac{m_j}{2}\left[\left(\frac{dx_j}{dt}\right)^2 + \left(\frac{dy_j}{dt}\right)^2 + \left(\frac{dz_j}{dt}\right)^2\right] = \frac{m_j v_j^2}{2}, \text{ where } v_j \text{ is the absolute value of}$$

the velocity of the jth particle is called the *kinetic energy* of that particle. Any increase or decrease in the total kinetic energy T is exactly balanced by a decrease or increase of the function V, which is accordingly called the *potential energy*. Notice finally that the differential equations of motion can be expressed in terms of the two functions T and V. Let $q_1, \cdots, q_{3n} = x_1,$

$y_1, z_1, \cdots, x_n, y_n, z_n, q_1, \cdots, \dot{q}_{3n} = \dfrac{dx_1}{dt}, \dfrac{dy_1}{dt}, \cdots, \dfrac{dz_n}{dt}$. Then T is a function of the $\dot{q}_j$ above and $\dfrac{\partial T}{\partial \dot{q}_j} = m_j \dot{q}_j$. Thus the equations of motion may be written in the form $\dfrac{d}{dt}\left(\dfrac{\partial T}{\partial \dot{q}_j}\right) = -\dfrac{\partial V}{\partial q_j}$.

I shall not give details here, but it is not difficult to write down an analogue of these equations when the system is a continuum so that a configuration is described by a scalar or vector-valued function on a portion of space instead of a $3n$-tuple of real numbers. There is no difficulty in defining the kinetic energy of a moving continuous distribution of matter. One simply integrates the local kinetic energy defined by the velocity distribution and mass density function. The equations of motion are thus determined as soon as the analogue of V is known. This is a numerical function defined on all possible configurations of the continuous matter in question, and it must be determined by suitable experiments in each case. Fortunately the possibilities are not as various as one might think. A fluid (liquid or gas), for example, is characterized for mechanical purposes by the fact that V depends only on the function ϱ that describes the (possibly variable) mass per unit volume and may be computed from ϱ by integrating $g(\varrho)$ over the space occupied by the fluid. Here g is a real function defined on the positive real axis and characteristic of the fluid in question. One is usually not given g directly but rather the function $\varrho \to -\varrho^2 \dfrac{dg}{d\varrho}$, whose value at any given ϱ is called the *pressure* at that density. Actually, g (or equivalently the pressure function) depends not only on the nature of the fluid but on its so-called "temperature" as well, and the whole of classical continuum mechanics is valid only insofar as temperature changes can be ignored. The beautifully subtle theory known as thermodynamics, which deals with the interplay between continuum mechanics and temperature changes, was not developed until around 1850. Indeed, the work of Joseph Black (1728-1799) clarifying the distinction between temperature and quantity of heat and introducing the concept of specific heat did not begin until 1764.

The theory of the possible potential energy functions for a solid (elasticity theory) is much more complicated than for a fluid, and even the linear approximation (valid for small displacements from equilibrium) was not adequately worked out until the 1820s. Eighteenth-century work on elasticity was by and large confined to doing special problems by ad hoc methods. On the other hand, the theory of a flexible string is like that of a one-dimensional fluid in that the potential energy is determined by a single real function relating "tension" to linear density. In terms of this function and the generalization of Newton's laws indicated above, it is not difficult to write down the equation of motion—a non-linear partial differential equation analogous to Euler's equations for nonviscous fluids. (A nonviscous fluid is one in which one can ignore the "dissipation of energy" as heat.) The equation studied by D'Alembert in 1747, $\dfrac{\partial^2 f}{\partial t^2} = \mu^2 \dfrac{\partial^2 f}{\partial x^2}$, is the approximation that results when one assumes f to be small and the tension linearly related to the density. D'Alembert discovered the rather easy argument leading to the conclusion that every solution may be written uniquely in the form

$$f(x,t) \equiv \phi(x - \mu t) + \psi(x + \mu t)$$

where ϕ and ψ are differentiable functions of one variable but are otherwise arbitrary. A year later Euler pointed out an important implication of D'Alembert's result. Since $f(x, 0) \equiv \phi(x) + \psi(x)$ and $\dfrac{\partial f}{\partial t}(x, 0) = \mu\psi(x) - \mu\phi(x)$, ϕ and ψ are uniquely determined by the configuration and its rate of change at $t = 0$. Thus the whole trajectory of the string is determined once its configuration and the rate of change of that configuration are known at $t = 0$. Five years later in 1753 Bernoulli considered the case of the string with fixed end points and length ℓ and for the first time emphasized the significance of the linearity of the equation in permitting the construction of solutions by "superposition," i.e., by taking arbitrary linear combinations of solutions already at hand. He saw in particular that $\displaystyle\sum_{n=0}^{\infty} a_n \sin\dfrac{\pi x n}{\ell}$ $\cos\dfrac{n\pi\mu}{\ell}(t - b_n)$ must be a solution for "all" choices of the real constants a_n and b_n and gave heuristic arguments indicating that every solution can be written in this form. Combined with D'Alembert's general solution, the validity of Bernoulli's argument would imply the possibility of expanding a more or less arbitrary function in the form $\displaystyle\sum_{n=0}^{\infty} a_n \sin\dfrac{\pi x n}{\ell}$. This seemed paradoxical to many mathematicians of the time; a controversy arose and Bernoulli's conclusion was rejected. As a result the systematic application of harmonic analysis to the solution of linear partial differential equations came over half a century later than it might have.

Of the linear partial differential equations arising in continuum mechanics, the first to be studied after D'Alembert's equation of the vibrating string were "Laplace's equation" $\dfrac{\partial^2 \psi}{\partial x^2} + \dfrac{\partial^2 \psi}{\partial y^2} + \dfrac{\partial^2 \psi}{\partial z^2} = 0$, and the "wave equation" $\dfrac{\partial^2 \varrho}{\partial t^2} = \mu^2\left(\dfrac{\partial^2 \varrho}{\partial x^2} + \dfrac{\partial^2 \varrho}{\partial y^2} + \dfrac{\partial^2 \varrho}{\partial z^2}\right)$. Both arise in studying special cases and approximate solutions of Euler's non-linear equations for nonviscous fluid flow and were written down by Euler in 1752 and 1759 respectively. When a fluid moves in such a way that the three components u, v, w of its velocity can be written in the form $\dfrac{\partial \psi}{\partial x}, \dfrac{\partial \psi}{\partial y}, \dfrac{\partial \psi}{\partial z}$ where ψ is a single scalar function, one says that the flow is irrotational and that ψ is the *velocity potential*. The concept and even the term occur in Bernoulli's *Hydrodynamica* of 1738. When the fluid is also incompressible so that $\dfrac{\partial u}{\partial x} + \dfrac{\partial v}{\partial y} + \dfrac{\partial w}{\partial z} \equiv 0$, one sees (as noted by Euler in 1752) that the velocity potential satisfies the equation $\dfrac{\partial^2 \psi}{\partial x^2} + \dfrac{\partial^2 \psi}{\partial y^2} + \dfrac{\partial^2 \psi}{\partial z^2} = 0$ and that many problems in fluid flow can be reduced to finding suitable solutions. It was well into the nineteenth century, however, before its theory began to be well understood. The same is true of the wave equation, which is a three-dimensional analogue of the vibrating string equation studied by D'Alembert in 1747. Euler used it to describe the small density oscillations of a fluid, such as occur for example in the propagation of sound.

In the later part of the eighteenth century, other aspects of the physical world began to be subjected to non-trivial mathematical analysis. We have already mentioned Black's work on heat and temperature. Black was also a pioneer in introducing quantitative methods into chemistry, making careful measurements of the masses involved when calcium carbonate decomposes into calcium oxide and carbon dioxide and preparing the way for the fundamental work of Priestley (1733-1804) and Lavoisier (1743-1794) between 1775 and 1785 on the nature of combustion and the law of conservation of matter. Modern chemistry is considered to have been founded with the publication of Lavoisier's book *Traité élémentaire de chimie* in 1789. In the 1750s and 1760s Michel (1724-1793), Priestley, and others began to study electrical and magnetic phenomena more quantitatively. They did experiments and made deductions that made it seem quite likely that magnetic poles and electric charges attract and repel one another with a force that is like gravitational attraction in varying inversely with the square of the distance. Then between 1785 and 1789 Coulomb published a series of memoirs reporting his own very careful measurements and convincing the scientific world of the validity of what is now known as Coulomb's law.

While Laplace's equation $\dfrac{\partial^2 \psi}{\partial x^2} + \dfrac{\partial^2 \psi}{\partial y^2} + \dfrac{\partial^2 \psi}{\partial z^2} = 0$ first appears in work of Euler in fluid mechanics, it was destined to be important in dealing with all attractive and repulsive forces varying according to an inverse square law. Lagrange in 1773 pointed out that the field of force F_x, F_y, F_z produced by the gravitational attraction of an arbitrary distribution of matter is like an irrotational velocity distribution in being realizable as the three partial derivatives of a single "potential function" V. Twelve years later in 1785 Laplace observed that it was also "solenoidal" in the sense that $\dfrac{\partial F_x}{\partial x} + \dfrac{\partial F_y}{\partial y} +$

$\dfrac{\partial F_z}{\partial z} = 0$, so that the potential V satisfies the equation that now bears his name. He overlooked the modification (later introduced by Poisson) that has to be made at points occupied by matter. Laplace (and his contemporary Legendre) were interested in computing the gravitational attraction due to bodies of various shapes and sizes and found it convenient to think in terms of the potential function V and its properties. It was in connection with this work that they introduced the functions on the surface of a sphere known as surface harmonics and expanded solutions of Laplace's equation (in polar coordinates) as a series of powers of r multiplied by surface harmonics. This expansion theory, when properly developed, permits one to show that an arbitrary continuous function on the surface of a sphere can be extended uniquely to satisfy Laplace's equation in the interior. Extending this theorem to closed surfaces of more general shape was to be a major theme of nineteenth-century mathematics and to lead to new developments of far-reaching importance. As I shall explain later, the use of expansions in surface harmonics to study Laplace's equation inside a sphere may be regarded as an early (and of course unconscious) application of non-commutative harmonic analysis.

Lagrange's observation of 1773 was only a small part of his contribution to the development of mathematical physics. He improved and extended the investigations of Bernoulli, Euler, and D'Alembert in many ways, and in competition with Laplace he made important inroads into celestial mechanics. He is credited above all, however, with completing the transition from a geometrical to an analytical point of view in dealing with mechanics and with basing the whole subject on simple general principles. His *Mécanique analytique*, published in 1787 exactly one hundred years after Newton's *Principia*, is a magnificent survey of the discoveries of Galileo, Newton, Bernoulli, Euler, D'Alembert, etc., all reworked into a coherent elegant scheme. His thoroughly analytical point of view is exemplified by his boast that the book contains no diagrams.

8. THE WORK OF FOURIER, POISSON, AND CAUCHY, AND EARLY APPLICATIONS OF HARMONIC ANALYSIS TO PHYSICS

The work of the eighteenth century on continuum mechanics was severely handicapped by a lack of systematic methods for solving or otherwise coping with the partial differential equations to which Euler and his contemporaries had been led. Only simple equations like $\dfrac{\partial^2 \psi}{\partial t^2} = \dfrac{1}{a^2} \dfrac{\partial^2 \psi}{\partial x^2}$ could be dealt with in a general way. The clue provided by the theory of this equation was missed — not only by Euler and D'Alembert as indicated in the last section, but again by Lagrange in a memoir on the theory of sound published in 1759.

The breakthrough came in 1807, when J. B. Fourier (1768-1830) submitted a long memoir on the conduction of heat to the French Academy of Sciences. Simple hypotheses and arguments led to the conclusion that the variable temperature T in a homogeneous body will satisfy a partial differential equation of the form $\frac{\partial T}{\partial t} = \mu \left(\frac{\partial^2 T}{\partial x^2} + \frac{\partial^2 T}{\partial y^2} + \frac{\partial^2 T}{\partial z^2} \right)$ where μ is a constant depending on the material of which the body is made, x, y, and z are spatial coordinates, and t is the time. With this equation as a starting point Fourier made a profound analysis of a number of problems in heat flow. Once the temperatures at the boundaries are specified, it turns out that the distribution of temperature over the body at time $t = 0$ determines this distribution at all later times, and this later distribution can be calculated from a simple algorithm.

The simplest case to discuss is that in which the body is a thin bar, insulated so that heat flows in and out only through the ends and T depends only on the coordinate x. Then the equation becomes $\frac{\partial T}{\partial t} = \mu \frac{\partial^2 T}{\partial x^2}$. If the temperatures are held fixed at A and B at the ends and these occur at $x = 0$ and $x = \ell$, we may write $T = \left(\frac{B-A}{\ell} \right) x + A + \tilde{T}$ where $\tilde{T}$ satisfies the same equation but $\tilde{T}(0, t) = \tilde{T}(\ell, t) = 0$. Now $\sin \frac{\pi n x}{\ell}$ is zero at $x = 0$ and $x = \ell$ for all $n = 1, 2, \cdots$. Thus if $g_t(x) = \tilde{T}(x, t)$ can be written as a linear combination of the functions $x \to \sin \frac{\pi n x}{\ell}$, the coefficients will depend upon t and we will have $\tilde{T}(x, t) = \sum_{n=0}^{\infty} c_n(t) \sin \frac{\pi n x}{\ell}$ where the $c_n(t)$ remain unknown. However, at least at a formal level one verifies at once that $\tilde{T}$ satisfies the partial differential equation of heat conduction if and only if $\frac{d}{dt} c_n(t) = \frac{-\mu \pi^2 n^2}{\ell^2} c_n(t)$ so that $c_n(t) = e^{(-\mu \pi^2 n^2 t)/\ell^2} c_n(0)$. This implies that $\tilde{T}(x,t) = \sum_{n=0}^{\infty} c_n(0) e^{(-\mu \pi^2 n^2 t)/\ell^2} \sin \frac{\pi n x}{\ell}$, and we need only know the constants $c_n(0)$ to know $\tilde{T}(x, t)$ for all x and t. But these constants are just the expansion coefficients of $\tilde{T}(x, 0)$ in the Fourier series expansion $\tilde{T}(x, 0) = \sum c_n(0) \sin \frac{\pi n x}{\ell}$ and can be computed from Fourier's formula

$$\int_0^\ell \tilde{T}(x, 0) \sin \frac{\pi n x}{\ell}\, dx = c_n(0) \int_0^\ell \sin^2 \frac{\pi n x}{\ell}\, dx \ .$$

This is not the way Fourier proceeded, but is perhaps the easiest way to understand the reason for the truth of Fourier's algorithm, which amounts simply to this: Expand the reduced temperature $\tilde{T}$ at $t = 0$ in a sine series, computing the coefficients as indicated. To get the temperature at time t, simply multiply the nth coefficient by $e^{(-\mu\pi^2 n^2 t)/\ell^2}$ and add up the modified sine series.

The key point of course is the assumption that a more or less arbitrary function on a finite interval can be written as the sum of a linear function and a sine series $\sum\limits_{n=1}^{\infty} c_n \sin \dfrac{n\pi x}{\ell}$. This is what Fourier insisted upon and what his eighteenth-century predecessors had refused to believe. As a matter of fact, Fourier's 1807 memoir was rejected by the French Academy as insufficiently rigorous — Lagrange was a member of the jury. On the other hand, he was encouraged to continue his admittedly very original researches and in 1812 won an academy prize for another version of the same memoir. This also was criticized for insufficient rigor and Fourier's work was not published in detail until the appearance in 1822 of his immensely influential classic *Théorie analytique de la chaleur*.

The importance of Fourier's treatise of course does not lie in its applications to solving problems in heat flow, but to the universality of the methods he employed (not to mention the influence on foundational questions and the development of set theory produced by the study of convergence and other points of rigor). There is no problem in extending the expansion technique to functions of several variables. One simply replaces $\sin\dfrac{n\pi x}{\ell}$ by $\sin\dfrac{n\pi x}{\ell_1}\sin\dfrac{m\pi y}{\ell_2}$ for two variables and the extension to more variables is obvious. More generally, the method is one that can be adapted to deal with the wave equation, with Laplace's equation, and in fact with any linear partial differential equation with constant coefficients. Not since Newton and Leibniz introduced the calculus well over a century earlier had mathematical physicists been provided with so powerful a tool. Now the partial differential equations that had accumulated (and were still accumulating) as the mathematical analysis of physical phenomena proceeded could be solved and their implications studied. It was an enormous advance.

Before saying more about what Fourier and his contemporaries accomplished, let us look briefly at how Fourier's method fits into the general scheme of harmonic analysis as outlined earlier. A function of x, y, and z which is periodic in each variable with periods ℓ_1, ℓ_2, and ℓ_3 respectively can be regarded as a function on the group obtained from the additive group of all triples of real numbers by factoring out the subgroup of all triples of the form $n_1 \ell_1$, $n_2 \ell_2$, $n_3 \ell_3$ where n_1, n_2, and n_3 are integers. The continuous characters on this group are just the functions $x, y, z \to e^{(2\pi i n_1 x)/\ell_1} e^{(2\pi i n_2 y)/\ell_2} e^{(2\pi i n_3 z)/\ell_3}$ where n_1, n_2, and n_3 are integers. Since the real and imaginary parts of $e^{(2\pi i n x)/\ell}$ are just $\cos\dfrac{2\pi n x}{\ell}$ and $\sin\dfrac{2\pi n x}{\ell}$, Fourier's theorem on expanding in a sine series is easily deducible from a theorem permitting the expansion of a more or less arbitrary functions on our quotient group in terms of characters. Functions on intervals, rectangles, etc., may be looked upon of course as restrictions of periodic functions on the line, the plane, etc., respectively. The expansion is useful in solving partial differential equations with constant coefficients for just the same reason that passing to the generating function is useful in solving the difference equations of probability theory. If one thinks of the set of coefficients as a function on the group of characters and thinks of this function as the primary unknown, the differential equation becomes an algebraic equation. This is because partial differentiation transforms into multiplication by a function. The chief difference between Fourier's application of this principle and the earlier applications in probability and number theory lie in the fact that Fourier was dealing with a continuous group.

In dealing with problems on all of space or on the whole real line, one can apply the same principles but not to a compact quotient group. One must deal with the full locally-compact group of all n-tuples of real numbers for various values of n. The most general continuous character on R^n is $x_1, \cdots, x_n \to e^{i(x_1 z_1 + \cdots + x_n z_n)}$ where the z_j are arbitrary complex numbers and these characters are bounded or equivalently of absolute value one when and only when the z_j are all real. In any event there is a whole continuum of possible characters, and infinite sums have to be replaced by integrals over this continuum. Instead of the formula $f(x) = \sum_{n=-\infty}^{\infty} c_n e^{inx}$ where $c_n = \frac{1}{2\pi} \int_0^{2\pi} f(x) e^{-inx} dx$, one has $f(x) = \int_{-\infty}^{\infty} c(y) e^{ixy} dy$ where $c(y) \equiv \frac{1}{2\pi} \int_{-\infty}^{\infty} f(x) e^{-ixy} dx$ and similarly for higher dimensions. Taking f to be real and even, these formulae reduce to $f(x) = \int_0^{\infty} 2c(y) \cos xy \, dy$ and $2c(y) = \frac{2}{\pi} \int_0^{\infty} f(x) \cos xy \, dx$, and in this form were known and used by Fourier in his 1812 prize paper.

The applications of Fourier's ideas to other branches of physics did not have to wait for the publication of his book in 1822. Poisson (1781-1840) and Cauchy (1789-1857) were twenty-six and eighteen years old respectively in 1807, when the thirty-nine-year-old Fourier submitted his memoir to the French Academy, and both began after a while to study the partial differential equations of physics and to apply the methods of harmonic analysis. Indeed a sort of three-cornered competition arose. Although Fourier had published no details, Poisson read the 1807 memoir in manuscript and published a five-page summary and review of it in 1808. Moreover, eight years later Fourier published his own summary — including his ideas on the Fourier integral. By 1816 both Poisson and Cauchy had written papers applying harmonic analysis to the solution of the wave equation in three dimensions, and in 1823 Cauchy published a paper explicitly pointing out how the Fourier transform made it possible to deal with an arbitrary linear partial differential equation with constant coefficients. In a paper published in 1817 Cauchy claims to have independently discovered the reciprocal formulae of the cosine transform

$$c(y) = \sqrt{\tfrac{2}{\pi}} \int_0^\infty f(x) \cos xy \, dx \quad f(x) = \sqrt{\tfrac{2}{\pi}} \int_0^\infty c(y) \cos xy \, dy,$$

but acknowledges the priority of Fourier. For a very full account of the whole story the reader is referred to the book *Joseph Fourier 1768-1830* [7] by I. Grattan-Guinness, which contains among other things the complete text of Fourier's 1807 memoir.

Simultaneously with this work on harmonic analysis and its applications to the wave equation, a new field of application was emerging in a revival of Huygens's ideas about the wave nature of light. Between 1801 and 1827 the work of Young (1773-1829) and Fresnel (1788-1827) with important contributions by Malus (1775-1812) and Arago (1786-1853) led to a complete overthrow of the corpuscular theory and the establishment of the wave theory as far superior in explaining known phenomena. Young's early work in explaining diffraction patterns and the colors of thin films as due to interfering waves was not well received in spite of the fact that he showed how to compute the wave length of the light from the diffraction pattern. However, in 1810 Malus accidentally passed light reflected from a window pane through a doubly refracting crystal and found that the two refracted rays were of radically different intensities. He had discovered the polarization of reflected light, and Malus investigated this phenomenon in detail. In 1816 Fresnel and Arago discovered that oppositely polarized light rays do not interfere, and Young offered an explanation based on the hypothesis that light waves are transverse. Transverse waves are waves in which the vibrations take place perpendicular to the direction of propagation. If one thinks of oppositely polarized waves as being waves in which the vibrations are perpendicular to one another as well as to the common direction of motion, one can understand the non-interference. Fresnel published an elaborate memoir based on these ideas in 1827. One great problem remained, however. If light is a wave, what is it that is waving and what are its properties? The only known examples of transverse waves occurred in the vibrations of elastic solids. Stimulated by the memoir of Fresnel, Poisson and Cauchy took up the study of the small vibrations of elastic solids. Cauchy was the first to write down the correct system of three linear second order differential equations. He did this in 1828, and in the same year Poisson analyzed this system and showed that there would be two kinds of waves, longitudinal and transverse, each with its own characteristic velocity. In the ensuing years Cauchy made three attempts to find a possible elastic solid whose transverse waves would behave like light, but none succeeded. Decades later the puzzle was solved by Maxwell's theory of the oscillations of an electromagnetic field.

9. HARMONIC ANALYSIS, SOLUTIONS BY DEFINITE INTEGRALS, AND THE THEORY OF FUNCTIONS OF A COMPLEX VARIABLE

The mapping set up by the Fourier transform $f(x_1, \cdots, x_n) \rightarrow \hat{f}(y_1, \cdots, y_n) = \int_{-\infty}^{\infty} \cdots \int_{-\infty}^{\infty} f(x_1, \cdots, x_n) e^{i(x_1 y_1 + \cdots + x_n y_n)} dx_1 \cdots dx_n$ has an important formal property which is an integrated counterpart of the fact that partial differentiation transforms into multiplication by $-i$ times the corresponding coordinate. It is the property that the product of two Fourier transforms $\hat{f}$ and $\hat{g}$ is the Fourier transform $\hat{h}$ of a function h which can be constructed from f and g by a simple integral formula and is called their *convolution.* $h(x_1, x_2, \cdots, x_n) = f * g(x_1, \cdots, x_n) = \int_{-\infty}^{\infty} \cdots \int_{-\infty}^{\infty} f(x_1 - t_1, x_2 - t, \cdots, x_n - t_n) g(t_1, t_2, \cdots, t_n) dt_1 dt_2 \cdots dt_n.$
If one solves a linear partial differential equation with constant coefficients by using the Fourier transform to turn it into an algebraic equation, it will often turn out that one has an explicit expression for the Fourier transform of the unknown function as the product of the Fourier transform of a given function and some other explicitly known function which is determined by the partial differential equation in question. Taking the inverse Fourier transform, the solution to the problem is exhibited as the convolution of the given function with the inverse Fourier transform of the function determined by the differential equation. For example, consider the problem of solving $\dfrac{\partial^2 V}{\partial x^2} + \dfrac{\partial^2 V}{\partial y^2} + \dfrac{\partial^2 V}{\partial z^2} = 4\pi\varrho$ where ϱ is known and V is to be determined. Taking Fourier transforms one has $-(u^2 + v^2 + w^2)\hat{V}(u,v,w) = 4\pi\hat{\varrho}$ so that $\hat{V} = -\dfrac{1}{u^2 + v^2 + w^2} 4\pi\hat{\varrho}$. Taking inverse Fourier transforms one finds that $V = 4\pi\varrho * g$ where g is the inverse Fourier transform of $-\dfrac{1}{u^2 + v^2 + w^2}$ and this can be computed to be $\dfrac{1}{4\pi \sqrt{x^2 + y^2 + z^2}}$. Thus

$$V(x,y,z) = \int_{-\infty}^{\infty} \int_{-\infty}^{\infty} \int_{-\infty}^{\infty} \frac{\varrho(x - x', y - y', z - z')}{\sqrt{(x')^2 + (y')^2 + (z')^2}} dx' \, dy' \, dz',$$

which may be recognized as the formula for computing the potential due to a charge or mass distribution of density ϱ. The equation it solves is Poisson's correction of Laplace's equation for points of space at which the density is not zero. Of course, these formal considerations are only valid when the functions satisfy suitable regularity and boundedness conditions. Other examples one can give include the general solution of the wave equation as a "superposition of plane waves" and the solution of the initial value problem for the heat equation as a convolution of the initial temperature distribution with the function $x,y,z \rightarrow \dfrac{1}{(2a\sqrt{\pi t})^3} e^{-(x^2 + y^2 + z^2)/4a^2 t}$.

Many of these explicit solutions can be (and originally were) obtained by other methods that do not involve the use of Fourier analysis. Poisson was particularly active in developing such integral formulae. He seems to have been the first to recognize that the existence of conductors of electricity combined with Coulomb's law presented a challenging mathematical problem: How does a given charge distribute itself on a conductor? And, more generally, given a system of a finite number of conductors with a given charge on each, how does the charge distribute itself? It is easy to reduce this problem to a purely mathematical one involving Laplace's equation, and Poisson created (mathematical) electrostatics with a long memoir on the subject published in 1812. Twelve years later he published an important memoir on magnetism showing how the inverse square law for magnetic poles and the apparent non-existence of isolated magnetic poles both follow from a theory in which the fundamental entity is a continuous distribution in space of so called "magnetic dipoles." Let $\alpha^2 + \beta^2 + \gamma^2 = 1$ and consider two magnetic poles of strengths $\dfrac{-m}{\epsilon}$ and $\dfrac{m}{\epsilon}$ located at $x_0 \mp \dfrac{\epsilon\alpha}{2}, y_0 \mp \dfrac{\epsilon\beta}{2}, z_0 \mp \dfrac{\epsilon\gamma}{2}$. Then the net magnetic field they produce has a limit as ϵ tends to zero. It is called the field of a dipole at x_0, y_0, z_0 whose dipole moment is the vector $m\alpha, m\beta, m\gamma$. From a strictly mathematical point of view, Poisson's assumption (when the field behaves suitably at ∞) is equivalent to the assertion that the magnetic field components H_x, H_y, H_z satisfy the partial differential equation $\dfrac{\partial H_x}{\partial x} + \dfrac{\partial H_y}{\partial y} + \dfrac{\partial H_z}{\partial z} = 0$ (in vector form div $\vec{H} = 0$). Indeed, introducing the vector notation $\mathrm{curl}(\vec{A}) = \left(\dfrac{\partial A_z}{\partial y} - \dfrac{\partial A_y}{\partial z}, \dfrac{\partial A_x}{\partial z} - \dfrac{\partial A_z}{\partial x}, \dfrac{\partial A_y}{\partial x} - \dfrac{\partial A_x}{\partial y}\right)$, one verifies easily that $\vec{A}$ is uniquely determined by div $\vec{A}$ and curl $\vec{A}$ provided that it goes to zero properly at ∞. Since div $\vec{H} = 0$ one can attempt to find a formula expressing H in terms of curl $\vec{H}$. This amounts to solving a system of partial differential equations of the form div $\vec{H} = 0$, curl $\vec{H} = \vec{I}$ and the method of Fourier transforms may be applied. It works and leads to a formula for $\vec{H}$ in terms of $\vec{I}$ of (vector) convolution type. This formula may be interpreted as asserting that $\vec{H}$ is the field due to a distribution of dipoles of density $\dfrac{\vec{I}}{4\pi}$. The relationship between $\vec{H}$ and $\vec{I}$ is quite analogous to that between V and ϱ in electrostatics. Indeed, just as any vector field vanishing suitably at ∞ and having zero divergence can be regarded as the magnetic field due to a continuous distribution of magnetic dipoles, any vector field with zero curl (and suitably vanishing at ∞) can be regarded as the electric field due to a continuous distribution of charge. We are dealing on the one hand with the problem of expressing $\vec{H}$ in terms of curl $\vec{H}$ given that div $H = 0$ and on the other with the problem of expressing $\vec{E}$ in terms of div $\vec{E}$ given that curl $\vec{E} = 0$.

The early nineteenth century was an exciting period in the development of physics and chemistry quite apart from the discovery of powerful mathematical methods. We have already mentioned the renaissance and development of the wave theory of light. Another advance of tremendous importance was the discovery by Oersted (1777-1851) in 1819 of a direct relationship between electricity and magnetism — a discovery made possible by Volta's invention of the electric battery in 1800. The work of Volta (1745-1827), based on still earlier work of Galvani (1737-1798) and others on "animal electricity," was of equal importance in its own right. It made steady electric currents available for the first time and through the work of Nicholson (1753-1816) and Carlisle (1768-1840) in the same year forged a fundamental link between electricity and chemistry. Nicholson and Carlisle showed that (impure) water can be decomposed into its elements by passing an electric current through it. What Oersted did in 1819 was to observe that an electric current in a wire deflects nearby compass needles. Almost immediately the quantitative aspects of this unexpected new phenomenon were under intensive investigation by a number of scientists, the most important being Biot (1774-1862), Savart (1791-1841), and Ampère (1775-1836). Ampère investigated the magnetic effects of one current on another and published a long and important memoir on electromagnetism in 1825. The basic quantitative law is usually attributed to Biot and Savart and was formulated by them in infinitesimal physical terms. An equivalent formulation in the spirit of our discussion of Poisson's theory of ordinary magnetism is that the magnetic field produced by any finite system of moving charges satisfies the partial differential equation div $\vec{H} = 0$, curl $\vec{H} = \dfrac{4\pi\vec{i}}{c}$ where $\vec{i}$ is the current density and c is a fundamental constant. One can recover the Biot-Savart law by integrating these equations and expressing $\vec{H}$ as a (vector) convolution of the appropriate function with $\vec{i}$. The mathematics is identical with that of Poisson's theory. One simply treats an "infinitesimal" current element $\vec{i}$ as a point dipole of (vector) dipole moment $\dfrac{i}{c}$. A few years later, in 1830, Faraday (1791-1867) and Henry (1797-1878) independently discovered that there is a converse relation between electricity and magnetism. A changing magnetic field produces an electric field. When this effect was quantitatively understood, one saw that the equation curl $\vec{E} = 0$ is valid only when $\vec{H}$ does not change with time, and more generally should read curl $\vec{E} = -\dfrac{1}{c}\dfrac{\partial\vec{H}}{\partial t}$. That the equation curl $\vec{H} = \dfrac{4\pi i}{c}$ has to be similarly corrected by adding $\dfrac{1}{c}\dfrac{\partial\vec{E}}{\partial t}$ to the right-hand side was recognized only a generation later (on theoretical grounds) by a man who was born just after Faraday and Henry made their discoveries. Maxwell (1831-1879) made this proposal in the 1860s, showed that it implied that both $\vec{E}$ and $\vec{H}$ satisfy the wave equation $\dfrac{1}{c^2}\dfrac{\partial^2\vec{A}}{\partial t^2} = \left(\dfrac{\partial^2\vec{A}}{\partial x^2} + \dfrac{\partial^2\vec{A}}{\partial y^2} + \dfrac{\partial^2\vec{A}}{\partial z^2}\right)$ at points of space free of charge, and was led thereby to his celebrated electromagnetic theory of light. The constant c is of course just the velocity of light — a fact which had been noted some years earlier as a curious and possibly significant coincidence.

The main point I have been trying to make in this section so far is that the use of formulas of convolution type is thinly disguised harmonic analysis and that in this disguise harmonic analysis was a key factor in the development of electricity and magnetism in the early nineteenth century as well as in the theory of heat conduction and wave propagation.

Another major development of the early nineteenth century was the founding of the theory of functions of a complex variable by Gauss and Cauchy. The use of calculations involving complex numbers, which were more or less equivalent to the Cauchy integral formula, goes back well into the eighteenth century, but it was mistrusted and not well understood. Complex numbers themselves were still regarded as rather mysterious entities in the early nineteenth century. The exact history is rather complicated and I shall not attempt to trace it. I shall rather content myself with stating that Gauss and Cauchy founded the theory in the sense that they systematized and rigorized earlier uses of the basic idea. Its formal birthdate is often taken to be 1825, the year in which Cauchy published the integral formula that bears his name, although Cauchy and Gauss both knew the result a decade earlier. Our main purpose here is to indicate briefly the very close connections of this theory with harmonic analysis.

Consider a function f which depends analytically on the complex variable z inside the disk $|z| < R$. For each r with $0 < r < R$ let f_r denote the restriction of f to the circle $|z| = r$. Expanding in a Fourier series one finds that $f_r(re^{i\theta}) = \sum_{n=-\infty}^{\infty} c_n(r) \, e^{in\theta}$, and the analyticity condition implies that $c_n(r) = c_n r^n$. Finally the continuity at the origin implies that $c_n = 0$ for $n < 0$. Thus $f(re^{i\theta}) = \sum_{n=0}^{\infty} c_n \, r^n \, e^{in\theta}$ so if $z = re^{i\theta}$, $f(z) = \sum_{n=0}^{\infty} c_n \, z^n$ and the expansibility of an analytic function in a power series is established. Of course the coefficients c_n may be determined from any f_r. Thus if $r_1 < r_2$ we have $f(r_1 e^{i\theta}) = \sum_{n=0}^{\infty} c_n \, r_1^n \, e^{in\theta}$ where $c_n r_2^n = \dfrac{1}{2\pi} \int_0^{2\pi} f(r_2 \, e^{i\phi}) \, e^{-in\phi} \, d\phi$. Hence $f(r_1 \, e^{i\theta}) = \dfrac{1}{2\pi} \sum_{n=0}^{\infty} \left(\dfrac{r_1}{r_2} e^{i\theta}\right)^n$

$$\int_0^{2\pi} f(r_2 \, e^{i\phi}) \, e^{-in\phi} \, d\phi = \frac{1}{2\pi} \int_0^{2\pi} \left(\sum_{n=0}^{\infty} \left(\frac{r_1}{r_2} e^{i(\theta-\phi)}\right)^n \right) f(r_2 \, e^{i\phi}) \, d\phi. \quad \text{But}$$

$$\sum_{n=0}^{\infty} \left(\frac{r_1}{r_2} e^{i(\theta-\phi)}\right)^n = \frac{1}{1 - \dfrac{r_1 e^{i\theta}}{r_2 e^{i\phi}}},$$

so that $f(r_1 \, e^{i\theta}) = \dfrac{1}{2\pi} \int_0^{2\pi} \dfrac{f(r_2 e^{i\phi}) \, d\phi}{1 - \dfrac{r_1 \, e^{i\theta}}{r_2 \, e^{i\phi}}}$, and writing $z = r_1 e^{i\theta}$, $\zeta = r_2 e^{i\phi}$ thus becomes $f(z) = \dfrac{1}{2\pi i} \displaystyle\int_{|\zeta| = r_2} \dfrac{f(\zeta)\,d\zeta}{\zeta - z}$, which is Cauchy's integral formula for the circle $|z| = r_2$.

With the Cauchy integral theorem on the circle and the expansibility of an analytic function in a power series as a starting point, one can deduce the basic theorems in elementary complex variable theory quite quickly and easily. Thus there is a sense in which the theory of functions of a complex variable is an aspect of Fourier analysis. From another point of view the theory of functions of a complex variable is the theory of the solutions of the Cauchy-Riemann equations $\dfrac{\partial u}{\partial x} = \dfrac{\partial v}{\partial y}$, $\dfrac{\partial u}{\partial y} = -\dfrac{\partial v}{\partial x}$, and the properties of this system can be deduced from Fourier analysis just as with the partial differential equations of electromagnetism. Indeed, thinking of v, u as a two vector, the Cauchy-Riemann equations are just the two-dimensional version of curl $\vec{E} = \operatorname{div} \vec{E} = 0$.

The argument used to deduce the Cauchy integral theorem for the circle from the theorem on expansibility of periodic functions in a Fourier series can be used in almost exactly the same way to deduce Poisson's formula expressing a harmonic function in a disk in terms of its boundary values. There is a similar formula (also due to Poisson) expressing a function harmonic in a ball in terms of its boundary values, and it is natural to ask whether this formula can be deduced in an analogous fashion. This would seem to require giving the surface of the sphere the structure of a commutative group and using an expansion theorem in terms of its characters. While it can be proved that this surface S cannot be made into a group at all in a manner consistent with its topology, there is a compact non-commutative group which acts transitively on S—namely $SO(3)$, the rotation group in three dimensions. Moreover, there is an analogue of the Fourier series proof of Poisson's formula. It differs from the Fourier series proof in using expansions in the surface harmonics of Legendre and Laplace (see section 7) instead of the $e^{in\theta}$. This suggests that there might be some relationship between surface harmonics on a sphere and complex exponentials on the circle. There is, and understanding this relationship is the key to understanding why the theory of group representations may be regarded as a natural generalization of Fourier analysis. The rotation group in two dimensions acts transitively on the unit circle with center 0, 0 and takes each function $\theta \to e^{in\theta}$ into a constant multiple of itself. Moreover, every one-dimensional invariant subspace of measurable functions on the circle is the one-dimensional subspace of all multiples of $e^{in\theta}$ for some n. The rotation group in three dimensions acts transitively on the unit sphere with center 0,0,0 but has *no* invariant one-dimensional subspace of measurable functions except for the space of constants. On the other hand it has finite-dimensional invariant subspaces, which are irreducible in the sense that no proper subspace is invariant. These are mutally orthogonal with respect to integration over the sphere and each is spanned by surface harmonics. Thus the expansion theorem in surface harmonics may be looked upon more fundamentally as an expansion theorem in terms of members of irreducible invariant subspaces. Each such subspace defines a representation of $SO(3)$ by linear transformations, and it is these irreducible representations which replace characters in dealing with a transitive action of a non-commutative group. A commutative group has only one-dimensional irreducible representations. In a sense the theory of surface harmonics of Legendre and Laplace is an anticipation of non-commutative harmonic analysis made over a century before the world was ready to appreciate it as such.

Poisson's formula expressing a harmonic function inside a ball in terms of its values on the spherical surface bounding the ball can be used to show that for every continuous function ϕ on this surface, there is a unique solution of Laplace's equation continuous at the boundary and having the values of ϕ as boundary values. The corresponding result for a rectangular parallelepiped can easily be proved using Fourier series in three variables, and it is natural to conjecture a general theorem applying to more or less arbitrary closed surfaces. The first person to state and seriously to attempt a proof of this important theorem was G. Green (1791-1841). Green did not get to a university until he was forty, but he read Poisson's 1812 paper on electrostatics and Fourier's book on heat conduction while helping his father run the family mill and bakery in Nottingham. Starting to think about things for himself, Green managed to go much further than Poisson. In 1828 he published a remarkable paper entitled "An essay on the application of mathematical analysis to the theories of electricity and magnetism." Unfortunately he was too much out of touch with the scientific world to think of sending it to a journal. He had it privately printed and circulated in Nottingham; it did not become widely known until Lord Kelvin (then William Thompson) stumbled upon it and arranged for its publication in *Crelle's Journal* in 1850. In the meantime much of its contents had been rediscovered by Gauss and others. It contained the well-known Green's identities from which one deduces the uniqueness of solutions of Laplace's equation with given boundary values and Green's famous "physical" argument for the existence of what is now called a Green's function. From its alleged existence Green could deduce a generalization of Poisson's formula and the existence of a solution of Laplace's equation with arbitrary continuous boundary values. Gauss later gave a different proof, which was unsatisfactory in another way, and what came to be known as the "Dirichlet problem" remained open for many years. The first satisfactory solution for convex regions was given by C. Neumann (1832-1925) in 1870. Work on the problem was an important stimulus to the development of analysis—especially the theory of integral equations. The theory of integral equations led in turn to the modern theory of operators in Hilbert space, which plays a central role in present day non-commutative harmonic analysis (see section 14).

10. ELLIPTIC FUNCTIONS AND EARLY APPLICATIONS OF THE THEORY OF FUNCTIONS OF A COMPLEX VARIABLE TO NUMBER THEORY

The usefulness of considering the functions of analysis for complex values of the argument became dramatically apparent in the years 1827-1829 when Abel (1802-1829) and Jacobi (1804-1851) published their memoirs, *Recherches sur les fonctions elliptiques* (in two parts) and *Fundamenta nova theoriae functionum ellipticarum* respectively. Independently (Abel slightly earlier) they had discovered and developed the consequences of the following important fact: The theory of the functions defined by the indefinite integrals that arise when one attempts to find the length of an arc of an ellipse becomes very much simpler if one a) concentrates attention on the inverse functions and b) considers complex as well as real values of the variables. These inverse functions are analytic except for poles in the whole complex plane, and moreover have two complex periods ω_1 and ω_2 which are not real multiples of one another: $f(z + \omega_1) = f(z) = f(z + \omega_2)$ for all z. Any function with these properties is called an elliptic function. Because of its two independent periods an elliptic function is uniquely determined by its values in a parallelogram with one vertex at 0, and because of its analyticity except for poles it is determined up to a multiplicative constant by the positions (and orders) of its (finitely many) zeros and poles in this parallelogram. It is thus possible to get a complete overview of all possible elliptic functions with given periods ω_1 and ω_2, and a very beautiful theory emerges.

Jacobi and Abel were interested in the dependence of their functions on ω_1 and ω_2 as well as on z and found innumerable elegant identities and relationships. In the course of such work Jacobi compared the power series expansions resulting from two different expressions for the same function and deduced the remarkable fact that $r_4(n)$, the number of representations of a positive integer n as the sum of four squares, is equal to eight multiplied by the sum of all divisors of n which are not divisible by 4. Since 1 and n count as divisors, it follows at once that *every* positive integer n can be written as a sum of four squares. This result, conjectured by Fermat, had been proved by Lagrange in 1773 after many years of unsuccessful attempts by Euler (Euler found his own proof a year later). The formula for the exact number of representations was completely new. Other identities in Jacobi's work led to similar formulae for r_2, r_6, and r_8. Of course that for r_2 was already known. Twenty years later Eisenstein (1823-1852) found purely arithmetical proofs of the formulae for r_6 and r_8 as well as similar formulae for r_5 and r_7, but the analytical proofs for r_4, r_6, and r_8 could be understood only as curious accidents until well into the twentieth century. Between 1925 and 1940 Hecke and Siegel fitted them into two (somewhat different) beautiful general theories which will be described later in this article. Both theories are developments of the theory of "modular forms" begun by Dedekind, Klein, and Hurwitz in the last quarter of the nineteenth century and, since it is not difficult, I shall give a brief account here of how modular forms relate to elliptic functions on the one hand and Jacobi's number-theoretical results on the other.

For each pair ω_1 and ω_2 of independent periods, there is a canonical elliptic function $\wp$ characterized by the fact that it has second order poles at all points $n\omega_1 + m\omega_2$ of the discrete group of all periods and no other poles, and that the principal part at each pole z_j is $\dfrac{1}{(z - z_j)^2}$. A key (and not difficult) result in the theory of elliptic functions is that every elliptic function with periods ω_1 and ω_2 is a rational function of $\wp$ and its derivative $\wp'$, and that $\wp$ satisfies a differential equation of the form $\wp'(z)^2 \equiv 4(\wp(z))^3 - g_2 \wp(z) - g_3$ where g_2 and g_3 are "constants" which of course depend upon ω_1 and ω_2. If one investigates g_2 and g_3 as functions of ω_1 and ω_2, two facts emerge very easily. Substitution of λz for z in the differential equation leads at once to the conclusion that g_2 and g_3 are homogeneous of orders -4 and -6 respectively; that is, $g_2(\lambda\omega_1, \lambda\omega_2) \equiv \dfrac{1}{\lambda^4} g_2(\omega_1, \omega_2)$ and $g_3(\lambda\omega_1, \lambda\omega_2) = \dfrac{1}{\lambda^6} g_3(\omega_1, \omega_2)$. It follows trivially that g_2 and g_3 may be written in the form $g_2(\omega_1, \omega_2) = \dfrac{1}{\omega_2^4} \phi_2\!\left(\dfrac{\omega_1}{\omega_2}\right)$ and $g_3(\omega_1, \omega_2) = \dfrac{1}{\omega_2^6} \phi_3\!\left(\dfrac{\omega_1}{\omega_2}\right)$ where ϕ_2 and ϕ_3 are functions of one variable. On the other hand it is clear that g_2 and g_3 depend only on the discrete group of all periods and not on ω_1 and ω_2 themselves. Suppose then that $\left(\begin{smallmatrix} a & b \\ c & d \end{smallmatrix}\right)$ is any two-by-two matrix of integers with $ad - bc = 1$. If $\omega'_1 = a\omega_1 + b\omega_2$ and $\omega'_2 = c\omega_1 + d\omega_2$, then ω'_1 and ω'_2 generate the same group that ω_1 and ω_2 do. Hence $g_2(\omega_1, \omega_2) = g_2(\omega'_1, \omega'_2)$ and $g_3(\omega_1, \omega_2) = g_3(\omega'_1, \omega'_2)$. Expressing g_2 and g_3 in terms of ϕ_2 and ϕ_3, one discovers immediately that ϕ_2 and ϕ_3 have the property that $\phi_2\!\left(\dfrac{az + b}{cz + d}\right) = (cz + d)^4 \phi_2(z)$ and $\phi_3\!\left(\dfrac{az + b}{cz + d}\right) = (cz + d)^6 \phi_3(z)$ for all integer matrices $\left(\begin{smallmatrix} a & b \\ c & d \end{smallmatrix}\right)$ of determinant one. The functions ϕ_2 and ϕ_3 can also be shown to be analytic throughout the upper half of the complex plane and to be bounded as z approaches infinity along the imaginary axis.

Quite generally, a function f analytic in the upper half-plane and bounded as just indicated is said to be a *modular form* of weight k (or dimension $-2k$) if $f\!\left(\dfrac{az + b}{cz + d}\right) = (cz + d)^{2k} f(z)$ for all z in the upper half-plane. The functions ϕ_2 and ϕ_3 defined by g_2 and g_3 as above were not only the first modular forms to be considered as such, but turn out to generate all the others. In the general theory of modular forms it is shown that $\phi_2{}^3$ and $\phi_3{}^2$ span the two-dimensional vector space of all modular forms of weight 6, and a member of the one-dimensional subspace vanishing at ∞ is singled out and called Δ. Given a positive integer k, the equation $2\alpha + 3\beta + 6\gamma = k$ has a finite number of solutions in non-negative integers α, β, γ. Moreover, for each solution, $\phi_2{}^\alpha \phi_3{}^\beta \Delta^\gamma$ is evidently a modular form of weight k. These particular modular forms of weight k turn out to span the vector space of all modular forms of weight k.

If f is any modular form of weight k and we choose $\left(\begin{smallmatrix} a & b \\ c & d \end{smallmatrix}\right) = \left(\begin{smallmatrix} 1 & 1 \\ 0 & 1 \end{smallmatrix}\right)$, the identity $f\left(\dfrac{az + b}{cz + d}\right) = (cz + d)^{2k}f(z)$ reduces to $f(z + 1) = f(z)$ so that f has a Fourier series expansion $\sum\limits_{n=-\infty}^{\infty} c_n e^{2\pi i n z}$ which converges in the upper half-plane. The condition on f at $i\infty$ implies that $c_n = 0$ for $n < 0$ so that $f(z)$ always has the form $\sum\limits_{n=0}^{\infty} c_n e^{2\pi i n z}$. The c_n are called the Fourier coefficients of the form and c_0 is called the constant term. The constant term c_0 may be thought of as the value of f at $i\infty$, and forms for which it is zero are called *cusp forms*. For $k = 2, 3, 4, \cdots$ there is a very simple and straightforward way of defining a modular form of weight k which is *not* a cusp form. One simply sums the so-called Eisenstein series $G_k(z) = \sum\limits_{n,m \neq 0,0}^{\infty} \dfrac{1}{(nz + m)^{2k}}$ where n and m are integers. This sum obviously satisfies the identity characteristic of modular forms of weight k, and it is not difficult to check that it has the necessary analyticity and boundedness properties. Simple arguments based on the easily established formula $\dfrac{1}{(\sin \pi z)^2} = \sum\limits_{n=-\infty}^{\infty} \dfrac{1}{(z + n)^2}$ allow one to compute the Fourier coefficients of G_k, to show that it is not a cusp form, and more significantly to show that these coefficients have simple and suggestive number theoretical properties. Indeed $G_k(z) = c_0 + \dfrac{2(-1)^k 2\pi^{2k}}{(2k - 1)!} \sum\limits_{n=1}^{\infty} a_n e^{2\pi i n z}$ where $c_0 = 2[1 + \dfrac{1}{2^{2k}} + \dfrac{1}{3^{2k}} + \dfrac{1}{4^{2k}} \cdots] = 2\zeta(2k)$ and $a_n = \Sigma\, d^{2k-1}$ where d ranges over all the divisors of n. Evidently every modular form of weight k is uniquely a sum of a cusp form and a constant multiple of the Eisenstein series of weight k. The cusp forms also have Fourier coefficients with number-theoretical properties, but these are more subtle and were not discovered until well into the twentieth century.

The genesis of Jacobi's results on sums of squares can now be understood, at least in principle, by confronting the facts about the Fourier coefficients of Eisenstein series with a remarkable connection between sums of squares and modular forms which emerges when one applies the Poisson summation formula to the function $x \to e^{-\alpha x^2}$. Let ϕ be any continuous function which goes to zero sufficiently rapidly at ∞ and let $\hat{\phi}(y) = \int_{-\infty}^{\infty} e^{ixy}\, \phi(x)\, dx$

be its Fourier transform. The Poisson summation formula is a simple consequence of elementary manipulations with Fourier series and asserts that $\sum\limits_{n=-\infty}^{\infty} \phi(n) = \sum\limits_{n=-\infty}^{\infty} \hat{\phi}(2\pi n)$. Now if $\phi(x) = e^{-ax^2}$, it is easy to compute that $\hat{\phi}(y) = \sqrt{\dfrac{\pi}{a}} e^{(-y^2)/4a}$ so that the Poisson summation formula yields $\sum\limits_{n=-\infty}^{\infty} e^{-an^2} = \sqrt{\dfrac{\pi}{a}} \sum\limits_{n=-\infty}^{\infty} e^{(-\pi^2 n^2)/a}$. This is valid for complex values of a having positive real part, and making the substitution $a = -\pi i z$ yields the identity

$$\sum_{n=-\infty}^{\infty} e^{\pi i n^2 z} = \sqrt{\frac{i}{z}} \sum_{n=-\infty}^{\infty} e^{(-\pi i n^2)/z}$$

for all z in the upper half-plane. Setting $\theta(z) = \sum\limits_{n=-\infty}^{\infty} e^{\pi i n^2 z}$, this identity may be written as $\theta(z) = \sqrt{\dfrac{i}{\pi}}\, \theta(-\dfrac{1}{z})$, a relation found by Jacobi and sometimes called the Jacobi inversion formula. Using the Jacobi inversion formula and the fact that $\theta(z+2) = \theta(z)$, it is possible to show that for $k = 1, 2, \cdots$, θ^{4k} is "almost" a modular form—specifically that it is a modular form of weight k not for the whole "modular group" of all 2×2 integer matrices $\left(\begin{smallmatrix} a & b \\ c & d \end{smallmatrix}\right)$ with $ad - bc = 1$ but for the subgroup Γ_2 of all $\left(\begin{smallmatrix} a & b \\ c & d \end{smallmatrix}\right)$ with a and d odd and b and c even. On the other hand it is an immediate consequence of the definitions that $\theta^{4k}(z) = \left(\sum\limits_{n=-\infty}^{\infty} e^{\pi i n^2 z}\right)^{4k} = \sum\limits_{n=1}^{\infty} r_{4k}(n) e^{\pi i n z}$, so that the Fourier coefficients of this "almost modular form" are the representation numbers for sums of squares. To get actual theorems like Jacobi's by this route, the theory of Eisenstein series and their Fourier coefficients has to be extended to modular forms of "higher level" as was done by Hecke in 1927.

It is interesting to compare the role of the Poisson summation formula in the above considerations with its use in establishing the quadratic reciprocity law via the determination of the sign of a Gauss sum (see section 6). It is also interesting to note that Jacobi and Dirichlet, who introduced analysis into number theory in such different ways, were almost exact contemporaries and also very good friends.

Thirty years after the appearance of Jacobi's 1829 memoir on elliptic functions, Riemann (1826-1866) published a short note giving a still different application of the theory of functions of a complex variable to number theory and providing an important connecting link between the ideas of Jacobi and those of Dirichlet. Recall that in the mid-eighteenth century Euler proved that the sum of the reciprocals of the primes diverges by considering the identity $\Pi_p \dfrac{1}{1 - \dfrac{1}{p^s}} \equiv \sum_{n=1}^{\infty} \dfrac{1}{n^s}$ for real values of the variable greater than 1, and that Dirichlet in 1837 extended Euler's ideas to the primes in an arithmetic progression. Riemann had the idea of obtaining more refined information about the distribution of the primes by considering Euler's function $\sum_{n=1}^{\infty} \dfrac{1}{n^s} = \zeta(s)$ for *complex* values of s. (The notation $\zeta(s)$ is due to Riemann.) It is evident that the series converges absolutely and uniformly in the right half-plane $\sigma > 1$ where $s = \sigma + i\tau$ and defines an analytic function there. Riemann went further, however, and showed that ζ could be continued to be analytic over the whole complex plane except for a first order pole at $s = 1$. Moreover, he found a simple functional equation connecting the values of ζ in the half-plane $\sigma > \dfrac{1}{2}$ with those in the half-plane $\sigma < \dfrac{1}{2}$. It reads $\dfrac{\zeta(s)\Gamma(s/2)}{\pi^{s/2}} = \dfrac{\zeta(1-s)\Gamma(\frac{1-s}{2})}{\pi^{(1-s)/2}}$ where Γ is Euler's well known gamma function. The gamma function is easily seen to be analytic in the entire complex plane except for simple poles at $-1, -2, -3, \cdots$ and to have no zeros. The significance of this identity is in the information it provides about the zeros and poles of ζ. The product formula $\zeta(s) = \Pi_p \dfrac{1}{1 - \dfrac{1}{p^s}}$ implies that there are no zeros or poles in the half plane $\sigma > 1$, and Riemann's functional equation allows one to conclude a) that the zeros and poles in the half plane $\sigma < 0$ are the poles and zeros of Γ and b) that the poles and zeros in the "critical strip" $0 < \sigma < 1$ are symmetrical about the center line. Thus ζ in addition to having a unique pole at $s = 1$ has no zeros outside the closure of the critical strip except for simple ones at $0, -1, -2, \cdots$. Since a function analytic except for poles (and well behaved at ∞) is almost determined by its zeros and poles, it is of interest to know more about the location of the unknown zeros inside and on the boundary of the strip. Riemann's famous unproved conjecture asserts that they all lie on the center line $\sigma = 1/2$. The much weaker result, that there are no zeros on the line $\sigma = 1$, was proved independently by Hadamard (1865-1963) and de La Vallée-Poussin (1866-1964) in 1896 and used by them to prove another of Riemann's conjectures.[2] This is the celebrated prime number theorem, which asserts that $\lim_{x \to \infty} \dfrac{\pi(x)\log x}{x} = 1$ where $\pi(x)$ is the number of primes less than or equal to x.

One of the two proofs that Riemann gave of his functional equation exhibits it as a corollary of Jacobi's inversion formula and hence of the Poisson summation formula. Moreover, the connection between the functional equation and Jacobi's formula is provided by the so-called Mellin transform. This in turn is just the analytic continuation of the Fourier transform applied to functions on the multiplicative group of all positive real numbers.

11. THE EMERGENCE OF THE GROUP CONCEPT

While a modern mathematician looking back can see the pervasive role of group theory in the mathematics of the nineteenth century, this role was quite invisible to the mathematicians themselves until very late in the century. Until the end of the 1860s, group theory was the theory of finite permutation groups and its only application was to the theory of equations. Although Feit [5] prefers to regard Cauchy as having founded group theory in 1815, one can make a case for Lagrange's having done so forty-five years earlier. While Lagrange did not have the group concept — not even that of a group of permutations — he was the first to realize the significance of the study of permutations of the roots for the theory of equations. Moreover, his long memoir on the theory of equations published in 1770 stimulated the later work of Cauchy and Galois and contained in essence the proof of what is known today as Lagrange's theorem. This is the theorem that the order of a subgroup of a finite group necessarily divides the order of the group. The main purpose of Lagrange's memoir was to study systematically the various methods that had been found for solving polynomial equations of the second, third, and fourth degrees, to understand why they worked and what stood in the way of extending these methods to equations of the fifth and higher degrees. He found that the known methods could be understood in a unified way by considering what happened to rational functions of the roots when they were permuted amongst themselves. Lagrange's analysis gave strong indications that there was a difficulty in principle in solving fifth-degree equations in the same sense that this was possible for equations of lower degree — namely by formulas involving only rational operations and the taking of roots. However, an actual proof of the impossibility of such a solution was first given by Ruffini (1765-1822) in 1813. Abel gave another proof in 1824 without knowledge of Ruffini's work. Both men had read Lagrange's paper and were influenced by his ideas. Cauchy also read Lagrange's paper but was influenced in another way. In 1815 he introduced the concept of a group of permutations and in a series of papers developed some of the elementary theory of such groups. He is responsible for the notions of subgroup, transitive group, and conjugate elements, and he made an attempt at classifying his groups.

The first man to make really fundamental progress using the group concept was E. Galois (1811-1832), who combined the ideas of Lagrange, Cauchy, and Abel to produce a beautiful general theory explaining in terms of group theory just why some equations are solvable in terms of radicals and others are not. He distinguished a certain subgroup of the group of all permutations of the roots of an equation by the symmetry properties of its elements. This group is now called the Galois group of the equation, and Galois showed that the solvability of the equation depends completely on the structure of the Galois group G. The equation is solvable if and only if there exists a family $N_0 \subset N_1 \subset N_2 \cdots N_k = G$ of subgroups such that N_j is normal in N_{j+1} for $j < k$ and the quotient group N_{j+1}/N_j is abelian. The concept of a normal subgroup and of a quotient group are due to Galois — and so is the word group. Galois wrote up his work for publication rather hurriedly just before engaging in the duel which was to take his life. His concise exposition of very original ideas was difficult for many to follow, and for this and other reasons publication was delayed for many years. The paper did not appear until 1846.

Kronecker (1823-1891) seems to have been the first to understand Galois's ideas thoroughly enough to carry them further. He published on the subject as early as 1853. His first really striking achievement, however, was his use of Galois theory to give a more perspicacious proof of an astonishing discovery of Hermite. Hermite (1822-1905) became interested in the theory of equations at an early age and, like Ruffini and Abel before him, succeeded in proving the impossibility of solving fifth-degree equations by radicals after reading Lagrange's memoir of 1770. Later he became a close student of the work of Abel and Jacobi on elliptic functions and in 1858 blended his two interests by showing that the roots of the general equation of the fifth degree could be expressed in terms of the coefficients if one used certain transcendental functions arising in the theory of elliptic functions and related to modular forms. Kronecker not only showed how to prove and understand this result, using Galois theory, but went on to study in depth the intricate relationships that exist between groups, polynomial equations, and elliptic functions.

The concept of an abstract group was formulated in 1854 by Cayley (1821-1895), but until around 1870 group theory remained a very specialized topic, known and understood by relatively few mathematicians and having the theory of equations as its only significant application. The first exposition of the theory to occur in a textbook appeared in 1866 as a section in Serret's *Cours d' Algebre Superieure*. The first book to be completely devoted to group theory was Jordan's (1838-1922) very influential *Traités des substitutions et des equations algébriques*, published in 1870. Almost simultaneously with the appearance of Jordan's comprehensive treatise, the scope of group theory began to increase dramatically as a consequence of the activities of two young mathematicians just beginning their careers. Sophus Lie (1842-1899) began in 1869 to study continuous groups with a view of doing for differential equations what Galois had done for algebraic equations. In 1872 Felix Klein (1849-1925) devoted his inaugural lecture as a professor at Erlangen to announcing his celebrated program for unifying geometry through group theory and soon was developing the ideas of Hermite and Kronecker into an elaborate study of the interplay between discrete groups (both finite and infinite) and the theory of functions of a complex variable. We have already mentioned (see section 10) Klein's connection with the theory of modular forms. The first systematic treatment of this theory occurs in the thesis (published in 1881) of Klein's student Hurwitz (1859-1919) and is based quite directly on ideas of Klein and Dedekind published in the late 1870s. The so-called modular group of all matrices $\left(\begin{smallmatrix} a & b \\ c & d \end{smallmatrix}\right)$ with integer coefficients and determinant one could now of course be recognized and dealt with as a group. Above all, however, Klein was enthusiastic about the importance, unifying power, and wide applicability of group theoretic ideas and viewpoints. Being an energetic organizer and proselytizer by temperament, he did as much as or more than Jordan's book in spreading and popularizing group theory and the group concept.

After 1872 group theory developed rapidly in many directions, and mathematicians slowly became more and more group-conscious. An early development in the latter direction was the realization that Gauss not only had been working with finite commutative groups in his composition of quadratic forms, but in effect had proved that every such group is a direct product of cyclic groups. Deeper structure theorems for finite groups began to be found, beginning with the important Sylow theorems in 1872, and by 1897 Burnside (1852-1927) was able to publish a book on finite groups alone going far beyond the 1870 book by Jordan. Lie worked out his ideas on continuous groups between 1869 and 1884, and (with the collaboration of Engel) presented them in a three-volume treatise, the final volume of which appeared in 1893. Lie's central concept — that of the Lie algebra of a continuous group — made it possible to reduce many questions about the group (more precisely about its local behavior) to purely algebraic questions about the Lie algebra. In his thesis of 1894 E. Cartan (1869-1951) corrected the work of Killing (1847-1923) published between 1888 and 1890, and thereby gave a complete classification of all "simple" Lie algebras over the complex numbers. Not quite a decade later he found the corresponding classification for real Lie algebras. Klein's earlier work on discrete groups and complex variable theory led to his celebrated book of 1884, *Vorlesungen über das Ikosadeder und die Auflösung der Gleichung vom fünften Grade,* and his two volume treatise with Fricke, *Vorlesungen über die theorie der elliptischer Modulfunktionen,* published in 1890 and 1892. In 1881 a formidable rival to Klein appeared in the person of a young Frenchman named Henri Poincaré (1854-1912) who at that time began to develop his ideas on what are now called automorphic functions. Inspired by earlier work of Fuchs (1833-1902) and Schwarz (1843-1921) on the inverse functions of solutions of second order differential equations, Poincaré was led to consider functions invariant under much more general discrete groups of two-by-two matrices than the subgroups of finite index of the modular group. Klein had been led in the same direction by his studies of Riemann's work on integrals of algebraic functions, and the final theory is a result of their combined efforts. Klein's version of the theory appeared in 1897 and 1901 in another two-volume treatise co-authored with Fricke. While the theory of automorphic functions was developing, infinite discrete groups appeared in a rather different connection. Their relationship to earlier studies in crystallography was realized, and with this application in mind Federov (1853-1919), Schoenflies (1853-1928), and Barlow (1845-1934) independently classified the so-called "space groups." A quarter of a century after Klein's famous inaugural lecture, group theory was a well established subject with lengthy treatises devoted to its various aspects.

12. INTRODUCTION TO SECTIONS 13-16

The group theoretical nature of expansions in Fourier series could not very well have been recognized until Jordan's book and Klein's missionary activities had had time to do their work, that is, until the very end of the nineteenth century. It was only in 1882 that Weber (1842-1913) formally defined the character notion for abstract finite commutative groups. Actually another three decades were required. The delay is not difficult to understand and may be attributed to the fact that (in spite of the efforts of Klein) analysts and mathematical physicists resisted thinking in group theoretical terms until well into the twentieth century. Moreover, noticing that the functions $x \rightarrow e^{inx}$ are group characters might have seemed of minor interest unless it also occurred to one to seek non-commutative analogues of these characters and so extend the scope of harmonic analysis. What actually happened was that a non-commutative extension of group characters was discovered in 1896 in a purely algebraic context. It was studied purely algebraically for over a quarter of a century and then extended to compact Lie groups. Specialization to the one-dimensional torus group made the connection obvious.

Fredholm's work on integral equations in 1900 and the introduction of the Lebesgue integral in 1902 inspired and made possible the modern L^2 theory of Fourier series and integrals as well as the spectral theory of self-adjoint and unitary operators in Hilbert space. This development took place more or less simultaneously with that described in the preceding paragraph and was destined to be blended with it in the general theory of unitary group representations, which emerged later. Although group theory played no role as such, one could see looking back that the work of Hilbert and his students on spectral theory (augmented by later work of Stone and von Neumann) was equivalent to a very thorough and complete reduction theory for unitary representations of the additive group of the real line. Moreover, when properly formulated, this reduction theory turned out to have an essentially verbatim generalization to arbitrary locally-compact commutative groups.

Before these two lines of development could be blended as indicated, they both found extensive applications to physics in connection with the marvelously effective and subtle refinement of classical mechanics known as quantum mechanics. Quantum mechanics emerged between 1924 and 1927 after a quarter-century of confused, inconsistent, and semi-successful attempts to deal with the anomalies that arose when one attempted to explain phenomena by applying classical mechanics to the atoms of which matter was supposed to be composed. It seems almost miraculous that two sets of necessary and appropriate mathematical tools (later to be seen as parts of one whole) were being independently forged just as the need for them was arising.

Quantum mechanics not only provided a rich source of applications for the mathematical developments of the three preceding decades but also inspired and strongly influenced the unified theory to which they led. Before giving details, I shall devote the next three sections to independent discussions of the three components, beginning with a section on physics and proceeding through spectral theory, etc., to non-commutative characters.

13. THERMODYNAMICS, ATOMS, STATISTICAL MECHANICS, AND THE OLD QUANTUM THEORY

One of the more striking aspects of nineteenth-century physical science is the extent to which profound relationships between apparently independent phenomena were discovered. In earlier sections I have already mentioned the relationship between chemistry and electricity, between electricity and magnetism, and between electromagnetism and light. Another and by no means the least important was suggested at the turn of the century by work of Rumford (1753-1814) and Davy (1778-1829), but only became worked up into an exact quantitative theory around 1850. This is the relationship between mechanics and heat, whose principal features constitute the first and second laws of thermodynamics. Black's distinction between temperature and quantity of heat and his concept of specific heat (see section 7) depended on the (approximately verifiable) notion that heat behaves in many respects like a fluid and that when a hot body cools down because of contact with a colder one, the quantity of heat lost by one equals the quantity gained by the other. There is in effect a law of conservation of heat. In mechanical processes where there is "friction," however, heat seems to be created out of nothing, and when one deals with expanding gases, heat seems to disappear without reappearing elsewhere. Moreover, the conservation law for mechanical energy implied by Newton's laws also seems to fail when bodies change temperature because of friction or the expansion and contraction of gases.

The first law of thermodynamics is in essence the statement that these two failures are mutually self-correcting. The quantity of heat that appears or disappears is directly proportional to the amount of mechanical energy that disappears or appears. There is a so-called "mechanical equivalent of heat" and, while neither mechanical energy nor quantity of heat is conserved by itself, there is a conservation law for the sum of the mechanical energy and the mechanical equivalent of the quantity of heat. While the existence of a well defined mechanical equivalent of heat was suggested by the experiments of Rumford and Davy, it did not become an accepted part of physics until Kelvin (1824-1907) drew attention to very careful measurements made by Joule (1818-1889) around 1840, and Helmholtz (1821-1894) published an influential paper in 1847. Helmholtz generalized the observation that heat and mechanical energy can be converted into one another, pointing out that they can be converted into other things involving electricity, chemistry, etc., as well, and proposing a universal energy conservation law. (Similar ideas were expounded five years earlier by J. R. Meyer (1814-1878), but Helmholtz was more detailed, specific, and persuasive.)

The second law of thermodynamics is more subtle and is concerned with certain limits on the possibility of converting energy in the form of heat back into mechanical energy. It has its origin in a paper on the efficiency of heat engines published in 1824 by S. Carnot (1796-1832). While Carnot had the necessary key ideas, he could not express them clearly because he did not think in terms of conservation of energy and along with Lavoisier thought of heat as an element. Carnot's paper was forgotten for a quarter of a century, but Kelvin called attention to it in 1848, and by 1850 Kelvin and Clausius (1822-1888) had combined its ideas with those of Joule and Helmholtz to formulate the second law as we understand it today.

In understanding this law it is useful to realize that it has two aspects. On the one hand it is a general principle which is a bit awkward to formulate in terms that are both sufficiently general and sufficiently precise. On the other hand it is a collection of exact laws about the properties of matter that can be deduced as consequences of this general principle. Consider for example unit mass of some gas (not assumed to be "perfect" but assumed incapable of chemical change) enclosed in a container of variable volume V. It follows from the laws of continuum mechanics (see section 7) that at each temperature T there is a function $V \rightarrow p(V, T)$ characteristic of the gas in question, and giving the pressure as a function of the variable volume. In addition it follows from the first law that there is a function U of V and T defined up to an additive constant by the formulae $C_v = \dfrac{1}{\lambda}\dfrac{\partial U}{\partial T}$ and C_p

$$= \frac{1}{\lambda}\left[\frac{\partial U}{\partial T} - \left(p + \frac{\partial U}{\partial v}\right)\left(\frac{\partial p}{\partial T} \Big/ \frac{\partial p}{\partial v}\right)\right]$$ which is called the *internal energy function*. Here C_v and C_p are the specific heats at constant volume and constant pressure respectively, and λ is the conversion factor from heat units to mechanical energy units. Until one takes the second law into account there are no restrictions on U and p. There is no *a priori* reason why there should not exist a gas having two more or less arbitrary functions as U and p. However, for every known gas there is a relationship between U and p which holds with great exactitude. This relationship consists in satisfying the partial differential equations equivalent to the existence of a function θ such that $\frac{dU + pdV}{\theta(T)}$ is an exact differential. The function θ is uniquely determined up to a multiplicative constant and is the same for all gases. One can redefine temperature so that $\theta(T) \equiv T$ and so obtain the "absolute" temperature scale introduced by Kelvin. The connection of this very useful fact about gases to the principle known as the second law of thermodynamics is as follows: If one could discover a single gas for which the relationship did not hold, one could use this gas in a manner described by Carnot to take heat energy from a lower to a higher temperature without at the same time turning any mechanical energy into heat. One could thereby turn heat energy back into mechanical energy without restriction and so construct a "perpetual motion machine of the second kind." The various formulations of the second law are equivalent ways of asserting the impossibility of such a machine.

In the case of our chemically stable gas, the exactness of $\frac{dU + pdV}{T}$ (where T is now the absolute temperature) implies the existence of a function S of V and T such that $dU + pdV = TdS$. S is determined up to an additive constant and is called the *entropy* function for the gas. Let $F = U - TS$. Then F is uniquely determined by the gas up to an arbitrary linear function of T. Moreover, straightforward calculations show that the functions U and p can be computed from F by the formulae $p = -\frac{\partial F}{\partial V}$, $U = F - T(\frac{\partial F}{\partial T})$. Thus it suffices to know the single function F to know both U and p and hence all mechanical and thermal properties of the gas. Its value at any V and T is called the *free energy* of the gas in the state defined by V and T.

Exactly the same ideas can be applied to liquids and solids and more importantly to relationships between different substances or substances in different states of aggregation; as in chemical reactions, evaporation, and melting. In every case the impossibility of a perpetual motion of the second kind can be shown to imply exact quantitative relationships between different experimentally measurable quantities and thus to cut down enormously on the amount of experimental work that has to be done. The pioneer in applying thermodynamics to chemistry was J. Willard Gibbs (1839-1903).

Throughout most of the nineteenth century it was a moot question whether matter is best conceived as a continuum or as built out of discrete point-like entities called atoms. In 1803 John Dalton (1766-1844) had shown how the laws of definite proportions and multiple proportions in chemistry could be explained in terms of atoms and molecules and introduced the concepts of atomic weight and molecular weight. While Dalton's view soon became the accepted one in chemistry, and the experiments of Faraday on electrolysis in the early 1830s suggested that there were also atoms of electricity, many scientists continued to doubt the real existence of atoms of any kind. They preferred to think of them as fictional entities useful in organizing the facts of chemistry, but otherwise not to be taken seriously. It was difficult to decide the matter experimentally because the laws of continuum mechanics were the same whether one postulated a continuum or a sufficiently fine particle structure. Indeed, a popular method for deriving the partial differential equations of motion of a continuum was to begin with an atomic model and pass to the limit as the atoms became lighter and more numerous.

The position changed radically, however, in the final decades of the nineteenth century, when serious quantitative efforts were made to explain heat as mechanical energy due to the motions of the atoms and an algorithm was found for computing the free energy function of an arbitrary gas (or other homogeneous matter) from the kinetic and potential energy functions for the mechanical system defined by its constituent atoms. This "statistical mechanics" evolved from work on the kinetic theory of gases published by Maxwell (1831-1879) in 1859 and Boltzmann (1844-1906) in 1868 and was extensively developed by Boltzmann and Gibbs. The fundamental result can be derived by several probabilistic arguments, none of which is entirely satisfactory, and can be stated very simply. Let Ω denote the "phase space" of the underlying atomic system, that is, the space of all possible position and momentum coordinates of the constituent atoms, and let H denote the real valued function on Ω giving the total energy of the system in terms of the positions and momenta of its atoms. Then the free energy of the substance as a function of V and T is $-kT \log \int \cdots_\Omega \cdot \int e^{-H/kT} dq_1 \cdots dq_{3n}\, dp_1 \cdots dp_{3n}$ where the q_j and the p_j are the position and momentum coordinates of the n atoms and k is a universal constant known as Boltzmann's constant. The dependence on V results from the dependence of H on V. One of the interesting elementary consequences of this algorithm is obtained by thinking of a solid as a set of n atoms vibrating about their equilibrium positions in some regular lattice. To the extent that the vibrations are small enough so that the equations of motion may be taken to be linear, one can conclude that the specific heat at constant volume is independent of the temperature and equal to $3nk$. This implies that the specific heat of unit mass of a substance is inversely proportional to the atomic weight — a fact discovered empirically by Dulong and Petit in 1819. It also implies that in the limit of infinitely many zero-mass atoms, the specific heat per unit mass will be infinite. The continuum limit of statistical mechanics gives absurd results it if can be said to exist at all.

While the result just described could be considered almost as an experimental verification of the existence of atoms, the disbelievers had a way out. It was possible to object to statistical mechanics on at least two grounds. First of all, the arguments leading to the algorithm for computing the free energy from H were far from compelling. Second and even more seriously, the results of this calculation often gave results in pronounced and inexplicable disagreement with experiment. For example, the specific heats of solids are not always independent of the temperature. Indeed, if one goes to low enough temperatures they *never* are. Instead, one finds a monotone increase to a constant asymptotic value and it is only at this high temperature limit that the law of Dulong and Petit applies. Similar difficulties were found with the specific heats of gases having polyatomic molecules. It was as though some degrees of freedom were frozen at low temperatures. Gibbs and Boltzmann fought a losing battle against their critics and became quite discouraged. Unfortunately they both died in the early twentieth century just before being vindicated by Einstein's application of quantum ideas to the theory of specific heats. It is rather ironic that J. J. Thomson's discovery of the atom of negative electricity at the very end of the century was particularly discouraging to Gibbs. It suggested that an atom of matter was a complex object with many degrees of freedom. Since none of these appeared to have any influence on specific heats, the ideas of Gibbs and Boltzmann seemed more inadequate than ever.

The immediate stimulus leading to the theory that was to rehabilitate statistical mechanics and revolutionize physics was an anomaly closely related to the failure of the Dulong and Petit law to hold at low temperatures. With the development of Maxwell's ideas about the identity of light waves with oscillating electric and magnetic fields (see section 9), it became clear that one could think of an electromagnetic field as a generalized oscillating continuum and apply the concepts of dynamics and even thermodynamics to its study. In particular one could try to understand the relationship between heat and light suggested by the fact that hot objects are "red hot" at one temperature and "white hot" at a higher temperature. The precise problem actually considered was that of a perfectly reflecting enclosure with walls at a fixed temperature and electromagnetic radiation being reflected back and forth across it. One could let some of the radiation escape through a small hole, pass it through a spectroscope, and study the distribution of energy across the spectrum. There is a characteristic distribution at each temperature that is independent of the material of which the enclosure is made. The problem was to explain this distribution theoretically.

Using the linearity of the equations of motion, it is not difficult to show that the equivalent dynamical system is equivalent in turn to a system consisting of a countable infinity of *independent* harmonic oscillators of natural frequencies ν_1, ν_2, $\cdots$ where the ν_j are uniquely determined by the shape and size of the container. In a state of the system in which the jth oscillator has energy E_j, the radiation with wave length between λ and $\lambda + \Delta\lambda$ will have an energy $\Sigma' E_j$ where the sum Σ' is extended over all j with ν_j between c/λ and $\dfrac{c}{\lambda + \Delta\lambda}$ and c is the velocity of light. The problem reduces to computing the E_j as a function of the temperature, and this can be dealt with (as Rayleigh [1842-1919] saw in 1900) by applying the fundamental algorithm of statistical mechanics. In fact, except for having infinitely many oscillators instead of $3n$, the dynamical system has the same structure as the one that arises in computing the specific heat of a solid made up of n atoms. One finds easily that the energy of the jth oscillator is independent of ν_j and equal to kT where k is Boltzmann's constant. Ignoring the fact that $\sum_j E_j$ is predicted to be infinite (classical statistical mechanics and classical continuum mechanics are incompatible as indicated above), one immediately deduces that the amount of energy in the part of the spectrum with wave length between λ and $\lambda + \Delta\lambda$ is just kT times the number of ν_j such that c/ν_j is between λ and $\lambda + \Delta\lambda$. When the container is a rectangular parallelepiped, a simple application of Fourier analysis permits an explicit determination of the possible ν_j and the conclusion that for a sufficiently large volume V there is an approximately continuous distribution with density $\dfrac{8\pi\nu^2}{c^3} V$. This immediately implies Rayleigh's law stating that the energy density in the spectrum at wave length λ is

$$\frac{8\pi(\frac{c}{\lambda})^2}{c^3} kTV \frac{d}{d\lambda}\left(\frac{-c}{\lambda}\right) = \frac{8\pi kTV}{\lambda^4} .$$

While Rayleigh's law was in gross disagreement[3] with experiment for small λ (experiments did *not* suggest an infinite total energy), agreement was good for large λ. More precisely agreement was good when $T\lambda$ was large. Thus the range of good agreement increased with temperature. Once again statistical mechanics seemed to be valid only at sufficiently high temperatures. A few years earlier Wien (1864-1928), with some theoretical justification, had found that one could fit the data well for small λT with an empirical formula of the form $\dfrac{A\, e^{-b/\lambda T}}{\lambda^5}$ where A and b are constants.

Quantum theory began in 1900 when Max Planck (1858-1947) found a startlingly simple but rather mysterious argument leading to a reconciliation of the contradictory formulae of Rayleigh and Wien. If one notices that Wien's formula differs from Rayleigh's only in replacing λT by $be^{-b/\lambda T}$, it is easy to guess Planck's formula. Indeed $\dfrac{1}{e^{1/x}-1}$ is very close to $e^{-1/x}$ when x is small, and very close to x when x is large. Planck's formula may be obtained from Rayleigh's by writing $\dfrac{8\pi kTV}{\lambda^4} = \dfrac{8\pi k\lambda TV}{\lambda^5}$ and replacing λT by $b/e^{b/\lambda T}-1$ where $b = \dfrac{hc}{k}$ and h is a new constant of nature introduced by Planck. This is not how Planck obtained his formula, however, and what is really significant and interesting is the physical hypothesis that led him to it.

Consider the fundamental algorithm of statistical mechanics. It may be written in the form $F = -kT \log P$ where $P = \int \cdots_\Omega \int e^{-H/kT}\, dq_1 \cdots dq_{3_n}\, dp_1 \cdots dp_{3_n}$ and $P = e^{-F/kT}$ is the so called *partition function*. The formula for P in terms of H may be rewritten as $\int_{-\infty}^{\infty} e^{-x/kT}\, d\beta(x)$ where β is the measure on the real line such that β of any interval I is the $dq_1, \cdots, dp_{3_n}$ measure of $H^{-1}(I)$. When the dynamical system is a harmonic oscillator of frequency ν, one computes easily (since Ω is two-dimensional) that $\beta(I)$ is just $\dfrac{1}{\nu}$ times the length of I when I lies in $x \geq 0$ and $\beta(-\infty,0) = 0$. Thus $P(T) = \int_0^\infty e^{-x/kT}\dfrac{dx}{\nu} = \dfrac{kT}{\nu}$. It follows at once from the thermodynamical relationship between energy and free energy that $E(T) = kT^2 \dfrac{P'(T)}{P(T)}$, so that in the case of a harmonic oscillator, $E(T) = kT$ as stated earlier. Suppose now that the measure $\beta = \dfrac{dx}{\nu}$ is replaced by a discrete approximation to it.

Choose an arbitrary number h and let us replace β by a measure β_h concentrated at and having the value h at each of the equally spaced points 0, a, $2a$, $3a$, $\cdots$. An interval with n such points will have β_h measure nh and β measure between $\dfrac{(n-1)a}{\nu}$ and $\dfrac{na}{\nu}$. Thus β and β_h will agree asymptotically if and only if $a = h\nu$. Using β_h instead of β, one finds $P_h(T) = \sum_{n=0}^{\infty} e^{-kn\nu/kT} = \dfrac{h}{1 - e^{-h\nu/kT}}$ and $E_h(T) = \dfrac{h\nu}{e^{k\nu/kT} - 1}$. Of course as h tends to zero so that β_h approaches β, $E_h(T)$ has kT as a limit as one would expect. However, if one stops short of zero at just the right value (Planck's constant) one gets a formula for $E_h(T)$ which, when substituted for kT in Rayleigh's derivation, gives the correct distribution law for all T and λ.

This is not of course how Planck proceeded.[4] He couched his argument in more physical terms. In essence, however, the argument was the one I have given. Planck did not intend to stop short at a non-zero value of h. He found it convenient to follow the time-honored custom of beginning his analysis with a discrete approximation and meant to pass to the limit of zero h. It is fortunate for physics that he was alert enough to notice that the right answer fell out before the limit was reached.

As Einstein (1879-1955) pointed out in 1907, the failure of the law of Dulong and Petit at low temperatures can be understood in exactly the same way. Replacing $\frac{kT}{\nu}$ by $1 - e^{-h\nu/kT}$ as the partition function of a harmonic oscillator of frequency ν, one finds that the specific heat of an n atom solid is not $3nk$ but $\sum_{j=1}^{3n} \frac{d}{dT} \frac{h\nu_j}{e^{h\nu_j/kT} - 1}$. Each term tends to k as T approaches ∞, but its distance from the limiting value k depends upon T/ν_j. Thus the high frequency components will be suppressed at low temperatures. Some degrees of freedom will indeed be "frozen."

In spite of these successes, the physical significance of the argument remained completely obscure. To make any sense at all out of it one had to assume that for some mysterious reason an oscillator of frequency ν could not occupy the whole continuum of energy states previously thought possible. It was restricted to energies of the form $nh\nu$ where n is a non-negative integer. It was to be a quarter of a century before this mysterious quantization was to be "understood" in the sense of being a consequence of the laws of a new mechanics — the subtle refinement of classical mechanics known as quantum mechanics. In the meantime a number of other cases were found in which it was possible to "explain" physical phenomena and derive formulae agreeing with experiment by arbitrarily "quantizing" energy or momentum. Perhaps the most striking examples are Einstein's explanation of the photo-electric effect in 1906 and Bohr's (1885-1962) explanation of the spectrum of the hydrogen atom in 1913. The body of results so obtained between 1900 and 1925 is sometimes referred to as "the old quantum theory."

14. THE LEBESGUE INTEGRAL, INTEGRAL EQUATIONS, AND THE DEVELOPMENT OF REAL AND ABSTRACT ANALYSIS

In 1900 I. Fredholm (1866-1927) announced an interesting new approach to the theory of certain linear integral equations, and in the winter of 1900-1901 this work was reported upon in the seminar of David Hilbert (1862-1943). In 1902 the thesis of H. Lebesgue (1875-1941) appeared. It contained a theory of measure and integration which came to be more or less universally accepted as the appropriate one for most purposes. Both events had their roots in the work of the early nineteenth century on methods for dealing with the partial differential equations of mathematical physics (see sections 8 and 9), and both were of fundamental importance for the future development of analysis.

Lebesgue's thesis was the culmination of a long development which I shall not attempt to trace in any detail. The failure of Fourier series to converge everywhere for continuous functions having insufficient smoothness had led to much debate and soul-searching about the true nature of functions and the best way of defining the definite integral. Well-known early studies of the question were made by Cauchy in 1823 and Riemann in 1854. The problems involved led Cantor (1845-1918) to his general theory of sets in 1884-85 and to various attempts to measure their "size." The correct way of going about this was perceived by E. Borel (1871-1956) and briefly indicated by him in 1898. Lebesgue developed Borel's ideas about measure and applied them to integration. For a full account, including the contributions of Jordan, Baire, and others, the reader may consult the recent book by Thomas Hawkins [8].

This new flexible and much more general theory of integration had a considerable impact not only on the theory of Fourier series (and integrals), but also on the theory of integral equations and probability theory. One highly unsatisfactory feature of the old theory of Fourier series was the lack of any necessary *and* sufficient conditions for a particular sequence of complex numbers to be the Fourier coefficients of a function in a certain class or vice versa. Already by 1907 this circumstance had been remedied by the beautiful and important Riesz-Fischer theorem. This asserts that the mapping $f \to \frac{1}{2\pi} \int_0^{2\pi} f(x)e^{-inx}dx$ sets up a one-to-one correspondence between *all* Lebesgue measurable complex valued functions f on the interval $[0, 2\pi]$ such that $\int_0^{2\pi} |f(x)|^2 dx < \infty$ and *all* sequences $\{c_n\}$ of complex numbers such that $\sum_{n=-\infty}^{\infty} |c_n|^2 < \infty$. In this correspondence $\sum_{n=-\infty}^{\infty} |c_n|^2 = \frac{1}{2\pi} \int_0^{\infty} |f(x)|^2 dx$, and one identifies two functions f when they differ only on a set of measure zero. The interesting part, of course, is that every square summable sequence actually arises as a sequence of Fourier coefficients for some function, and this of course could not be true without including the quite general Lebesgue integrable functions. Actually Riesz and Fischer (who wrote independent papers) proved a considerably more general theorem applying to orthogonal functions in general. Moreover, they made it clear that their result was an easy consequence of the central fact that the set of all square integrable functions is "complete" with respect to "mean convergence." Every sequence which is a Cauchy sequence in the sense of mean convergence has a square integrable limit. The analogue for Fourier integrals was formulated and proved by Plancherel and published in 1910.

A necessary and sufficient condition of equal importance but of a rather different character is due to G. Herglotz (1881-1953). If μ is any finite nonnegative measure on the interval $0 \le x \le 2\pi$ (and one identifies 0 and 2π), then one calls the complex numbers $c_n = \int_0^{2\pi} e^{-inx}d\mu(x)$ the Fourier coefficients of μ. It is trivial to verify that the sequence $n \to c_n$ is *positive definite* in the sense that $\sum_{n,m} c_{n-m}Z_n\overline{Z}_m \ge 0$ for every finite sequence $Z_{-\ell}, Z_{-\ell+1}, \cdots, Z_0, Z_1, \cdots, Z_{\ell'}$ of complex numbers. Conversely, as shown by Herglotz in 1911, every positive definite sequence is the sequence of Fourier coefficients for a unique positive measure μ. The analogue for Fourier integrals was first stated and proved twenty-one years later by Bochner (1899—).

Fredholm's work on integral equations was stimulated by slightly earlier work of Volterra (1860-1940), and this work in turn can be understood as the result of looking at Neumann's solution of the Dirichlet problem in 1870 (see section 9) from a new point of view. This new point of view is of central importance in modern abstract analysis (functional analysis) and has become so familiar that it is now hard to believe that it was startlingly new in 1887. In that year Volterra began to publish a series of papers systematically developing a theory of functions in which the arguments need not be k-tuples of numbers but may be other functions. In particular, he introduced and stressed the point of view that when one performs an operation on a function which leads to a number (e.g., when one evaluates a direct integral), this is analogous to substituting a number in a formula to get another number. He called such functions "functions of lines" — a term soon replaced by "functional." A closely related step, which became part of the same program, was to think of the operations of analysis such as differentiation as again analogues of functions. In other words, one can have generalized functions in which the ranges as well as the domains are sets of functions. These Volterra called *operations* or *operators*.

Neumann had solved the Dirichlet problem by way of a preliminary reduction to a linear integral equation — for functions on the surface of the region in question. Then in 1896 Volterra studied a class of integral equations that could be solved by the same method. Moreover, he interpreted and clarified the method using his operator point of view. Let $T_K(f)(y) = \int_a^y K(x, y)f(x)dx$ where K is a continuous function of two variables and a is a fixed constant. Then to solve the integral equation

$$f(y) = \varphi(y) - \int_a^y \varphi(x)K(x, y)dx$$

is to solve the operator equation $f = \varphi - T_K(\varphi) = (I - T_K)\varphi$, and one has $(I - T_K)(I + T_K + T_K^2 + \cdots) = I$ where I is the identity. T_K^n is an integral operator defined by a kernel K_n, and $T_K + T_K^2 + \cdots$ makes sense as the integral operator defined by $K_1 + K_2 + \cdots$. Thus $\varphi = (I + T_K + T_K^2 + \cdots)f = f + T_{K_1}(f) + T_{K_2}(f) + \cdots$. Fredholm showed how to use analytic continuation to deal with more general cases in which $I + T_K + T_K^2 + \cdots$ does not converge. The idea is that $I - \lambda T_K$ will fail to have an inverse only at the zeros of a certain entire function δ, and $(I - \lambda T_K)^{-1}\delta(\lambda)$ can be developed as an everywhere convergent power series in λ.

Fredholm's results were quite beautiful and interesting, but there is little doubt that their greatest importance lies in the fact that they inspired Hilbert to spend a decade making an intensive study of linear integral operators. Hilbert's work on the subject was first published in a series of six articles between 1904 and 1910, and then again as a book *Grundzüge einer allgemeinen Theorie der Linearen Integralgleichungen,* which appeared in 1912. Hilbert concerned himself above all with the existence of eigenfunctions and eigenvalues for his integral operators, that is, with functions f and constants λ such that $T(f) = \lambda f$ for the operators T. Moreover, he took an abstract algebraic approach, thinking of an integral operator as an infinite analogue of a matrix, and introduced the "Hilbert space" of all infinite sequences $c_1, c_2, \cdots$ of complex numbers such that $|c_1|^2 + |c_2|^2 + \cdots < \infty$ as the corresponding analogue of the space of all n-tuples of complex numbers. He made the connection with function spaces via expansions in systems of orthogonal functions and thereby inspired the work of Riesz and Fischer discussed above.

In the purely algebraic case, an $n \times n$ matrix a_{ij} with $a_{ij} = \overline{a}_{ji}$ was known to be "diagonalizable" in the sense that there exists a basis of mutually orthogonal eigenvectors for the operator defined by the matrix. Hilbert set out to generalize this result to infinite matrices in Hilbert spaces working with the quadratic form $\Sigma \, a_{ij} x_i \overline{x}_j$ instead of the matrix itself. Assuming $a_{ij} = \overline{a}_{ji}$ and a strong continuity property called complete continuity, he was able to prove the obvious analogue of the algebraic result. Since many integral operators can be shown to be completely continuous, Hilbert's theorem had wide applicability to the differential operators that occur in mathematical physics. On the other hand, complete continuity is too strong a requirement for many purposes, and Hilbert's most striking achievement in this area is his formulation and proof (in the special case of a bounded self-adjoint operator) of what is today called the spectral theorem. From Hilbert's point of view, he found the analogue for a bounded quadratic form in infinitely many variables of the classical theorem asserting that a linear change of coordinates reduces every quadratic form to a linear combination of squares. The difficulty produced by dropping the hypothesis of complete continuity is that sums have to be replaced by integrals. Anticipating later concepts one can say that, in effect, Hilbert showed that every bounded self-adjoint operator in Hilbert space can be decomposed as a "continuous direct sum" or "direct integral" of constant operators. The replacement of sums by integrals makes it difficult to attach a meaning to such concepts as the "multiplicity of occurrence" of an eigenvector and so to discuss "equivalence" of operators or forms. This difficulty can be met, and a preliminary version of the resulting "spectral multiplicity theory" appeared in the 1907 thesis of Hilbert's student Hellinger (1883-1950). Hellinger published an improved version in 1909, and Hahn (1879-1934) showed in 1911 that further improvements and simplifications could be made by making systematic use of the theory of the Lebesgue integral. One usually speaks today of the Hahn-Hellinger theory. The modern theory of unitary group representations, which contains classical harmonic analysis as a very special case, may be regarded as a sort of blend of the spectral theorem combined with the

Hahn-Hellinger theory on the one hand and the group representation theory of Frobenius and Schur (see section 15) on the other. Thus this work of Hilbert, Hellinger, and Hahn is of the greatest importance for our main theme. I shall formulate their results more precisely in section 16.

The work of Hilbert and Lebesgue described above led naturally to a very fruitful approach closely allied to the operator point of view advocated and developed by Volterra twenty years earlier. This approach consists in looking at sets of functions as infinite-dimensional analogues of Euclidean space with individual functions as "points." One can then think of convergence of functions in geometrical terms and apply one's geometric intuition to get new insights. The concept of an abstract metric space was introduced by Fréchet (1878-1973) in 1906. The following year Fréchet and E. Schmidt introduced geometric language into the study of Hilbert's sequence space, and Fréchet and F. Riesz (1880-1956) observed that the same language could be applied to the space of square summable functions. Indeed, the Riesz-Fischer theorem implied that the two spaces were isomorphic. Between 1907 and 1918 this geometric point of view was applied to a number of concrete function spaces different from Hilbert space by several mathematicians, among whom the undisputed leader was F. Riesz. The modern axiomatic approach was started in 1922 by a paper of S. Banach (1892-1945).

Concurrently with the discovery of the Riesz-Fischer theorem and the rise of the function space concept, another less abstract branch of analysis was developing out of applications of "summability" methods to the study of divergent Fourier series. The main initial stimulus was a paper by Fejér (1880-1957) published in 1904. Let S_n be the sum of the first n terms of the Fourier series of a function f. Fejér proved that $\dfrac{S_1 + \cdots S_n}{n}$ converges uniformly to f whenever f is continuous, and in 1905 Lebesgue proved that if f is measurable and such that $\int |f(x)|\, dx < \infty$, then $\dfrac{S_1(x) + \cdots + S_n(x)}{n}$ converges to $f(x)$ for all x except for a set of measure zero. Two years after that, Fatou (1878-1929) proved an analogue of Lebesgue's theorem using "Abel summability" instead of the "Cesarò summability" of Fejér and Lebesgue. In Abel summability one replaces

$$\frac{S_1 + S_2 + \cdots + S_n}{n} = \frac{na_1 + (n-1)a_2 + \cdots + a_n}{n}$$

where $S_n = a_1 + \cdots + a_n$ by $a_1 + ra_2 + \cdots$ where $0 < r < 1$ and calls $\lim_{r \to 1} (a_1 + ra_2 + \cdots)$ the Abel sum if it exists. For Fourier series of the form $\sum_{n=0}^{\infty} c_n e^{in\theta}$, the Abel sum is $\lim_{r \to 1} \sum_{n=0}^{\infty} c_n r^n e^{in\theta} = \lim_{r \to 1} \sum_{n=0}^{\infty} c_n z^n$ where $z = re^{i\theta}$. Thus Fatou's theorem implies that a measurable function on the circle whose absolute value has a finite Lebesgue integral and whose negative Fourier coefficients are zero is the set of boundary values of a function of a complex variable analytic in $|z| < 1$. A closely related corollary connects harmonic functions in the unit circle with arbitrary measurable, absolutely integrable boundary functions. An important new element was introduced in 1909 and 1910 when Hardy (1877-1947) and Littlewood (1885-1977) began to study what came to be called Tauberian theorems. These are theorems allowing one to deduce the convergence of a series from its summability in various senses when appropriate auxiliary conditions are satisfied. Such a theorem was proved by Tauber in 1897 under rather strong conditions. In the decade or so following 1910, Hardy and Littlewood collaborated in a series of papers that improved Tauber's theorem in a number of non-trivial ways.

These Tauberian theorems proved by Hardy and Littlewood (with contributions from other mathematicians as well) turned out to be of central importance in applying Fourier analysis to number theory in a new way. Let $n \to \varphi(n)$ be a complex-valued function on the integers. The most general character on the additive group of all the integers is $n \to z^n = r^n e^{in\theta}$ where $z \neq 0$ and the "Fourier transform" of φ is $\sum_{n=-\infty}^{\infty} \varphi(n)z^n$. When $|z| = 1$ so that the corresponding character is unitary, this reduces to $\sum_{n=-\infty}^{\infty} \varphi(n)e^{in\theta}$, the function whose Fourier coefficients are the $\varphi(n)$. Of course this latter function will not exist unless $\varphi(n) \to 0$ as $|n| \to \infty$. On the other hand, if $\varphi(n) = 0$ for $n < 0$ and $\varphi(n)$ is not too badly unbounded as $n \to \infty$, then $\sum_{n=-\infty}^{\infty} \varphi(n)z^n = \sum_{n=0}^{\infty} \varphi(n)r^n e^{in\theta}$ will be defined and analytic for all z with $|z| < 1$, and this analytic function may be regarded as an analytic continuation of the "non-existent" function $\sum_{n=0}^{\infty} \varphi(n)e^{in\theta}$. Moreover, when $\sum_{n=0}^{\infty} \varphi(n)e^{int}$ does exist, studying Abel summability and Tauberian theorems amounts to studying the relationship of $z \to \sum_{n=0}^{\infty} \varphi(n)z^n$ to its boundary values. More generally, one can use similar methods to relate the asymptotic behavior of $\sum_{n=0}^{\infty} \varphi(n)z^n$ as $|z| \to 1$ to the behavior of $\varphi(n)$ for large n. For various functions φ of number-theoretical interest, such as the number of partitions of n, one can show by direct arguments that the "dominant" singularities of $z \to \sum_{n=0}^{\infty} \varphi(n)z^n$ as $|z| \to 1$ are at points of the form $e^{(2\pi ip)/q}$ where p and q are integers. Moreover, one can obtain quite precise information about the way in which $\sum_{n=0}^{\infty} \varphi(n)r^n e^{(2\pi ipn)/q}$ approaches ∞ as r tends to 1. Using this information, Hardy and his collaborators were able to apply Cauchy's theorem, Tauberian theorems, and delicate estimates to obtain useful asymptotic formulae for $\varphi(n)$. The method is now known as the circle method. It was used first by Hardy and Ramanujan (1887-1920) to study the number of partitions of n—the results being published between 1917 and 1919. Slightly later Hardy and Littlewood used it to study Waring's problem and in particular to obtain asymptotic formulae for the number of representations of n as a sum of a fixed number r of integer kth powers. They also showed that the prime number theorem of Hadamard and de la Vallée Poussin (see section 10) is deducible from a Tauberian theorem.

In 1914 Hardy introduced and studied a new class of functions—called H^p functions in his honor. If $p \geq 1$, the class H^p consists of all analytic functions in $|z| < 1$ have the property that $\int |f(re^{i\theta})|^p d\theta$ is bounded as a function of r. The study of this class was continued by F. and M. Riesz and led to a certain blending of function space ideas with those of the more concrete analysts such as Hardy and Littlewood.

15. GROUP REPRESENTATIONS AND THEIR CHARACTERS

In 1881 Weber defined a character of a finite commutative group G to be a complex valued function χ on G such that $\chi(xy) = \chi(x)\chi(y)$ for all x and y in G. This definition was an abstract generalization of one given three years earlier by Dedekind in connection with his work on algebraic number theory, which was inspired in turn by early work of Gauss and Dirichlet (see sections 6 and 12). While Weber's definition makes sense for arbitrary finite groups, it is more or less vacuous except insofar as the group has commutative aspects. Specifically, every character is identically one on the commutator subgroup and consequently the only characters not identically one are derived trivially from characters of commutative quotient groups. Group theory acquired a powerful new tool that was soon to become almost indispensable when G. Frobenius (1849-1917) published a paper in 1896 showing that there is a natural generalization of the character notion that involves the whole group G in a significant and interesting way—even when G is non-commutative. Considering the impact that this generalization was to have on group theory, it is interesting to note that Frobenius and Klein were born in the same year. It is even more interesting that the new definition was more or less directly inspired by Dedekind and the needs of this work on algebraic number theory.

After Dirichlet's work of 1837-1840, the theory of binary quadratic forms was generalized in two different directions. On the one hand, the preliminary work of Legendre and Gauss on ternary quadratic forms developed into a general theory of quadratic forms in n variables in the hands of Eisenstein (1823-1852), Hermite (1822-1905), and H. J. S. Smith (1826-1883). On the other hand, attempts to prove Fermat's "last theorem" and to generalize the quadratic reciprocity law led first Kummer (1810-1893) and then Kronecker (1823-1891) and Dedekind (1831-1916) to develop the theory of algebraic number fields. The theory of binary quadratic forms is more or less equivalent to the special case of the latter theory in which the field is generated over the rationals by a root of a quadratic equation with integer coefficients. In that case the group of automorphisms of the field (the Galois group) is of order two and one can deal with it without thinking in group theoretic terms. More generally, a theory as complete as that of Gauss and Dirichlet is not available even today (except when the Galois group is commutative). It was in working out aspects of this still incomplete theory (see section 19) that Dedekind was led to the problem that inspired Frobenius to introduce his "higher dimensional characters." Let G be a finite group of order h and let $g_1 \cdots g_h$ be the elements. Let $x_{g_1} \cdots x_{g_h}$ be h independent variables parameterized by the elements of G and let $\theta(x_1 \cdots x_h)$ denote the determinant of the matrix $||x_{g_i g_j^{-1}}||$. Then θ is a polynomial in h variables, which Dedekind called the group determinant. In a letter written to Frobenius in 1896 (and published in Dedekind's collected works), Dedekind states that many years earlier (around 1880) he had been led to study the group determinant through a consideration of the discriminant of an algebraic number field. He had soon discovered the interesting fact that θ factorizes into linear factors parameterized by the characters of G whenever G is commutative, and he had also factorized θ for various special non-commutative groups. But his attempts to generalize his theorem about commutative groups to general non-commutative groups had failed, and one purpose of his letter was to interest Frobenius in the problem (see Thomas Hawkins's article in this volume).

Frobenius's response was prompt and effective. A correspondence ensued, and before the end of the year, Frobenius had published a paper on the theory of his new characters and another applying them to the solution of Dedekind's problem. Each Frobenius character χ has a *degree* equal to its value at the identity element e, and it turns out that the *distinct* irreducible factors of θ are parameterized by the Frobenius characters. Each factor has a degree equal to the degree of the corresponding Frobenius character and occurs with a multiplicity equal to this degree.

Frobenius's original definition of character was a complicated one, which emerged from his analysis of Dedekind's problem. A year later, however, he showed that his definition is equivalent to another that is much simpler and more natural. Let us define an n-dimensional *representation* of the group G to be a homomorphism L of G into the group of all $n \times n$ complex matrices of non-zero determinant. Let us define L to be reducible if a change of basis can be made which throws all matrices L_x simultaneously into the form $\left(\begin{pmatrix} A_x & 0 \\ B_x & C_x \end{pmatrix}\right)$ and let us define L to be irreducible if it is not reducible. For each representation L of G one obtains a complex valued function χ^L on G by setting $\chi^L(x) = \text{Trace}(L_x)$. χ^L is called the *character* of L. The classical characters of Dedekind and Weber are just the characters of the *one-dimensional* representations of G, and the new characters introduced by Frobenius are just the characters of the other irreducible representations of G. When the representation L is reducible as explained above, it is clear that $x \to A_x$ and $x \to C_x$ are also representations, and that $\chi^L(x) = \chi^A(x) + \chi^C(x)$. It follows that the character χ^L of any representation L is a sum of a finite number of characters of irreducible representations and hence of characters in the new sense introduced by Frobenius. It will be convenient to adopt the following more or less standard terminology. A character in the sense of Dedekind and Weber is a *one-dimensional character*. The character of an irreducible representation is an *irreducible character*. A finite sum of irreducible characters, or equivalently the character of a (possibly reducible) representation, is a *character*. It follows at once from the definition that the characters of two representations are equal whenever one representation can be obtained from the other by a change of basis, and that every character is a constant on the conjugate classes of the group. Since Frobenius was able to prove that any two distinct irreducible characters χ_1 and χ_2 are *orthogonal* in the sense that $\sum_{x \epsilon G} \chi_1(x)\overline{\chi_2(x)} = 0$, it follows that the irreducible characters are linearly independent and hence that there can be only finitely many of them. Indeed, it follows that there can be no more than h where h is the number of conjugate classes. Frobenius proved further that there are exactly h, and succeeded in his first paper in determining all of them for several different non-commutative groups.

Burnside (1852-1927) became interested in Frobenius's new theory almost immediately and began to publish papers on the subject in 1898. He found different proofs of Frobenius's main results and was a pioneer in emphasizing the advantages of taking representations rather than their characters as the basic objects of the theory. Frobenius's student Schur (1875-1941) saw things in this way also. In 1905 he published a systematic account of the whole theory from the representation theory point of view. In the theory of representations a key role is played by a theorem discovered by Maschke in a slightly different context and published by him in 1899. It asserts that every reducible representation is actually *completely reducible* in the sense that the basis may be chosen so that the matrices take the form $\left(\begin{pmatrix} A_x & 0 \\ 0 & C_x \end{pmatrix}\right)$.

The theory of representations takes a more perspicuous form if one avoids a choice of basis and thinks in terms of abstract linear transformations. A representation L of G is then a homomorphism $x \to L_x$ of G into the group of all non-singular linear transformations of some finite-dimensional complex vector space $V(L)$, and two representations L and M are equivalent if there exists a non-singular linear transformation T from $V(L)$ onto $V(M)$ such that $TL_xT^{-1} = M_x$ for all x. L is reducible if there exists a proper subspace V_1 of $V(L)$ such that $L_x(V_1) = V_1$ for all x. The restriction of the L_x to V_1 defines a *subrepresentation* L^{V_1} whose space is V_1. Maschke's argument shows that for every such subspace V_1 there is another V_2 with $V_1 + V_2 = V(L)$ and $V_1 \cap V_2 = 0$. Thus in an obvious sense L is the "direct sum" of L^{V_1} and L^{V_2}. Iterating this procedure, one shows that every representation L is equivalent to a direct sum $M^1 \oplus M^2 \oplus \cdots \oplus M^k$ where the M^j are irreducible. Moreover, it is not hard to show that this direct sum decomposition is essentially unique in the sense that if also $L \simeq N^1 \oplus N^2 \oplus \cdots \oplus N^\ell$ where the N^j are irreducible, then $\ell = k$ and there exists a permutation π such that M^j and $N^{\pi(j)}$ are equivalent. Thus one knows the most general representation of G to within equivalence when one knows the most general irreducible representation of G to within equivalence. Since it can be shown that two representations are equivalent if and only if their characters are equal, it follows that there are only finitely many equivalence classes of irreducible representations. A representation of particular interest is the so-called regular representation. Its space is the space of all complex-valued functions on G, and one defines the representation by translation: $L_x(f)(y) = f(yx)$. A fundamental theorem asserts that the regular representation is equivalent to a direct sum in which each equivalence class of irreducibles occurs, and occurs with a multiplicity equal to the degree of its character; that is, the dimension of the representation space. It is suggestive to compare this fact with Frobenius's theorem on the factorization of group determinants.

There is an alternative route to the decomposition theory of group representations which takes its origin in the theory of algebras (or hypercomplex number systems as they used to be called). Influenced by the work of Killing cited in section 11, Molien (1861-1941) published two papers in 1893 giving the first deep and general theorems about the structure of associative algebras over the complex field. He introduced the notions of simplicity and semi-simplicity, and more or less proved the celebrated Wedderburn structure theorems in the complex case. (Similar results were obtained independently but slightly later by Cartan.) Then in 1897 (and apparently without knowledge of Frobenius's work) Molien applied his ideas to a particular algebra that one can associate with a finite group—the so-called group algebra—and obtained a number of Frobenius's more important results. For more particulars about the relationship between the work of Frobenius, Molien, and Burnside, as well as a detailed analysis of the correspondence between Dedekind and Frobenius and how Frobenius was led to invent characters, the reader is referred to three excellent articles by Thomas Hawkins [*9, 10,* and *11*]. I am indebted to these articles for many of the historical facts stated in this section.

In addition to solving Dedekind's problems and opening the door to a far-reaching extension of the method of harmonic analysis, Frobenius's discovery of non-one-dimensional irreducible characters provided group theory itself both with a powerful new tool and with a fascinating and difficult new problem. In 1900 Burnside published two papers using characters to prove new theorems about the structure of finite groups, and shortly thereafter Frobenius did likewise. Then in 1904 Burnside used characters to prove the very striking theorem that any group whose order is of the form $p^\alpha q^\beta$ where p and q are primes is necessarily solvable. For well over half a century no proof not using characters was known, and even today the character proof is by far the simplest. The new problem is that of actually finding the irreducible representations and their characters for particular finite groups. This problem is easily solved in some cases but quite difficult and challenging in others. In the special case of the so-called "Chevalley groups" (analogues of semi-simple Lie groups in which the real and complex fields are replaced by finite fields), study of this problem is a field of research of considerable current interest and one in which important progress has recently been made.

In studying the problem of finding the irreducible representations and characters of a finite group G, it is useful to consider the relationship of the representations of G to those of its various subgroups. Already by 1898 Frobenius had published a paper on the subject. In this paper he introduced the very important concept of an *induced character*. Let χ be an arbitrary character of the subgroup H of the group G and let χ^0 be the function on G which agrees with χ on H and is otherwise zero. Then define χ^* on G by the formula $\chi^*(x) = \dfrac{1}{o(H)} \sum_{y \epsilon G} \chi^0(yxy^{-1})$. Frobenius proved that χ^* is always a character and called it the character of G induced by the character χ of H. While χ^* need not be irreducible when χ is, it is irreducible in many cases, and inducing is one of the most important ways of constructing non-one-dimensional irreducible characters. For nilpotent groups *every* irreducible character which is not one-dimensional is induced by a one-dimensional character of a suitable subgroup, and for many non-commutative groups a significant fraction of their irreducible characters may be so obtained. Further insight into the nature of the inducing process may be obtained by considering the celebrated Frobenius reciprocity theorem. Let χ_1 and χ_2 be irreducible characters of G and a subgroup H respectively. Then χ_2^* expressed in terms of the irreducible characters of G contains χ_1 exactly as many times as the restriction of χ_1 to H contains χ_2. One verifies easily that the character of the regular representation of G is equal to the character induced by the unique irreducible character of the subgroup $\{e\}$ consisting of the identity alone. The Frobenius reciprocity theorem applied to this case yields at once the facts about the structure of the regular representation stated earlier. Another important elementary fact relating characters of subgroups to characters of groups is concerned with product groups. If $G = G_1 \times G_2$ and

χ_1 and χ_2 are characters of G_1 and G_2 respectively, then $x, y \rightarrow \chi_1(x)\chi_2(y)$ is a character of G which is irreducible if and only if χ_1 and χ_2 are both irreducible. Moreover, every irreducible character of G can be so obtained by composition from characters of G_1 and G_2 respectively.

That characters and group representations might have something to do with Fourier analysis seems to have first been recognized by Hermann Weyl (1885-1955) in 1927. But an essential first step was taken by Schur in 1924. Because of connections with the branch of algebraic geometry known as "invariant theory," Schur became interested in studying representations of the rotation group in n dimensions and discovered that he could carry over the main features of the character and representation theory of finite groups if he replaced summation over the elements of a finite group by a suitable integration over the compact manifold constituted by the elements of the rotation group. Hurwitz had made use of such an integration in 1897 in a method he discovered for constructing invariants. Schur adapted Hurwitz's integral to his needs. From a modern point of view, Schur and Hurwitz made use of the fact (proved by A. Haar in 1933) that every separable locally-compact group admits a measure (unique up to a multiplicative constant) that is defined on all Borel sets, is finite on compact sets, is invariant under right translation, and is not identically zero. When the group is a Lie group, the existence of this measure can be established easily using concepts from differential geometry. Using integration with respect to "Haar measure" to replace sums over the group elements, Schur was able to carry over Maschke's argument and prove the decomposability into irreducibles of an arbitrary representation of the rotation group. He was able to show also that an irreducible representation is determined to within equivalence by its character and to find all irreducible representations together with their characters for the groups with which he concerned himself.

Weyl had been informed of Schur's results in advance of publication. In the same year he published excerpts of a letter to Schur explaining how his results could be generalized to arbitrary semi-simple Lie groups by making use of work of Cartan that had appeared eleven years earlier in 1913. Cartan, working with Lie algebras, had solved the analogous infinitesimal problem. Weyl worked out his ideas in detail and published them in three papers which appeared in 1925 and 1926. Then in 1927 Weyl took the crucial step toward relating characters and representations to Fourier analysis.[5] In that year, in collaboration with his student Peter, he published a proof of the celebrated Peter-Weyl theorem. If L is an irreducible representation of any group, then the vector space spanned by the matrix elements with respect to a basis is independent of the basis chosen and is invariant under right and left translations. Moreover, the vector spaces so obtained from inequivalent representations are mutually orthogonal with respect to "Haar measure" when the group is a compact Lie group. In one formulation the

Peter-Weyl theorem asserts that for every compact Lie group, the linear span of these finite-dimensional orthogonal subspaces is uniformly dense in the space of continuous functions on the group. In another it asserts that one obtains a complete system of orthogonal functions for the group by choosing an orthogonal basis in each subspace. Specialized to the case of the one-dimensional torus group, the Peter-Weyl theorem is just the completeness of the functions $x \to e^{inx}$ with its implications for expansibility in Fourier series. The proof of the Peter-Weyl theorem is closely related to the proof (see section 14) that a completely continuous self-adjoint operator has a basis of eigenvectors. In fact, it is possible to deduce the Peter-Weyl theorem from the latter result. More generally, let the compact Lie group G act on the Riemannian space S so as to preserve the underlying metric and let it act transitively in the sense that for each s_1 and s_2 in S there exists an x in G with $(s_1)x = s_2$. For each x in G and each complex-valued function f on S, let $V_x(f)$ denote the function $s \to f((s)x)$. Then each V_x is a linear transformation and $x \to V_x$ defines a representation of G in any finite-dimensional vector space M of complex-valued functions on S which happens to be mapped onto itself by all V_x. Let us call this subspace irreducible if the corresponding representation is irreducible. In 1929 Cartan published a paper (admittedly inspired by the paper of Peter and Weyl) proving that M_1 and M_2 are orthogonal whenever the corresponding irreducible representations of G are inequivalent, and moreover that there exists a complete orthogonal set of continuous functions on S whose members belong to irreducible invariant subspaces M. The Peter-Weyl theorem is the special case in which $S = G$ and the action is by group multiplication. In the special case in which S is "symmetric," Cartan proved in addition that any two invariant irreducible subspaces which are not identical define inequivalent representations of G. In the subspecial case in which S is the surface of a sphere in three spaces and G is the rotation group, Cartan's result implies the completeness of the surface harmonics of Legendre and Laplace (see section 7). The connection between group representations and surface harmonics was recognized by Weyl in his book on group representation and quantum mechanics published in 1928.

16. GROUP REPRESENTATIONS IN HILBERT SPACE AND THE DISCOVERY OF QUANTUM MECHANICS

The extension of the theory of group representations and characters from finite groups to compact Lie groups does not produce many significant changes. One of the few is that the number of irreducible characters is countably infinite rather than finite. This implies of course that no representation can behave like the regular representation of a finite group in containing a representative of every equivalence class of irreducibles unless one permits infinite-dimensional representations in some sense. Nowadays, nothing seems more natural than to define the regular representation of a compact Lie group G to be a representation L whose space is the Hilbert space of all complex-valued functions on G which are square integrable with respect to Haar measure and where $L_x(f)(y) = f(yx)$. If one does so (and suitably generalizes the direct sum notion to apply to infinite sums) one finds that the Peter-Weyl theorem implies a very straightforward generalization of the structure theorem for the regular representation of a finite group. This generalization states that the regular representation of a compact Lie group is a direct sum of finite-dimensional irreducible subrepresentations and that each equivalence class occurs with a multiplicity equal to its dimension. Similarly, Cartan's theorem about functions on S can be stated in terms of the decomposition of the representation L of G in $\mathcal{L}^2 (S, \mu)$ defined by setting $L_x(f)(s) = f((s)x)$. The Lebesgue integral and Hilbert spaces of square integrable functions were still strange and unfamiliar objects to most mathematicians in the 1920s, however, and a systematic theory of group representations in an infinite-dimensional Hilbert space was slow to develop. When it did, this development was directly inspired by the discovery of quantum mechanics in the period between 1924 and 1927, especially by von Neumann's success in putting this theory into a rigorous, mathematically coherent form based on the theory of operators in Hilbert space.

The "old quantum theory" initiated by Planck's paper of 1900 was replaced by the much more satisfactory quantum mechanics during a period of about three years beginning at the end of 1924. In late 1924 and early 1925, Heisenberg (1901-1976) and Schrödinger (1887-1961) respectively published two apparently very different methods for deducing the spectrum of the hydrogen atom without imposing arbitrary quantization rules. These methods were later shown to be equivalent. More importantly, they turned out to provide the key to the puzzle. After a few years of intensive activity, difficult to trace in detail, physicists were in possession of a subtle refinement of classical mechanics which had classical mechanics as a limiting case and from which the quantum rules of the old quantum theory followed in a logical and consistent manner. Besides Heisenberg and Schrödinger, the chief architects of this new quantum mechanics were Born (1882-1970), Jordan (1902—), and Dirac (1902—).

The key idea in the finished theory is that one must give up the naïve notion that the state of a physical system at a given time can be described by the positions and velocities of its particles. Measurements interfere with one another in a manner that becomes more pronounced the lighter the particles are, and in the case of electrons it is very pronounced indeed. It turns out, however, that it makes sense to assign simultaneous probability distributions to all positions, velocities, and various functions of these and that such a collection of probability distributions may be considered to be a state of the system. Indeed, the laws of quantum mechanics permit one to calculate all probability distributions at time t_1 when they are known at time $t_2 < t_1$.

The physicists' formulations of the laws permitting one to make these calculations and draw various conclusions from them were somewhat vague and unsatisfactory from the standpoint of a pure mathematician, and von Neumann (1903-1956) became interested in clarifying them. He succeeded admirably, and in 1927 published a remarkable paper showing how a subtle generalization of Hilbert's spectral theorem was the key to the whole question. Altering slightly von Neumann's definition and terminology, let us define a *projection-valued measure* on the real line to be a mapping $E \to P_E$ assigning a projection operator P_E in a separable Hilbert space $\mathscr{H}(P)$ to each Borel set E in the line R in such a fashion that 1) $P_{E \cap F} = P_E P_F$ for all E and F, 2) $P_\emptyset = 0$ and $P_R = I$ where $\emptyset$ is the empty set and I is the identity operator, and 3) $P_{\cup E_j} = \Sigma P_{E_j}$ whenever the E_j are pairwise disjoint. Let us say that the projection-valued measure P has *bounded support* if $P_{[a,b]} = I$ for some finite interval $[a,b]$ and *countable support* if there exists a countable subset Λ of R such that $P_\Lambda = I$. Given a vector φ in the Hilbert space $\mathscr{H}(P)$ it is easy to check that the function $E \to (P_E(\varphi) \cdot \varphi)$ is a measure on the line which is finite when P has bounded support. When this is the case, one can form the integral $\int_{-\infty}^{\infty} xd(P_x(\varphi) \cdot \varphi)$, and it is not difficult to show that there exists a unique bounded linear operator A such that $(A(\varphi) \cdot \varphi) = \int_{-\infty}^{\infty} xd(P_x(\varphi) \cdot \varphi)$ for all vectors φ. Moreover, the bounded linear operator A is self-adjoint in the sense that $(A(\varphi) \cdot \psi) = (\varphi \cdot A(\psi))$ for all φ and ψ in $\mathscr{H}(P)$. In terms of projection-valued measures, Hilbert's spectral theorem can now be stated very simply. It is just a converse of the result just stated. For every bounded self-adjoint operator A there exists a unique projection-valued measure P^A defined on the real line and having bounded support such that $(A(\varphi) \cdot \varphi) = \int_{-\infty}^{\infty} xd(P_x(\varphi) \cdot \varphi)$ for all φ in $\mathscr{H}(P)$. To understand the one-to-one correspondence between bounded self-adjoint operators and projection-valued measures on R with bounded support set up by the spectral theorem, it is useful to consider the special case in which A has a basis of eigenvectors φ_j, $A(\varphi_j) = \lambda_j \varphi_j$. In that case, P has the set $\bigcup_{j=1}^{\infty} \{\lambda_j\}$ of eigenvalues as countable support. $P_E = \sum_{\lambda_j \in E} P_{\{\lambda_j\}}$ and $P_{\{\lambda_j\}}$ is the projection on the vector space of all φ with $A(\varphi) = \lambda_j \varphi$.

While the spectral theorem was in a sense just what was needed, it did not go far enough. Von Neumann's first task was to extend it so that all projection-valued measures were involved and not just those of bounded support. The problem was to find a corresponding extension of the class of bounded self-adjoint operators. Extending an idea of E. Schmidt, von Neumann found the answer in a certain class of operators that are not necessarily bounded and are defined on a dense subspace $\mathscr{D}$ of $\mathscr{H}(P)$. Given such an operator A, let A^* be defined as follows: $A^*(\psi)$ is defined if and only if $(A(\varphi) \cdot \psi)$ is continuous as a function of φ and then is the unique vector θ in $\mathscr{H}(P)$ such that $(A(\varphi) \cdot \psi) \equiv (\varphi \cdot \theta)$. One says that A is self-adjoint if A^* is defined on $\mathscr{D}$ and only on $\mathscr{D}$, and there agrees with A. With the concept of self-adjointness extended in this way to possibly unbounded operators, it became possible to extend the spectral theorem so that it set up a one-to-one correspondence between all self-adjoint operators on the one hand and all projection-valued measures on R on the other.

With the spectral theorem so extended, von Neumann was now able to unify the various probabilistic statements of the physicists in one simple general principle. This may be stated as follows: To each physical system there corresponds a complex Hilbert space (usually separable) whose one-dimensional subspaces define the states of the system. To every observable (position coordinates, momentum coordinates, etc.) there corresponds a self-adjoint operator. Given the self-adjoint operator A corresponding to a particular observable, let P^A denote the associated projection-valued measure and let φ denote any unit vector. Then $E \rightarrow (P_E^A(\varphi) \cdot \varphi)$ is a probability measure on the real line which depends only on the one-dimensional subspace to which φ belongs. This is the probability distribution assigned to the observable corresponding to A by the state corresponding to the one-dimensional subspace of complex multiples of φ.

Notice that in every state the observable corresponding to A takes a value outside of the spectrum of A with probability zero. When the spectrum of A is discrete or partially discrete, the values taken on by the observable are correspondingly restricted. This is the source of the mysterious "quantization rules" of the old quantum theory and explains why some observables are quantized and others are not. Similarly, the impossibility of finding states in which two different observables have highly concentrated probability distributions at the same time may be traced to the lack of commutativity of the corresponding self-adjoint operators. In fact, the celebrated Heisenberg uncertainty principle may be formulated as an inequality on the product of the dispersions of two probability distributions where the lower bound involves the commutator of the corresponding operators.

In the special case of a system of n "spinless," "distinguishable" particles it is possible to make these abstract statements more concrete. Let the masses of the n particles be $m_1, m_2, \cdots, m_n$ and let their coordinates $x_1 \, y_1 \, z_1 \cdots x_n \, y_n \, z_n$ be relabeled as $q_1, q_2, \cdots, q_{3n}$. Let $p_1, \cdots, p_{3n}$ be a relabeling of $m_1(dx_1/dt), m_1(dy_1/dt), \cdots, m_n(dz_n/dt)$. Then the Hilbert space of the system may be taken to be the space of all complex-valued functions on Euclidean $3n$ space which are square integrable with respect to Lebesgue measure, and when this is done, the operators Q_j and P_j corresponding to the q_j and p_j are the following: $Q_j(\varphi)$ is defined whenever $q_j\varphi$ and φ are both square integrable and $Q_j(\varphi)(q_1 \cdots q_{3n}) = q_j\varphi(q_1 \cdots q_{3n})$. $P_j(\varphi)$ is defined whenever φ is absolutely continuous in the jth variable and φ and $\dfrac{\partial\varphi}{\partial q_j}$ are both square integrable and $P_j(\varphi)(q_1 \cdots q_{3n}) = \dfrac{h}{2\pi i}\dfrac{\partial\varphi}{\partial q_j}(q_1 \cdots q_{3n})$ where h is Planck's constant (see section 13).

Since the Q_j commute with one another there is no problem in discussing the joint probability distribution of the q_j and it is easy to deduce from von Neumann's general principle that in the state described by the function φ the probability that the coordinates q_j be in the region R of $3n$ space is $\int_R \cdots \int |\varphi(q_1 \cdots q_{3n})|^2 dq_1 \cdots dq_{3n}$. While not every classical observable has a quantum counterpart, there are many that do, and those that do are usually ones whose classical definition may be written in such a form in terms of the q_j and p_j that the same formula applied to the Q_j and P_j makes sense as a self-adjoint operator. For example, the classical angular momentum of a particle about the z axis is $m(x\dfrac{dy}{dt} - y\dfrac{dx}{dt}) = q_1 p_2 - p_1 q_2$ and $Q_1 P_2 - P_2 Q_1$ is i times a self-adjoint operator which defines the quantum analogue of angular momentum about the z axis. This operator (when its domain is suitably defined) has a discrete spectrum with the integer multiples of $\dfrac{h}{2\pi}$ as eigenvalues. In 1913, Bohr derived the spectrum of the hydrogen atom from the *assumption* that angular momentum could only take on the values $0, \pm\dfrac{h}{2\pi}, \pm 2\dfrac{h}{2\pi}$, etc.

Thus far we have discussed only quantum statics and said nothing about how the state of a system varies with the time. The physicists' answer to this question carries over immediately to von Neumann's formulation. If $t \to \varphi_t$ is a one parameter family of unit vectors describing the state of the system at any time t, then this vector-valued function of t must satisfy a differential equation of the form $\dfrac{d\varphi_t}{dt} = \dfrac{2\pi i}{h} H(\varphi_t)$ where H is the self-adjoint operator defining the total energy of the system. Up to possible ambiguities of domain this operator H is well defined whenever the classical expression for the energy is the sum of a potential energy term which is a function V of the q_j alone and the usual kinetic energy term $\displaystyle\sum_{j=1}^{n} \dfrac{m_j}{2}\left[\left(\dfrac{dx_j}{dt}\right)^2 + \left(\dfrac{dy_j}{dt}\right)^2 + \left(\dfrac{dz_j}{dt}\right)^2\right]$.

One simply rewrites the kinetic energy in terms of the p_j, substitutes $\dfrac{h}{2\pi i}\dfrac{\partial}{\partial q_j}$ for each p_j and adds on the operator of multiplication by $V(q_1 \cdots q_{3n})$. Written out in concrete form, $\dfrac{d\varphi_t}{dt} = \dfrac{2\pi i}{h} H(\varphi_t)$ becomes the celebrated Schrödinger equation — a linear partial differential equation in $3n + 1$ variables. An immediate consequence of Schrödinger's equation is that the eigenvectors of the self-adjoint operator H corresponding to the energy play a special role. Indeed, if $H(\varphi) = \lambda\varphi$, then $e^{i\lambda t}\varphi$ satisfies Schrödinger's equation, and since φ and $e^{i\lambda t}\varphi$ define the same state, φ defines a so-called stationary state — a state which does not change with time. Conversely, every stationary state is easily shown to have this form. In other words, the eigenvectors of H are on the one hand the stationary states and on the other the states in which the energy has a definite sharp value. It turns out that when some external perturbation causes an atom to shift from one stationary state to another of lower energy, the energy difference manifests itself as a quantum of electromagnetic radiation of frequency equal to $\frac{\Delta E}{h}$. Thus the problem of predicting the spectral lines emitted by an atom reduces to finding the eigenvalues of the appropriate operator H. (Reduces is perhaps too strong a word. When one investigates the mechanism more closely, one finds that some lines occur with zero intensity and so are not observed.) While Bohr's simple idea of quantizing angular momentum was sufficient to predict the eigenvalues of the H for the hydrogen atom, the old quantum theory was quite incapable of dealing with atoms having more than one electron. Even with quantum mechanics the problem is difficult, because it is far from trivial to find the eigenvalues of the appropriate H. Indeed, one has to resort to approximate methods of various kinds.

Von Neumann did not content himself with clarifying and rigorizing the conceptions of the physicists. In two further papers published in 1927 he made a basic contribution to physics in that he showed how to combine the ideas of quantum mechanics with those of statistical mechanics and produce a "quantum statistical mechanics." This seems difficult if not impossible at first, because classical statistical mechanics is based on the consideration of joint probability distributions for all the dynamical variables; that is, on the consideration of probability measures in the $6n$-dimensional "phase space" of all possible $6n$-tuples of position and momentum coordinates. The Heisenberg uncertainty principle seems to preclude a quantum mechanical analogue of phase space. However, von Neumann found an analogue of probability measures in phase space in what are now called von Neumann density "matrices" (more accurately density operators). We already have seen that a unit vector φ in the underlying Hilbert space assigns a probability measure on the line to each self-adjoint operator and hence to each observable. More generally, let T be a self-adjoint operator with a basis $\varphi_1, \varphi_2, \cdots$ of eigenvectors and let the eigenvalues $\gamma_1, \gamma_2, \cdots$ be non-negative and such that $\gamma_1 + \gamma_2 + \cdots = 1$. Then for each self-adjoint operator A we may consider the set function $E \to \mathrm{Trace}(TP_E^A) = \sum_{j=1}^{\infty} \gamma_j (P_E^A(\varphi_j) \cdot \varphi_j)$ and verify at once that it is an infinite convex combination of the probability measures $E \to (P_E^A(\varphi_j) \cdot \varphi_j)$ and hence a probability measure itself. Here, of course, P^A is the projection-valued measure corresponding to A by the spectral theorem. The generalized states so defined by non-negative self-adjoint trace operators with Trace 1 are related to the states defined by unit vectors just as probability measures in phase space are related to points in phase space. In each case, it is a question of comparing a convex set with its set of extreme points. To distinguish them one now speaks of *mixed states* and *pure states*. The mixed states (or *mixtures* as von Neumann called them) are von Neumann's substitutes for probability measures in phase space. His substitute for $\int_\Omega e^{-H/kT}d\zeta$ is Trace $(e^{-H/kT})$ where in the second expression H is the self adjoint operator corresponding to the energy observable. Just as $\int_\Omega^{-H/kT}d\zeta$ may be written in the form $\int_{-\infty}^{\infty} e^{-x/kT}d\beta(x)$ where β is the image of ζ by H (see section 13), so Trace$(e^{-H/kT})$ may be written in the form $\sum_{j=1}^{\infty} e^{-E_j/kT}$ where $E_1, E_2, \cdots$ are the (not necessarily distinct) eigenvalues of H. In the systems to which statistical mechanics applies, H has a pure point spectrum.) This may be rewritten as $\int_{-\infty}^{\infty} e^{-x/kT}d\beta_q(x)$, where β_q is the measure that counts the number of eigenvalues in each set. In other words, as suggested by Planck's discovery, quantum mechanics tells us that in forming the partition function $T \to \int e^{-x/kT}d\beta(x)$, the continuous measure β must be replaced by a discrete measure. In addition, it tells us what discrete measure to choose. Known facts about the relationship of measures on the line to their Laplace transforms and the observation that classical and quantum statistical mechanics agree at high temperatures suggest that β_q and β should be asymptotically equal to constant multiples of one another. This suggestion, formulated as a theorem, turns out to include as special cases various theorems on the asymptotic distribution of eigenvalues proved earlier by Weyl and Courant and later by a number of mathematicians, among whom Titchmarsh (1899-1963) may be especially mentioned.

Group-theoretical ideas were introduced into quantum mechanics in quite different ways in two papers published in 1927 by Weyl and Wigner (1902—) respectively. Part I of Weyl's paper is devoted to a discussion (independent of von Neumann's) of the concept of a mixed state. Parts II and III contain the group theory. Let A be any bounded self-adjoint operator in a Hilbert space $\mathcal{H}$. For all real numbers t, let $U_t = e^{iAt} = I + iAt + \dfrac{(iAt)^2}{2!}$ $+ \dfrac{(iAt)^3}{3!} \cdots$. The series converges for all t, and it is not hard to verify that each U_t is unitary and that $t \to U_t$ is a unitary representation of the additive group R of the real line. Moreover, this representation is continuous in the sense that $t \to U_t(\varphi)$ is a continuous function from the real line to the Hilbert space for all φ in the Hilbert space. It is not difficult to verify that A is uniquely determined by the representation U so that there is a natural one-to-one correspondence between bounded self-adjoint operators on the one hand and *certain* continuous unitary representations of R on the other. Without actually formulating a theorem, Weyl suggested that von Neumann's extension of the spectral theorem to unbounded operators should make it possible to extend the correspondence just described to one between all continuous unitary representations of R and all self-adjoint operators. At the same time, he pointed out that diagonalizing A, when this is possible, is equivalent to decomposing the group representation as an infinite direct sum and that the spectral theorem correspondingly must be equivalent to some sort of continuous decomposition theorem for U.

These suggestions of Weyl were given a solid mathematical foundation by Stone (1903—). In a short note published in 1930, Stone sketched a proof of the following theorem: Let $t \to U_t$ be an arbitrary continuous unitary representation of R. Then there exists a unique projection-valued measure P on R such that for all φ in the (separable) Hilbert space $\mathscr{H}(U)$, one has $(U_x(\varphi) \cdot \varphi) = \int e^{ixy} d(P_y(\varphi) \cdot \varphi)$ identically in x. Conversely, every projection-valued measure P on R arises in this way from some continuous unitary representation U of R. Combining the one-to-one correspondence between representations and projection-valued measures produced by Stone's theorem with the one between self-adjoint operators and projection-valued measures produced by von Neumann's extension of Hilbert's spectral theorem, one has a natural one-to-one correspondence between representations and self-adjoint operators, which reduces in the case of bounded operators to that defined above. Actually, one can even make sense of the formula e^{iAt} in the general case by using the "operational calculus" implied by the spectral theorem. Stone's note was the third in a series devoted to the theory of operators in Hilbert space. Von Neumann's paper formulating quantum statics did not actually contain a proof of the generalized spectral theorem, and Stone found a different proof which he sketched in an earlier note. Von Neumann's proof appeared in a long and famous paper published in 1929.

In Schrödinger's equation as formulated by the physicists, the domain of the operator H is left vague. This is an important gap in the theory because, as emphasized in Weyl's paper, the time evolution of the system is determined by the representation $t \to e^{(2\pi i/h)Ht}$. This representation is not known until H is precisely defined as a self-adjoint operator with a definite domain. In the language of the theory of partial differential equations, the initial value problem does not have a unique solution unless appropriate boundary conditions are imposed. It is interesting that the same condition on an operator that makes it suitable for assigning probability distributions to states also makes it suitable for defining a dynamics in which the present state uniquely determines all future states.

In his 1927 paper on group theory and quantum mechanics, Weyl also pointed out that unitary representations of the real line are technically easier to deal with than unbounded self-adjoint operators that are not everywhere defined. Moreover, he showed that the unitary representations associated with the position and momentum operators for a set of n particles form a system with very simple and natural group-theoretical properties. Specifically, let $U_t^j = e^{iQ_j t}$ and $V_s^j = e^{iP_j s}$. Then $(U_t^j)(f)(q_1, \cdots, q_{3n}) = e^{itq_j} f(q_1, \cdots, q_{3n})$ and $V_s^j(f)(q_1, \cdots, q_{3n}) = f(q_1, q_2, \cdots, q_{j-1}, q_j + \dfrac{sh}{2\pi}, q_{j+1}, \cdots, q_{3n})$. From this one computes easily that the following commutation relations are satisfied: $U_{t_1}^j U_{t_2}^k - U_{t_2}^k U_{t_1}^j = V_{s_1}^j V_{s_2}^k - V_{s_2}^k V_{s_1}^j = 0$ for all $j, k, s_1, s_2,$ and $U_t^j V_s^k = e^{-(isth\delta_k^j)/2\pi} V_s^k U_t^j$ for all $j, k, s,$ and t. Using the first set of relations, one obtains a continuous unitary representation U of the additive group of $3n$-dimensional vector space of all n-tuples of real numbers by setting $U_{t_1, t_2, \cdots, t_{3n}} = U_{t_1}^1 U_{t_2}^2 \cdots U_{t_{3n}}^{3n}$, and another V by setting $V_{s_1, s_2, \cdots, s_{3n}} = V_{s_1}^1 V_{s_2}^2 \cdots V_{s_{3n}}^{3n}$. Moreover, the second set of relations is equivalent to the following simple commutation relation between U and V: $U_{t_1, \cdots, t_{3n}} V_{s_1, s_2, \cdots, s_{3n}} = e^{-(ih/2\pi)(s_1 t_1 + \cdots + s_{3n} t_{3n})} V_{s_1, \cdots, s_{3n}} U_{t_1, \cdots, t_{3n}}$. If we recall that the most general continuous character (of absolute value one) on the group of $3n$-tuples can be written uniquely in the form $s_1, s_2, \cdots, s_{3n} \rightarrow e^{-(ih/2\pi)(s_1 t_1 + \cdots + s_{3n} t_{3n})}$, this last relation can be written more simply and suggestively in the form $U_s V_x = \chi(s) V_x U_s$, where s stands for the general $3n$-tuple $s_1, \cdots, s_{3n}$ and χ for the general character $s = s_1, s_2, \cdots, s_{3n} \rightarrow e^{-(ih/2\pi)(s_1 t_1 + \cdots + s_{3n} t_{3n})}$. Let G now denote the additive group of all $3n$-tuples of real numbers and let $\hat{G}$ denote its (isomorphic) group of all continuous characters of absolute value 1 (unitary characters). If we define $W_{s,x} = U_s V_x$, we "almost" obtain a continuous unitary representation of the commutative product group $G \times \hat{G}$. However, $W_{(s_1, x_1)(s_2, x_2)} = W_{s_1 s_2, x_1 x_2} = U_{s_1 s_2} V_{x_1 x_2} = U_{s_1} U_{s_2} V_{x_1} V_{x_2} = U_{s_1} \chi_1(s_2) V_{x_1} U_{s_2} V_{x_2} = \chi_1(s_2) W_{s_1, x_1} W_{s_2, x_2}$ so that $W_{(s_1, x_1)(s_2, x_2)}$ is equal to $W_{s_1, x_1} W_{s_2, x_2}$ only up to multiplication by a complex number of modulus one. W is what is known as a *projective* or *ray* representation of $G \times \hat{G}$. Such representations for finite groups were studied by Schur beginning in 1904. Weyl pointed out that they occur naturally in quantum mechanics because two vectors in Hilbert space determine the same state when one is a constant multiple of the other. He also pointed out that while an irreducible ordinary representation of a finite commutative group is necessarily one-dimensional, an irreducible projective representation of such a group can be multi-dimensional.

Weyl regarded it as highly significant that the position and momentum operators for a quantum mechanical system of n interacting particles are related in such a simple way to a projective unitary representation of a $6n$-dimensional vector group, which, as he suggested, turned out to be irreducible. In fact, as stated by Stone in the paper cited above and as proved shortly afterwards by von Neumann, the projective representation W is, to within unitary equivalence, the only irreducible projective unitary representation of $G \times \hat{G}$ having $s_1, \chi_1, s_2, \chi_2 \to \chi_2(s_2)$ as its corresponding multiplier. Some years later, Weyl's views were shown to be essentially correct in that methods related to his made it possible to go a long way toward deducing Schrödinger's equation and the form of the position and momentum operators from plausible assumptions about invariance and symmetry. Of course, the commutation relations satisfied by U and V are just the celebrated Heisenberg commutation relations for the Q_j and P_j in global form. It is perhaps not too far from the truth to assert that the essential idea of Weyl is that one does not have to assume the truth of the Heisenberg commutation rules — rather that they may be deduced from plausible *a priori* symmetry considerations.

Wigner's paper, as already suggested, applied the theory of group representations in a completely different way. In the first place, it was concerned with non-commutative finite groups rather than with continuous commutative groups, and in the second place, it dealt with the technique of finding approximate eigenvalues of the energy operator rather than with foundational questions. Let H be a self-adjoint operator whose eigenvalues are to be found. Suppose that H can be written in the form $H_0 + J$ where H_0 has known eigenvalues and eigenvectors and J is in some sense "small." One can then attempt to estimate the eigenvalues of H as follows: Replace $H_0 + J$ by $H_0 + \epsilon J$ where ϵ is a variable parameter, *assume* that the eigenvalues of $H_0 + \epsilon J$ vary smoothly with ϵ and can in fact be expanded in power series in ϵ, compute the first few terms and set $\epsilon = 1$. Suppose that λ^0 is an eigenvalue of H_0 whose corresponding eigenspace M is finite-dimensional, and let ψ_1, ψ_2, $\cdots$, ψ_n be an orthonormal basis for M. One cannot expect that the n occurrences of λ^0 as an eigenvalue will all change in the same way as ϵ varies from zero to one. Instead, one must seek n different functions $\lambda_1^0(\epsilon), \cdots,$ $\lambda_n^0(\epsilon)$, each of which reduces to λ^0 when $\epsilon = 0$ and is an eigenvalue of $H_0 + \epsilon J$ for small ϵ. If these n functions can be expanded in powers of ϵ, $\lambda_j^0(\epsilon) = \lambda^0 + \lambda_j^1 \epsilon + \lambda_j^2 \epsilon^2 + \cdots$, then a simple analysis allows one to conclude that the $\lambda_1^1, \lambda_2^1, \cdots, \lambda_n^1$ are the n eigenvalues of the matrix $((J(\psi_i) \cdot \psi_j))$. This is the fundamental theorem of so called "first order perturbation theory," and in many problems one gets a useful approximation to the eigenvalues of $H_0 + J$ by using only the first two terms of the series and setting $\epsilon = 1$.

Now, whatever one may think about the validity of the assumptions leading to this approximation, finding the eigenvalues of $((J(\psi_i) \cdot \psi_j))$ is a well-defined mathematical problem. Moreover, it is usually rather non-trivial because eigenvalues of high multiplicity are the rule rather than the exception. They occur whenever the underlying mechanical system possesses symmetries. These are reflected in automorphisms of the state space which are implemented by unitary operators which commute with the energy operator.

On the other hand, let U be a unitary representation of a group G in a Hilbert space $\mathcal{H}(U)$ which decomposes as a direct sum of finite-dimensional irreducible representations L^1, L^2, $\cdots$ in finite-dimensional orthogonal invariant subspaces $\mathcal{H}^1, \mathcal{H}^2, \cdots$. Suppose that T is any self-adjoint operator which commutes with all U_x and has a pure point spectrum. It follows from elementary considerations that each eigenspace of T is an invariant subspace for U and hence that the $\mathcal{H}^j$ may be chosen to be inside the eigenspaces. This implies, however, that for every L^j whose dimension is greater than one there will be an eigenvalue whose multiplicity is at least equal to this dimension.

It would take us too far afield to give all the details, so let it suffice to say that the high order matrices $((J(\psi_i) \bullet \psi_j))$ which one finds it necessary to diagonalize turn out to commute with all the operators L_x of some representation $x \to L_x$ of a finite or compact group G. The matrices of this representation are explicitly known with respect to the basis ψ_j, and one can exploit these facts to simplify rather considerably the problem of diagonalizing $((J_{ij})) = ((J(\psi_i) \bullet \psi_j))$. The easiest case is that in which no irreducible constituent of L occurs more than once. In that case, the decomposition of L into irreducibles is necessarily a decomposition of the operator defined by $((J_{ij}))$ as a direct sum of constants. Moreover, one can compute these constants directly from the characters of G, the J_{ij}, and the matrix elements of the L_x without solving any equations of higher degree. More generally, when multiplicities occur in the decomposition of L, the same methods may be used to reduce the problem to diagonalizing matrices of lower dimension. The dimensions that occur are the multiplicities.

In the generality described in the preceding paragraphs, the method emerged gradually in a sequence of papers by Wigner and by Wigner and von Neumann in collaboration, published in 1927 and 1928. In Wigner's first paper, he considered only the symmetric group on n objects. This arises because of the identity of the electrons in an n-electron atom. In an earlier paper he had managed the three-electron case without using the theory of group representations as such. Moreover, he credits von Neumann with having called his attention to the existence and applicability of the latter theory.

Weyl followed up his 1927 paper with a remarkable book published in 1928. Based on a course of lectures Weyl gave at the Eidgenossische Technische Hochschule in Zurich in the winter semester of 1927-28, this book, *Gruppentheorie und Quantenmechanik,* was destined to become one of the great classics of mathematical physics. In addition to a presentation in developed form of the group-theoretical ideas of Wigner, von Neumann, and himself, it contained an astonishingly complete, coherent account of the conceptual structure of quantum mechanics, together with its application to concrete physical problems. Perhaps for pedagogical reasons Weyl had little to say about von Neumann's use of the spectral theorem to deal with observables whose operators are not discretely decomposable. Thus for a full appreciation of the extent to which the physicists' discoveries could be incorporated into a beautiful and rigorous mathematical model, one had to read Weyl's book in conjunction with von Neumann's presentation of his own ideas in book form. This book, *Mathematische Grundlagen der Quantenmechanik,* appeared in 1932. On the other hand, it must be emphasized that both in physical content and in the extent to which it integrated group representations with physics, the book went far beyond the brief indications which I have given here. In particular, Weyl added considerably to what could be found in the literature of the time, and more than once it has turned out that some "new" idea in physics could be found hidden away in some little-understood part of Weyl's book. A considerably revised second edition appeared in 1930 and an English translation in 1931.

It is difficult to overemphasize the importance for physics (and chemistry) of the discovery of quantum mechanics. It did much more than explain away the inconsistencies between classical mechanics and the mysterious quantum rules of Planck, Einstein, and Bohr. Now it was possible (at least in principle—and subject to certain qualifications to be indicated below) to deduce all the properties of matter from its atomic constitution and Rutherford's hypothesis of 1911 that an atom of atomic number n consists of n "electrons" of charge $-e$ interacting with one another and with a much heavier nucleus of charge ne according to Coulomb's law. Of course in applying Coulomb's law one has to replace classical mechanics by quantum mechanics. Moreover, one has to determine the fundamental charge e and the relevant masses by suitable experiments. After that, the theory (as modified by the discovery of electron "spin" and the Pauli exclusion principle) provides a mathematically well-defined procedure (which may be extremely difficult to carry out in practice) for computing the free energy function, the electric and magnetic properties, etc., of any piece of matter whose atomic constitution is known. In the same sense, the theory allows one to compute the binding energies of all molecules and in other ways to deduce the laws of chemistry from first principles. As stated by Dirac in the introduction to a paper published in 1929 (in volume 123, series A of the Proceedings of the Royal Society of London),

> The general theory of quantum mechanics is now almost complete, the imperfections that still remain being in connection with the exact fitting in of the theory with relativity ideas. These give rise to difficulties only when high speed particles are involved, and are therefore of no importance in the consideration of atomic and molecular structure and ordinary chemical reactions, in which it is, indeed, usually sufficiently accurate if one neglects relativity variation of mass with velocity and assumes only Coulomb forces between the various electrons and atomic nucleii. The underlying physical laws necessary

for the mathematical theory of a large part of physics and the whole of chemistry are
thus completely known and the difficulty is only that the exact application of these laws
leads to equations much too complicated to be soluble.

It was Dirac himself who showed how to reconcile Einstein's light quanta
with Maxwell's equations. In a paper published in 1927 he pointed out that
Maxwell's equations could be looked upon as the equations of motion of a
dynamical system with an infinite number of degrees of freedom. When the
standard procedures of quantum mechanics are applied to this dynamical
system, one obtains a quantum mechanical system which can be reinterpreted as a system of particles (photons or light quanta). Dirac was perhaps the
first physicist to see the new quantum mechanics as a logically coherent system, and his book on the subject is another great classic. While less satisfactory from the standpoint of a pure mathematician than the books of Weyl
and von Neumann, it was more acceptable to physicists and had an enormous influence. First published in 1930, it has gone through many editions.

The key to using quantum mechanics to explain the formation of molecules from atoms was found by Heitler (1904—) and London (1900-1954),
who applied it to the hydrogen molecule in a paper published in that magic
year 1927. Shortly thereafter they showed in independent papers that in
dealing with molecules with more than two electrons, one could apply the
representation theory of the symmetric group in much the same way that
Wigner had done. The fifth chapter of Weyl's book contains an exposition
of their ideas as interpreted by him and includes a beautiful group-theoretical explanation of chemical valence.

In 1931 Wigner published a book on the application of the theory of
group representations to the analysis of atomic spectra. He went on in the
1930s to publish a series of fundamental and influential papers showing
how to apply that same theory to a wide variety of quantum mechanical
problems.

17. THE DEVELOPMENT OF THE THEORY OF UNITARY GROUP REPRESENTATIONS BETWEEN 1930 AND 1945

It follows at once from Stone's theorem of 1930 connecting self-adjoint
operators, unitary representations of the real line and projection-valued
measures on the real line (see section 16), that the work of Hilbert and his
students and coworkers on self-adjoint operators in Hilbert space, as well as
the later work of Stone and von Neumann on the unbounded case, can be
reinterpreted as work on the problem of analyzing the unitary representations of R, the additive group of the real line. The spectral theorem itself is
the analogue of the theorem stating that any unitary representation of a
compact Lie group is a direct sum of irreducible representations, and the
spectral multiplicity theory of Hahn and Hellinger is the analogue of the

theorem stating that two direct sums of irreducible representations are equivalent if and only if the same irreducibles occur with the same multiplicities. Hahn and Hellinger dealt only with bounded self-adjoint operators, but in 1932 Stone published a now classic book on the theory of linear transformations in Hilbert space, which among other things contains the details of his own approach to the spectral theorem. Chapter VII of Stone's book is devoted to an improved exposition of the Hahn-Hellinger theory generalized to the unbounded case. Although it has a reputation for being complicated and difficult, the Hahn-Hellinger theory is actually quite easy to explain. I shall do so below in a more general context.

As matters stood at the end of 1932, one had all the ingredients of a complete theory of the unitary representations of the compact Lie groups, of certain finite groups, and of two non-compact groups. The two non-compact groups were the additive group R of all real numbers and the additive group Z of all integers. In the case of Z, the most general unitary representation is of course $n \to U^n$ where U is an arbitrary unitary operator. Since $H \to (H + i)(H - i)^{-1}$ can be shown to set up a one-to-one correspondence between all self-adjoint operators and all unitary operators, the spectral theorem, etc., for self-adjoint operators takes care of both R and Z. It must be confessed, however, that although the ingredients were there, their consequences for unitary group representations had not yet been spelled out. In particular, although it is an easy consequence of the Peter-Weyl theorem that every unitary representation of a compact Lie group is a discrete direct sum of irreducibles, this was not stated or proved in the literature until 1943. Moreover, the connection of the Hahn-Hellinger spectral multiplicity theory with the problem of classifying unitary representations of commutative groups was not explicitly pointed out until the 1950s.

An important stimulus to developing the theory of unitary group representations for infinite groups in a framework more general than that of compact Lie groups was a remarkable paper by A. Haar (1885-1933) published in 1933 and already mentioned in section 15. The notion of a topological group had been introduced by Schreier (1901-1929) only a few years before and Haar proved that whenever such a group is separable and locally compact, it admits a measure invariant under right translation which is defined on all Borel sets, is finite on compact sets, and is non-zero on open sets. (That such a measure is unique up to multiplication by a multiplicative constant was shown a few years later by von Neumann). Of course a left invariant measure with the same properties must also exist, but the two need not be equal. Groups for which the left and right invariant measures coincide are called unimodular and include the compact groups. Thus every compact separable group admits a unique left and right invariant Haar measure which assigns the measure one to the whole group. As pointed out by Haar, the arguments of Peter and Weyl can be applied to the most general compact separable groups once one has Haar measure to use to replace the Hurwitz integration process.

Another important stimulus was provided in 1934 when L. Pontrjagin (1908—) (inspired by the needs of duality theorems in topology) published a paper extending the known duality between finite abelian groups and their character groups to one between arbitrary discrete countable commutative groups on the one hand and separable compact commutative groups on the other. Let G be an arbitrary countable commutative group, and let us define a character on G to be a function χ from G to the complex numbers of modulus one such that $\chi(xy) = \chi(x)\chi(y)$ for all x and y in G. Then just as in the finite case, the pointwise product of two characters is again a character, and the set $\hat{G}$ of all characters is a group with multiplication as the composition law. Again as in the finite case, for each x in G the mapping $\chi \to \chi(x)$ is a character f_x of $\hat{G}$. If one gives $\hat{G}$ the weakest topology which makes all characters f_x continuous, it is not difficult to show that G becomes a topological group which moreover is compact and separable. Conversely, let A be any compact separable commutative topological group and let $\hat{A}$ denote the group of all *continuous* homomorphisms from A to the complex numbers of modulus one. Then $\hat{A}$ is countable and discrete, and as before each member a of A defines a character f_a of $\hat{A}$ by way of the definition $f_a(\chi) = \chi(a)$. Using Haar's extension of the Peter-Weyl theorem applied to A and $\hat{G}$, Pontrjagin was able to show that the maps $x \to f_x \to$ and $a \to f_a$ from G to $\hat{\hat{G}}$ and from A to $\hat{\hat{A}}$ are one-to-one and onto, and that in the case of A and $\hat{\hat{A}}$ that $x \to f_x$ is an isomorphism of topological groups. Thus every separable compact commutative group arises as the character group or dual of some countable discrete commutative group and conversely. There is a natural one-to-one correspondence between separable compact commutative groups on the one hand and countable discrete commutative groups on the other.

In the following year (1935), E. R. van Kampen (1908-1942) extended Pontrjagin's duality to a more general and more symmetrical one involving arbitrary locally-compact commutative groups. In particular, he was able to dispense with the hypothesis of separability. If G is an arbitrary locally-compact commutative group, one defines $\hat{G}$ just as before as the set of all continuous functions χ from G to the complex numbers of modulus one such that $\chi(xy) = \chi(x)\chi(y)$ for all x and y. $\hat{G}$ becomes a locally-compact topological group if one declares a set $\mathcal{O}$ of characters to be open whenever for each χ_0 in $\mathcal{O}$ there exists a compact subset C of G and $\epsilon > 0$ such that $|\chi(x) - \chi_0(x)| < \epsilon$ for all x in C implies $\chi \in \mathcal{O}$. Just as before, one defines $f_x(\chi) = \chi(x)$, and van Kampen's duality theorem asserts that $x \to f_x$ is simultaneously a homeomorphism and a group isomorphism of G on $\hat{\hat{G}}$. Every locally-compact commutative group is the dual of its dual. In the particular case in which G is an n-dimensional real vector group, G and $\hat{G}$ are isomorphic and the theorem is obvious.

Among the many properties of this duality relation, the following are particularly elegant and useful: a) If H is a closed subgroup of the locally-compact commutative group G, and $H^{\perp}$ is the subgroup of all χ in $\hat{G}$ such that $\chi(h) = 1$ for all h in H, then $H^{\perp\perp} = H$ and restricting χ to H sets up a one-to-one map of the quotient group $\hat{G}/H^{\perp}$ onto $\hat{H}$. This map is both a homeomorphism and an algebraic isomorphism—an isomorphism of topological groups. In particular, a character of a closed subgroup can always be extended to a character of the whole group. b) Let ψ be a continuous homomorphism from G_1 to G_2 where G_1 and G_2 are both locally compact commutative groups. Then for each χ in $\hat{G}_2$, the function $x \to \chi(\psi(x))$ is a character χ^* on G_1 and it is obvious that $\chi \to \chi^*$ is a homomorphism. It is actually a continuous homomorphism called the dual of ψ, which we may denote by ψ^*. One can prove that $\psi^{**} \equiv \psi$ and that ψ has a dense range if and only if ψ^* is one-to-one. More generally, if N^* is the subgroup on which ψ^* reduces to the identity, then $(N^*)^{\perp}$ is the closure of the range of ψ.

This last fact is the basis of an interesting application of the duality theorems to a topic in the harmonic analysis of functions on the real line. As already noted, the most general continuous character on R is $x \to e^{ixy} = \chi_y(x)$ so that $\hat{R}$ may be identified with R. However, it will be convenient here not to identify R and $\hat{R}$. Let D be $\hat{R}$ made into a discrete group by ignoring the topology, and let ψ be the identity map of D onto $\hat{R}$. Then ψ is a continuous homomorphism which is one-to-one. Hence ψ^* is a continuous homomorphism of $\hat{\hat{R}} = R$ into a dense subgroup of the *compact* character group $\hat{D}$ of D. Moreover, since ψ has all of $\hat{R}$ for its range, ψ^* is one-to-one. Thus the device of making $\hat{R}$ discrete yields a natural imbedding of the real line onto a dense subgroup of a compact group. Evidently the continuous complex-valued functions on the compact group $\hat{D}$ are in natural one-to-one correspondence with the members of a certain subclass of the bounded continuous complex-valued functions on the real line. These turn out to be precisely the so-called "almost periodic" functions introduced in 1924 by Harald Bohr (1887-1951).

Bohr's doctoral thesis of 1910 was on the summability of Dirichlet series, and his early work was primarily concerned with the further study of such series with particular emphasis on the Riemann zeta function and the location of its zeros. In collaboration with Landau (1877-1938), he published a paper in 1914 showing that the non-trivial zeros, if not on the central line, must at least be clustered about it. He made studies of the value distribution of the zeta function and other Dirichlet series on vertical lines and ultimately was led to ask for a characterization of those complex-valued functions on the line which can be represented in the form $t \to \sum_{n=1}^{\infty} a_n e^{-\lambda_n(s+it)}$, where s

is fixed real number and the a_j and λ_j are suitable sequences of complex and real numbers respectively. Writing $c_n = a_n e^{-\lambda_n s}$, this becomes $t \to \sum\limits_{n=1}^{\infty} c_n e^{-i\lambda_n t}$, which reduces to a Fourier series when the λ_n are integer multiples of a fixed real number. More generally, the function $t \to \sum\limits_{n=1}^{\infty} c_n e^{-i\lambda_n t}$ will not be periodic but will be "almost periodic" in the sense of having many "approximate" periods. Bohr gave a precise definition and in three long papers published between 1924 and 1926 presented a detailed theory of this new class of functions. One key theorem states that a bounded continuous function on the real line is almost periodic (in the sense of having enough approximate periods) if and only if it is a uniform limit of functions of the form $c_1 e^{i\lambda_1 x} + \cdots + c_n e^{i\lambda_n x}$ where the λ_j are real. Another asserts that every almost-periodic function f has a well defined "mean value" $M(f)$. Using $M(f)$ instead of $\frac{1}{2A} \int_{-A}^{A} f(x)dx$, one can develop an analogue of Fourier series expansions assigning to each almost-periodic function f the formal series $\sum c_n e^{i\lambda_n x}$ where $c_n = M(fe^{-i\lambda_n x})$ and $M(fe^{-i\lambda x}) = 0$, for all but countably many λ.

Bohr's theory aroused considerable interest at first, but, as I have already indicated, it was destined to be subsumed under the rubric of the duality theory of locally-compact commutative groups. In fact, the mean value $M(f)$ turns out to be just the Haar integral of the extension of f to the compact completion $\hat{D}$ of the real line and the Fourier series expansions to be reducible to the Peter-Weyl expansion on $\hat{D}$.

In 1927 Bochner showed that Bohr's definition in terms of approximate periods can be reformulated in a way that makes sense for arbitrary groups. A function is almost periodic in Bochner's sense if it is bounded and continuous, and if the set of all functions which can be uniformly approximated by its translates is compact in the topology of uniform convergence. For any locally-compact commutative group G, one has a theory of almost-periodic functions analogous to that of Bohr, and this theory can be reduced to the theory of functions on a compact group by taking the dual or adjoint of the natural map of the discretization of $\hat{G}$ into $\hat{G}$ just as for R. Von Neumann published a more general paper in 1934 developing a theory of almost-periodic functions on an *arbitrary* non-commutative group G. Of course the theory could be vacuous, and in 1935 A. Weil (1906—) showed that whenever G has sufficiently many almost-periodic functions, one can imbed it densely in a compact group K in such a fashion that the theory reduces to the Peter-Weyl theory on K.

The theory of almost-periodic functions on the group Z of all integers under addition turns out to have interesting connections with number theory. Rather than considering the whole of $\hat{Z}$, let Λ be the dense subgroup of $\hat{Z}$ consisting of all elements of finite order. The same argument as before shows that Z has a natural dense imbedding as a subgroup of the *separable* compact group $\hat{\Lambda}$. Moreover, if Λ_p is the subgroup of Λ consisting of all elements whose order is p^k for some k, then $\hat{\Lambda}$ is isomorphic to the direct product of all the compact groups $\hat{\Lambda}_p$. The restrictions to Z of the continuous functions on $\hat{\Lambda}$ are just the almost-periodic functions on Z whose non-zero Bohr-Fourier coefficients are those corresponding to the characters in Λ ($n \rightarrow e^{2\pi i r n}$ where r is rational). Moreover, such a restriction φ is multiplicative in the sense that $\varphi(nm) = \varphi(n)\varphi(m)$ for n and m relatively prime if and only if the original function on $\hat{\Lambda} = \prod_p \hat{\Lambda}_p$ is a product of continuous functions on the $\hat{\Lambda}_p$. Since each Λ_p is dense in $\hat{Z}$, there is a natural dense imbedding of Z in each $\hat{\Lambda}_p$, and the multiplication in Z has a unique continuous extension to a multiplication in $\hat{\Lambda}_p$. The ring which $\hat{\Lambda}_p$ thus becomes is isomorphic (algebraically and topologically) to the ring of all p-adic integers and its "field of quotients" is the field of p-adic numbers (see section 20). As will be explained more fully later, the Hardy-Littlewood results on Waring's problem (see section 14) have illuminating interpretations in terms of almost-periodic functions on Z.

The coherent theory of locally-compact groups which are either compact or commutative made possible by the results of Peter and Weyl, Haar, Pontrjagin, and van Kampen published between 1927 and 1935, inspired the publication of two very influential books a few years later. Pontrjagin's *Topological groups* was published in Russian in 1938, and an English translation appeared in 1939. *L'intégration dans les groupes topologiques et ses applications à l'analyse* by André Weil appeared in 1940. Although there is a large overlap, the two books are quite different in emphasis and to some extent in content. The key to the difference is revealed in the extra words in Weil's title. Pontrjagin is concerned above all with structure theorems for locally-compact groups and barely mentions harmonic analysis. Weil emphasizes harmonic analysis and shows in detail how one can define a Fourier transform for suitably restricted functions on any locally-compact commutative group and so include Fourier series and Fourier transforms in one and several variables in one unified theory. Specifically, if μ is a choice of Haar measure in G, then Weil defines the Fourier transform of a μ-integrable function f to be the function $\hat{f}$ on $\hat{G}$ such that $\hat{f}(\chi) = \int \chi(x)f(x)d\mu(x)$ for all χ in $\hat{G}$. When f is both integrable and square integrable, he shows that $\hat{f}$ is both continuous and square integrable and that the arbitrary constant in the Haar measure $\hat{\mu}$ in $\hat{G}$ can be so chosen that $f \rightarrow \hat{f}$ preserves the Hilbert space norms: $\int |f(x)|^2 d\mu(x) = \int |\hat{f}(\chi)|^2 d\hat{\mu}(\chi)$. Since the domain and range can be shown to be dense in $\mathscr{L}^2(G, \mu)$ and $\mathscr{L}^2(\hat{G}, \hat{\mu})$ respectively, it follows that $f \rightarrow \hat{f}$ has a unique extension to be a norm-preserving map of $\mathscr{L}^2(G, \mu)$ onto $\mathscr{L}^2(\hat{G}, \hat{\mu})$ just as in Plancherel's theorem about the classical Fourier transform. In addition to a proof of the generalized Plancherel theorem, Weil's book contains a statement and proof of a generalized Bochner-Herglotz theorem, a study of how Fourier transforms turn convolution and multiplication into one another, and a number of theorems about summability and pointwise convergence of generalized Fourier transforms. Weil's book makes it quite clear that the natural domain of classical commutative Fourier analysis is the study of the Fourier transform on general locally-compact commutative groups. The extra generality so provided was to prove of importance in applications of harmonic analysis to both number theory and probability. We have already hinted at one application to number theory, and there were to be many others. It is perhaps worth noting that Chevalley's "idèles," which were to play a central role in these applications, were introduced in 1936 — almost at the same time as the duality theory itself.

Let μ be a finite measure in the dual $\hat{G}$ of the locally-compact commutative group G, and let us define $\hat{\mu}$, the Fourier transform of μ, to be the continuous function on G defined by the equation $\hat{\mu}(x) = \int \chi(x)d\mu(\chi)$. One verifies at once that this function is "positive definite" in the sense that $\Sigma\, c_i\bar{c}_j\hat{\mu}(x_ix_j^{-1}) \geq 0$ for all pairs $x_1, x_2, \cdots, x_n, c_1, c_2, \cdots, c_n$ of n-tuples where the x_j are group elements and the c_j are complex numbers. The generalized Bochner-Herglotz theorem of Weil asserts conversely that given any continuous positive definite function f on G, there exists a unique finite ("Radon") measure μ on $\hat{G}$ such that $\hat{\mu} = f$. We shall see in the next section that this theorem is important in applications to probability theory. It is also interesting in that it turns out to be equivalent to the spectral theorem for unitary representations of locally-compact commutative groups. In 1929 Wintner (1903-1958) published a paper showing that the Herglotz theorem itself implied the spectral theorem for unitary operators (and so for unitary representations of Z, the additive group of the integers). Then in 1933 Bochner and Khinchin (1894—) independently proved that Stone's spectral theorem for unitary representations of the real line can be derived in the same way from Bochner's real line analogue of Herglotz's theorem. The argument can easily be extended to the general case. This was done in 1943 and 1944 in independent papers of Ambrose (1914—), Godement (1921—), and Naimark (1909—). The key point in connecting unitary representations with positive definite functions can be explained very easily: Let $x \rightarrow U_x$ be a continuous unitary representation of any topological group, and let φ be any vector in $\mathcal{H}(U)$, the space of U. Then it is trivial to verify that $x \rightarrow (U_x(\varphi) \cdot \varphi)$ is a continuous positive definite function on G. Indeed, $\Sigma\, c_i\bar{c}_j(U_{x_ix_j}{}^{-1}(\varphi) \cdot \varphi) = (\Sigma_j\, c_jU_{x_j}(\varphi) \cdot \Sigma_j\, c_jU_{x_j}(\varphi)) \geq 0$.

As one might guess immediately from the case of the real line, the spectral theorem for locally compact commutative groups G asserts the existence of a one-to-one correspondence between continous unitary representations U of G and projection-valued measures P on G such that $(U_x(\varphi) \cdot \varphi) =$

$\int \chi(x)d(P_x(\varphi) \cdot \varphi)$ for all φ in $\mathcal{H}(U)$. It immediately implies the generalized Bochner-Herglotz theorem for all continuous positive definite functions on G which may be put into the form $x \to (U_x(\varphi) \cdot \varphi)$. We need only take $\mu(E) = (P_E(\varphi) \cdot \varphi)$. Conversely, assuming the truth of the Bochner-Herglotz theorem, one can write $(U_x(\varphi) \cdot \varphi) = \int \chi(x)d\mu_\varphi(x)$ and obtain the spectral theorem by studying the dependence of μ_φ on φ. Actually, as observed by Gelfand and Raikov in 1943, there is a very simple argument showing that every continuous positive definite function on a topological group can indeed be thrown into the form $x \to (U_x(\varphi) \cdot \varphi)$. Thus the spectral theorem and the generalized Bochner-Herglotz theorem are equivalent results.

As remarked earlier, the spectral theorem for locally-compact commutative groups is the analogue of the theorem that any unitary representation of a compact group is a direct sum of irreducible representations. When combined with the essential uniqueness of the decomposition, this last theorem tells us that to classify all unitary representations it suffices to classify the irreducible ones. The uniqueness theorem for unitary representations of locally-compact commutative groups has a more subtle formulation and is less easy to prove but is still not very difficult. Moreover, it is essentially contained in the Hahn-Hellinger spectral multiplicity theory for self-adjoint operators mentioned in section 14 and developed between 1907 and 1911. Because of the spectral theorem, the problem of determining all continuous unitary representations of a locally-compact commutative group G to within equivalence is the same problem as determining all projection-valued measures P on $\hat{G}$ to within equivalence. Moreover, the solution of this second problem is independent of the group structure of $\hat{G}$ and depends only on its measure-theoretic structure—more precisely on the system consisting of the set $\hat{G}$ and its σ Boolean algebra of Borel sets. When G (and hence $\hat{G}$) is separable, then $\hat{G}$ is either finite, countable, or isomorphic as a Borel space to the real line. Hence when the solution of the problem is not trivial it may be deduced at once from the solution given by Hahn and Hellinger for projection-valued measures on the line.

I shall present this solution in a modernized form. Let S be a Borel space which is either countable or such that for each finite measure μ on S there exists a Borel set N of measure zero such that $S - N$ is Borel isomorphic to a Borel subset of a separable complete metric space. Such an S is said to be metrically standard. For each finite measure μ defined on all Borel subsets of S, let us define a projection-valued measure P^μ on S whose values are projections in the Hilbert space $\mathcal{L}^2(S, \mu)$ by setting $P^\mu_E(f)(s) = \varphi_E(s)f(s)$, where $\varphi_E(s)$ is 1 for s in E and zero for s not in E. Using the Radon-Nikodym theorem on the existence of "densities" for measures having the same sets of measure zero, it is almost immediate that there exists a unitary map V of $\mathcal{L}^2(S, \mu_1)$ on $\mathcal{L}^2(S, \mu_2)$ such that $VP^{\mu_1}_E V^{-1} = P^{\mu_2}_E$ for all E if and only if μ_1 and

μ_2 have the same sets of measure zero. To get insight into how general projection-valued measures of the form P^μ can be, it is useful to look at the special case in which P is supported by a countable set $\Lambda \subseteq S$. It is easy to see in that case that P is equivalent to some P^γ if and only if $P_{\{\gamma\}}$ has a zero- or one-dimensional range for all $\gamma \in \Lambda$. It is almost trivial that P is uniformly one-dimensional in this sense if and only if P has a commutative commuting algebra; that is, that $P_E T = T P_E$ and $P_E S = S P_E$ for all E implies $ST = TS$. This result suggests that in the general case the P^μ are precisely the P's with commutative commuting algebras. In fact, this theorem is true and moreover not difficult to prove. Let us say that P is *multiplicity free* if its commuting algebra is commutative. Moreover, let us say that P^1 and P^2 are disjoint if $TP_E^1 = P_E^2 T$ for all E implies that $T = 0$. If we define direct sums of projection-valued measures in the obvious way, then the rest of the Hahn-Hellinger spectral multiplicity theory can be summed up in the following propositions:

1) P^{μ_1} and P^{μ_2} are disjoint if and only if μ_1 and μ_2 are supported by disjoint Borel subsets of S; i.e., are mutually singular measures.

2) If P and P^1 are multiplicity free and $\overset{n \text{ times}}{P \oplus P \oplus \cdots}$ is equivalent to $\overset{m \text{ times}}{P^1 \oplus P^1 \oplus \cdots}$, then $n = m$ and P and P^1 are equivalent. (Here m and n may take on the value ∞.)

3) An arbitrary projection-valued measure on S in a separable Hilbert space may be written in the form $\infty P^\infty \oplus P^1 \oplus 2P^2 \oplus 3P^3 \oplus \cdots$ where some terms may be missing and the P^j are disjoint, multiplicity free, and uniquely determined up to equivalence.

Let μ be a finite measure defined in the dual $\hat{G}$ of the locally compact commutative group G. For each x in G and each f in $\mathscr{L}^2(\hat{G}, \mu)$ let $U_x^\mu(f)(\chi) = \chi(x)f(\chi)$. The reader can easily check that $x \to U_x^\mu$ is a unitary representation of G and that P^μ is the projection-valued measure on G canonically associated to U^μ by the spectral theorem. On the other hand, U^μ has an obvious interpretation as the "direct integral" with respect to μ of irreducible (one-dimensional) representations of G.

The results described so far have little to say about the unitary representations of groups that are neither compact nor commutative. The systematic theory of the unitary representations of such groups began to be developed rather abruptly in 1946, a year after the end of World War II. Three important contributions were made earlier, however, and I shall conclude this section with brief descriptions of these. Every irreducible unitary representation of a commutative group is one-dimensional and every continuous unitary irreducible unitary representation of a compact group is finite-dimensional unitary representations at all—except for the identity. In order to have a sufficiency of irreducible representations, one has to allow them to be infinite-dimensional. Also in order to have a reasonable theory, one has to define irreducibility in a topological rather than in an algebraic manner. The continuous unitary representation L of G is said to be irreducible if there exist no *closed* invariant subspaces. That there are in some sense "enough" infinite-dimensional irreducible unitary representations was proved by Gelfand and Raikov in 1943, in the same paper in which they proved that every continuous positive definite function is of the form $x \rightarrow (U_x(\varphi) \cdot \varphi)$. They did so by using the fact that the positive definite functions defined by irreducible continuous unitary representations are the extreme points in the convex set of all positive definite functions.

For a large (but far from exhaustive) class of locally compact groups, new phenomena appear which have to do with the possible structures of the commuting algebras of unitary representations. When the representation is a discrete direct sum of irreducibles, their commuting algebras are direct sums of algebras, each of which is isomorphic to the algebra of all bounded operators in a Hilbert space. One might hope in general for a direct integral of such algebras, but as first pointed out by Murray (1911—) and von Neumann in 1936, it is possible to have a commuting algebra with a trivial center that is not isomorphic to the algebra of all bounded operators in any Hilbert space. In a series of four papers published between 1936 and 1943, Murray and von Neumann made a detailed study of these strange new generalizations of full matrix algebras. They called them factors. After 1950 it became important to classify unitary representations according to the nature of the factors associated with their commuting algebras.

The third contribution was a now famous paper published by Wigner in 1939, containing an analysis of the possible irreducible unitary representations of the so-called inhomogeneous Lorentz group. This is the group generated by the translations in space-time and the celebrated Lorentz group of special relativity. According to the latter theory (advanced by Einstein in 1905), this group is the true group of symmetries of space-time. Because of the principles enunciated by Weyl and described in section 16, one expects a close relationship between the possible relativistic Schrödinger equations and the irreducible unitary representations of the inhomogeneous Lorentz group. It will be easier to describe Wigner's results in a later section. Here it will suffice to remark that Wigner's paper, while incomplete in certain respects, was the first to obtain a genuine classification of the irreducible unitary representations of a group having no non-trivial finite-dimensional unitary representations. Moreover, completing Wigner's analysis was to be a major stimulus to the systematic development which began in 1946.

18. HARMONIC ANALYSIS IN PROBABILITY; ERGODIC THEORY AND THE GENERALIZED HARMONIC ANALYSIS OF NORBERT WIENER

One of the several major consequences of the introduction of the Lebesgue integral (see section 14) was to make possible a convenient and rigorous mathematical framework in which to discuss the many mathematical problems that arose as probability theory found more and more applications in science, engineering, and human affairs. This framework emerged gradually between 1909, when E. Borel published a paper emphasizing the importance of countable additivity in probabilistic considerations, and 1933, when Kolmogorov's (1903—) book *Grundbegriffe der Wahrscheinlichkeitsrechnung* appeared showing how naturally all the main ideas of probability theory could be formulated in measure-theoretic terms. Important landmarks along the way were the papers of Wiener (1894-1964) on "Brownian motion" which appeared between 1920 and 1923, and a paper of Steinhaus (1887-1972) published in 1923. Among other things, Steinhaus's paper related the peculiar convergence properties of expansions in terms of the so-called "Rademacher functions" to the fact that these functions are not only orthogonal but are "independent" as "random variables."

The measure-theoretic framework for probability may be described very simply. One starts with a suitably restricted measure space Ω, μ such that $\mu(\Omega) = 1$, and thinks of the points ω in Ω as "events" in the "universe" Ω of all possible events. If E is a measurable subset of Ω, then $\mu(E)$ is the probability that the event that actually occurs is in the set E. The space Ω, μ is given and fixed once and for all, and probability theory concerns itself with measurable functions defined on Ω. These are given the suggestive name of "random variables" and are usually (but not necessarily) real valued. If g is a real-valued random variable, then the probability that g takes on a value in the Borel set F of real numbers is just $\mu(g^{-1}(F))$. Thus the behavior of the random variable g taken in isolation is completely determined by the measure α on the real line $F \rightarrow \mu(g^{-1}(F))$. This is a *probability measure* in the sense that $\alpha([-\infty, \infty]) = 1$ and is what is called the *distribution* of the random variable g. When it exists, $\int_\Omega g(x)d\mu(x) = \int_{-\infty}^{\infty} x d\alpha(x)$ is called the *expected value* or *expectation* of the random variable g, and if this is denoted by $\bar{g}$, then $\int(g(x) - \bar{g})^2 d\mu(x) = \int_{-\infty}^{\infty} (x - \bar{g})^2 \, d\alpha(x)$ is called the *variance* of g. It is clear that g has an expectation *and* a finite variance if and only if g is in $\mathscr{L}^2(\Omega, \mu)$. Of course, as long as only one random variable is involved, there is little point in introducing the "universe" Ω. Everything can be expressed in terms of the measure α. It is in dealing with relationships between different random variables that the usefulness of Ω becomes clear — especially when there are infinitely many.

Let g_1 and g_2 be two real-valued random variables and let α_1 and α_2 be the probability measures in the real line R that define their distributions. Then $\omega \to g_1(\omega), g_2(\omega)$ is a measurable function ψ from Ω to $R \times R$ and setting $\beta(F) = \mu(\psi^{-1}(F))$ defines a probability measure β in $R \times R$ which is far from uniquely determined by α_1 and α_2. It is called the joint distribution of the random variables g_1 and g_2. If $g_2 = g_1^2$, then β is supported by the curve $y = x^2$. On the other hand if g_1 and g_2 are so-called *independent* random variables, then β is the product measure $\alpha_1 \times \alpha_2$, i.e., $\beta(F_1 \times F_2) = \alpha_1(F_1)\alpha_2(F_2)$. In fact, this is the definition of independence, and this definition seems to capture quite completely the intuitive notion of what it means for random variables to be "independent." Of course, there are many possibilities intermediate between independence on the one hand and functional dependence $(g_2 = F \circ g_1)$ on the other. Random variables may be more or less "correlated." If g_1 and g_2 are in $\mathcal{L}^2(\Omega, \mu)$ and have expectations a_1 and a_2 respectively, then it is easy to verify that $g_1 - a_1$ and $g_2 - a_2$ are orthogonal functions whenever g_1 and g_2 are independent. The converse is not true, but $\int (g_1(x) - a_1)(g_2(x) - a_2)d\mu(x)$ can be used as a rough index of the extent to which g_1 and g_2 are not independent. It is called the *correlation coefficient* of the two random variables g_1 and g_2.

The idea that one could have a mathematical theory of a dependency relation weaker than strict functional dependency was a new and exciting one in the late nineteenth century. It is due to Sir Francis Galton (1822-1911), who became interested in continuous aspects of human inheritance (arm length, height, etc.) more or less at the same time as his exact contemporary Gregor Mendel (1822-1884) was concerning himself with the discrete aspects of inheritance in plants. He introduced the correlation coefficient in the 1870s (after much reflection), but being no mathematician defined it incorrectly. His book *Natural Inheritance,* published in 1889, caught the imagination of a young applied mathematician named Karl Pearson (1857-1936), who corrected Galton's concept of correlation and devoted the rest of his career to founding mathematical statistics.

Apart from modern refinements, the idea of thinking in terms of random variables and their expectations, variances, and higher moments goes back to work of Tchebycheff (1821-1894), as already mentioned in section 3. One of Tchebycheff's achievements was to use these concepts together with a famous inequality that bears his name to give a very simple proof of a generalized form of Bernoulli's "weak law of large numbers." This proof can be particularly elegantly formulated if one uses the modern definition of a random variable. I shall present it here as an example of the convenience of the modern framework. Let $f_1, f_2, \cdots$ be a sequence of independent random variables and let each f_j have the expected value a. Let $\varphi_n(\omega) = \dfrac{f_1(\omega) + f_2(\omega) + \cdots + f_n(\omega)}{n}$ for $n = 1, 2, \cdots$. Then $\varphi_n(\omega) - a$ has zero as expected value and $\int (\varphi_n(\omega) - a)^2 d\mu(\omega) =$

$$\int \frac{(f_1(\omega) - a + f_2(\omega) - a + \cdots + f_n(\omega) - a)^2}{n^2} d\omega = \frac{1}{n^2} \sum_{k=1}^{n} \int (f_k(\omega) - a)^2 d\mu(\omega).$$

If each f_k has finite variance less than or equal to δ, it follows that $\int (\varphi_n(\omega) - a)^2 d\mu(\omega) \leq \dfrac{n\delta}{n^2} = \delta/n$. Now let E_ϵ^n be the set in which $|\varphi_n(\omega) - a| \geq \epsilon$. Then $\int (\varphi_n(\omega) - a)^2 d\mu(\omega) \geq \epsilon^2 \mu(E_\epsilon^n)$ so $\mu(E_\epsilon^n) \leq \dfrac{\delta}{n\epsilon^2}$. Thus for each $\epsilon > 0$ and each $\eta > 0$, one can find n_0 so that $n > n_0$ implies $\mu(E_\epsilon^n) < \eta$. Now $\mu(E_\epsilon^n)$ is the probability that $\dfrac{f_1(\omega) + \cdots + f_n(\omega)}{n}$ differs from a by less than ϵ. We have proved that this probability can be made arbitrarily small by choosing n sufficiently large. Bernoulli's weak law of large numbers is the very special case in which the distributions of the f_j are all identical and are concentrated in a finite number of points. His proof was much more complicated.

Let $f_1, f_2, \cdots$ be a sequence of independent random variables with a common distribution α, and suppose that the f_j have zero expected value and finite variance ν. One verifies at once that $\varphi_n = \dfrac{f_1 + f_2 + \cdots + f_n}{n}$ has zero expected value and variance $\dfrac{\nu}{n}$. Thus $\varphi_n \sqrt{n}$ has zero expected value and variance ν. The central limit theorem of de Moivre and Laplace, in its simplest form, says that the distribution α_n of $\varphi_n \sqrt{n}$ converges as n tends to ∞ to a probability measure of the form $ce^{-x^2/a^2} dx$. Here a^2 and c are uniquely determined by the fact that the variance must be ν and the measure a probability measure. The first rigorous proof was given by Tchebycheff's student Markov (1856-1922) along lines suggested by Tchebycheff (see section 3). Shortly thereafter, a simpler proof was given by Liapunov (1858-1918), another student of Tchebycheff. Liapunov's proof makes use of Fourier transforms and is an interesting example of the application of harmonic analysis to probability. The idea is very simple. It depends on the easily checked fact that if f_1 and f_2 are independent random variables with distributions α_1 and α_2, and α is the distribution of $f_1 + f_2$, then α is the "convolution" of α_1 and α_2. It then follows from the general properties of Fourier transforms that the continuous positive definite functions $\hat{\alpha}, \hat{\alpha}_1, \hat{\alpha}_2$ obtained by taking the Fourier transforms of the measures are related by the equation $\hat{\alpha} = \hat{\alpha}_1 \hat{\alpha}_2$. (I may mention in passing that specialists in probability theory refer to the Fourier transform of a probability measure as its *characteristic function*.) Now if $f_1, f_2, \cdots, f_n, \cdots$ is a sequence of independent identically distributed random variables, the sum $f_1 + f_2 + \cdots + f_n$ will have a distribution whose characteristic function is g^n where g is the characteristic function of the common distribution α of the f_j. Thus $\dfrac{f_1 + \cdots + f_n}{\sqrt{n}}$ will have a distribution whose characteristic function is $y \to (g(y\sqrt{n}))^n$. To prove the central limit theorem one need only investigate the limit of $(g(y\sqrt{n}))^n$ (which turns out to be easy), verify that it is the characteristic function of a distribution of the form $ce^{-x^2/a^2}dx$, and prove an appropriate continuity theorem for the Fourier transform and its inverse. Liapunov's proof is not explicitly of this form, but in 1935 Levy (1886-1971) proved a theorem stating that the one-to-one map from probability measures to positive definite functions defined by the Fourier transform is a homeomorphism when the two spaces are given natural topologies. The central limit theorem in the simple form stated above is essentially a corollary of this theorem in harmonic analysis.

The so-called strong law of large numbers could not even be formulated until the concept of a set of measure zero was available. One says that the strong law of large numbers holds for a sequence $f_1, f_2, \cdots$ of random variables having a common expectation a if $\displaystyle\lim_{n \to \infty} \frac{f_1(\omega) + \cdots + f_n(\omega)}{n} = a$ for almost all ω; that is, if the sequence $\dfrac{f_1(\omega) + \cdots + f_n(\omega)}{n}$ converges to a with probability one. That this is so for independent random variables with a common distribution and finite variance was proved by Cantelli (1875-1966) in 1917, a less general result having been proved earlier by E. Borel.

Quite generally, a singly or doubly infinite sequence of random variables $f_1, f_2, \cdots$ or $\cdots, f_{-1}, f_0, f_1, f_2, \cdots$ is called a *discrete parameter stochastic process*. The various sequences $f_1(\omega), f_2(\omega), \cdots$ or $\cdots, f_{-1}(\omega), f_0(\omega), \cdots$ arising from the points ω of Ω are called the *sample sequences* of the process. When the random variables are not independent, one classifies processes by the nature of the dependency relations that exist. These are determined by the joint probability distributions in $R \times R \times \overset{n \text{ times}}{\cdots} \times R = R^n$ associated with the finite subsequences $f_k, f_{k+1}, \cdots, f_{k+n-1}$. For example, if the distribution in R^n is independent of k, the process is said to be *stationary*. Perhaps the most important and widely studied class of all is the class of *Markov processes*. To define this class one needs the notion of "conditional probability." Let α be a probability measure in $R \times R$ and let $\tilde{\alpha}$ be the projection on the first factor, i.e., $\tilde{\alpha}(F) = \alpha(F \times R)$ for each Borel subset F of R. By a simple and fundamental theorem in measure theory, there exists for each x in R a probability measure β_x in R such that $\alpha(E \times F) = \int_E \beta_x(F)d\tilde{\alpha}(x)$ for all Borel sets E and F. The measures β_x are uniquely determined (mod $\tilde{\alpha}$ sets of measure zero). If two random variables f and g have α as joint probability distribution, one says that $\beta_x(E)$ is the *conditional prob-*

ability that g is in E given that f takes the value x. The generalization to several variables is obvious. Thus if $f_k, f_{k+1}, \cdots, f_{k+n-1}$ are random variables in a process, one can introduce $\beta_{x_k, x_{k+1}, \cdots, x_{k+n-2}}$ the conditional probability distribution for f_{k+n-1} given that $f_k = x_k, f_{k+1} = x_{k+1}, \cdots, f_{n+n-2} = x_{k+n-2}$. The process is said to be a *Markov process* if $\beta_{x_k, \cdots, x_{k+n-2}}$ is independent of n and of $x_k, \cdots, x_{k+n-3}$ and depends only on x_{k+n-2}. It is said to be a Markov process with *stationary transition probabilities* if $\beta_{x_k, \cdots, x_{k+n-2}} = \beta_{x_{k+n-2}}$ is also independent of $k+n$; that is, if there exists a map $x \to \beta_x'$ of real numbers into probability distributions such that $\beta_{x_k, \cdots, x_{k+n-2}} = \beta_{x_{k+n-2}}'$. One can think of the sample sequences of such a process as generated by a "random walk." Choose an arbitrary starting point x_1. Move to x_2 "at random" using the probability measure β_{x_1}'. Then move to x_3 "at random" using β_{x_2}', etc. In the special case in which the probability distribution of the f_j are all supported by a fixed finite set — say $\{1, 2, \cdots, k\}$—one says that there is a "finite state space" and $x \to \beta_x'$ is defined by a $k \times k$ matrix. Markov began the theory in 1906 with an analysis of the finite state case.

Instead of a sequence of random variables $f_1, f_2, \cdots$, it is convenient in many problems to consider a family $t \to f_t$ of random variables parameterized by a real number t (interpreted as the time in most applications). While there are additional technical complications (of a highly non-trivial nature) and one brand new problem, in broad outline the theory of continuous parameter Markov processes is similar to the theory of the discrete parameter case. Of course, one has sample functions instead of sample sequences, and the measure μ in the universe Ω leads to a measure in the space of all sample functions. The celebrated Wiener measure is of this character. In fact, Wiener's theory of Brownian motion, developed in the early 1920s, is the theory of the most important special case of a continuous parameter Markov process with stationary probabilities and continuous state space. The general theory of such processes was inaugurated by Kolmogorov in an important paper published in 1931. The new problem arises out of a difference between discrete and continuous parameter Markov processes that is analogous to that between unitary representations of the groups Z and R. In the first case, the representation $n \to U_n$ is uniquely determined by the single unitary operator U_1. In the second case, since R has no least element, one must differentiate and express $t \to U_t = e^{iHt}$ in terms of its "infinitesimal generator" iH in order to describe the representation by a single operator. In a continuous parameter Markov process, the transition probabilities are described by a probability measure β_{x,t_1,t_2}' which depends in general on *three* real variables and has the following interpretation: $\beta_{x,t_1,t_2}'(F)$ is the probability that f_{t_2} will have a value in F given that f_{t_1} has the value x ($t_1 < t_2$). In the stationary case $\beta_{x,t_1,t_2}' = \beta_{x,0,t_2-t_1}'$, so that there are effectively only two variables. Quite generally, let $x \to \gamma_x^1$ and $x \to \gamma_x^2$ be two mappings of R into probability measures in R, and suppose that $x \to \gamma_x^j(E)$ is measurable in x for $j = 1, 2$, and all E. Then the two step random walk using first $x \to \gamma_x^1$ and then $x \to \gamma_x^2$ leads to a mapping $x \to \gamma_x^3$ of R into probability measures which one can think of as the composite $\gamma^2 \circ \gamma^1$ of γ^1 and γ^2. It follows from the definition of a continuous parameter Markov process that $\beta'_{t_3,t_2} \circ \beta'_{t_2,t_1} = \beta'_{t_3,t_1}$ when $t_1 < t_2 < t_3$, a relationship known as the Chapman-Kolmogorov equation. When the transition probabilities are stationary so that $\beta'_{t_1,t_2} = \beta'_{0,t_2-t_1}$, this reduces to $\beta'_{0,t_1+t_2} = \beta'_{0,t_1} \circ \beta'_{0,t_2}$ so that the $\beta'_{0,t}$ constitute a one-parameter semi-group under composition. This has an infinitesimal generator that is defined by a linear operator in a suitable function space. The new problem lies in choosing this function space in such a way that one can recover the $\beta'_{0,t}$ from the infinitesimal generator. In the case of an infinite discrete state space, one has a one parameter semi-group of infinite matrices and one can differentiate to get an infinite matrix. This would seem to be the infinitesimal generator, but unfortunately it does not determine the semi-group. One needs "boundary conditions" in addition. Papers published by Doob (1910—) in 1942 and 1945 and by Lévy in 1951 did much to clarify the situation, but there has been room for much subsequent work by other mathematicians. In the case of a continuous state space, the problems are even more difficult. In Wiener's theory of Brownian motion, the formal infinitesimal generator is a constant times d^2/dx^2, and more generally one defines a diffusion process in such a way that its formal infinitesimal generator is a second order ordinary differential operator (with rather general coefficients). W. Feller (1906-1970) devoted a large fraction of a distinguished career to a study of the determination of the process from its (suitably defined) infinitesimal generator in this case. Many applications of probability theory involve Markov processes, and as in physics one starts by knowing the infinitesimal generator.

The other much studied and extensively applied class of stochastic processes is the class of stationary processes — both discrete and continuous parameter. This class is also the most relevant to our main theme because its theory is related in an intimate way both to harmonic analysis and to the branch of mathematical analysis known as ergodic theory. Ergodic theory is one of the newer branches of mathematics, essentially non-existent before 1931. Moreover, it turns out to have a number of significant connections with the theory of unitary group representations and increases the scope of harmonic analysis in a manner which is only now beginning to be explored. Let us begin with a brief account of the nature and history of ergodic theory.

Let Γ denote the phase space of a classical dynamical system and let U_t be the one-to-one transformation of Γ into itself which is such that $U_t(\gamma)$ is the point of Γ that describes the state of the system t time units after it was described by γ. (We consider only systems that are "reversible" in the sense that the U_t exist.) Then $U_{t_1+t_2} = U_{t_1} U_{t_2}$ and we have an action of the additive group of the real line on Γ. Let H be the real-valued function on Γ that defines the energy, and for each real number E for which $H^{-1}(E)$ is not empty let $\Gamma_E = H^{-1}(E)$. Then the sets Γ_E are carried into themselves by the U_t, and moreover, each of them admits a natural measure ϱ_E which is U_t invariant. Furthermore, for the systems of interest in statistical mechanics (see section 13), $\varrho_E(\Gamma_E)$ is finite. As part of the program for deducing the fundamental algorithm of statistical mechanics, Boltzmann wanted to prove the following theorem: Let g be a bounded continuous real-valued function on Γ_E. Then the "time average" $\lim_{T\to\infty} \frac{1}{T} \int_0^T g(U_t(\gamma))dt$ is equal to the "space average" $\frac{1}{\varrho_E(\Gamma_E)} \int_{\Gamma_E} g(\gamma)d\varrho_E(\gamma)$ for all choices of γ. Boltzmann naïvely hoped to base a proof of this theorem on the hypothesis that each trajectory or path $t \to U_t(\gamma)$ goes through every point of the constant energy hypersurface Γ_E. Putting together the Greek words *ergon* 'work' and *odos* 'path,' he called his hypothesis the *ergodic hypothesis*. The validity of this hypothesis was much debated, and it was ultimately shown to be untenable for topological reasons that now seem obvious. While various alternative hypotheses were proposed, nothing much could be deduced from them, and the subject remained in a state of uncertainty and confusion for over half a century.

Ergodic theory in its modern sense came into being in 1932 as a consequence of the following sequence of events: In May 1931, B. O. Koopman (1900—), inspired by Stone's paper of 1930 on unitary representations of the real line and by von Neumann's 1929 work on the spectral theorem, published a short note making an observation that today seems quite obvious. He pointed out that if one forms the Hilbert space $\mathscr{L}^2(\Gamma_E, \varrho_E)$, one can obtain a unitary representation V of the real line by setting $V_t(f)(\gamma) = f(U_t(\gamma))$. He also pointed out that Stone's theorem permits one to assign a projection-valued measure to the system, and suggested that it might be fruitful to relate the properties of this projection-valued measure to the properties of the system. Koopman discussed his work with von Neumann before it was published, and this conversation suggested to von Neumann the possibility of applying operator-theoretic methods to prove the equality of space and time averages in statistical mechanics. That Koopman's work might be so applied was also suggested to von Neumann by Weil[6] a short time after Koopman's paper appeared. These facts are stated in the introduction to a short note by von Neumann published in early 1932. In this note von Neumann proves what is now called the mean ergodic theorem. This asserts that for any f in $\mathscr{L}^2(\Gamma_E, \varrho_E)$, the time averages $\frac{1}{T} \int_0^T f(U_t(\gamma))dt = f_T$ converge in the Hilbert space metric to a function $\bar{f}$ which is a constant on the U_t orbits. If the action of U_t on Γ_E is "metrically transitive" in the sense that there are no measurable invariant subsets (except sets of measure zero and their complements), then it is easy to show that $\bar{f}$ is almost everywhere equal to the constant $\dfrac{1}{\varrho_E(\Gamma_E)} \int f(\gamma)d\varrho_E(\gamma)$. Except for replacing transitivity by metric transitivity and using convergence "in the mean" instead of pointwise convergence, this is just what Boltzmann had wanted to do. In fact, whenever the hypothesis of metric transitivity can be verified, the physical conclusions desired by Boltzmann follow. Pointwise convergence is not really needed, but it too can be proved. Shortly after learning of von Neumann's result, G. D. Birkhoff (1884-1944) proved the much more difficult pointwise ergodic theorem. For f integrable with respect to ϱ_E, the time averages $\dfrac{1}{T}\int_0^T f(U_t(\gamma))dt$ converge for almost all γ to an integrable function $\bar{f}$ which is constant on the U_t orbits. As before, if the action is metrically transitive, $\bar{f}$ is almost everywhere equal to $\dfrac{\int f(\gamma)d\varrho_E(\gamma)}{\varrho_E(\Gamma_E)}$.

Let us look more closely at the hypothesis of metric transitivity. It is very close to the original ergodic hypothesis of Boltzmann, but differs from it in one important respect. The ergodic hypothesis may be reformulated as the hypothesis that Γ_E has no proper subsets that are invariant under the U_t. The hypothesis of metric transitivity seems to be only slightly weaker. It excludes proper invariant subsets, but only those which are measurable and not measure-theoretically improper. One might suppose at first that this weakening made little difference — that under reasonable regularity conditions metric transitivity could be reduced to ordinary transitivity by discarding an invariant set of measure zero. This naïve supposition is false. Consider the action of the infinite cyclic group Z on the circle $|z| = 1$ defined by setting $(z)n = ze^{in\theta}$ where θ is an irrational multiple of π. The ordinary arc length measure in $|z| = 1$ is preserved, and every Z trajectory is countable and hence of measure zero. Cannot some of these trajectories be gathered together into a measurable set other than by taking almost all of them or almost none? A simple argument using Fourier analysis shows that they cannot. One need only study the Fourier coefficients of the characteristic function of a Z-invariant measurable set to see that every such set is either of measure zero or the complement of a set of measure zero. Metric transitivity, though analogous to ordinary transitivity, is *not* reducible to it and in fact turns out to be much more general.

That non-transitive metric transitivity can exist at all was noted for the first time only a few years earlier. Following the lead of G. W. Hill (1838-1914) and Poincaré (1854-1912), G. D. Birkhoff had been studying the qualitative properties of low-dimensional dynamical systems whose equations of motion could not be integrated. Using Poincaré's idea of "surfaces of section," he could reduce problems about actions of the real line on three-dimensional manifolds to problems about actions of the integers on two-dimensional surfaces. In this connection and in collaboration with Paul Smith (1900—), he published a long paper in 1928 on the structure of such surface transformation groups. This paper contains a definition of metric transitivity and the example sketched above.

It seems to have been von Neumann, however, who first realized the far-reaching significance of the new concept. A slight modification of the example of Birkhoff and Smith shows that it is possible for the real line to act in a metrically transitive manner on a compact manifold of arbitrarily high dimension. Thus (though the question seems difficult to settle in concrete cases) it is at least conceivable that many, and even most, dynamical systems have the property that the action defined by the time evolution of the system is metrically transitive on the constant energy hypersurfaces. As von Neumann emphasized, the hypothesis of metric transitivity is exactly the right substitute for Boltzmann's untenable ergodic hypothesis. Because of this and because "ergodic" is shorter than "metric transitivity," it has become customary to follow von Neumann's lead and call a metrically transitive action an *ergodic* action. Of course, a transitive action or even one reducible to a transitive action by neglecting an invariant set of measure zero is automatically ergodic. It will be convenient to distinguish between the two possibilities by using the terms *essentially transitive* and *properly ergodic* depending on whether there is or is not an orbit of positive measure.

The existence of proper ergodicity raises a host of interesting questions, and modern ergodic theory may be loosely defined as the branch of mathematics that attempts to answer them. The first detailed paper on the subject was published by von Neumann in 1932 and is entitled "Zur Operatorenmethode in der klassischen Mechanik." Theorem 1 of the paper is von Neumann's mean ergodic theorem, which is proved in detail. Theorem 2 is a fundamental decomposition theorem underlining the fact that the ergodic actions are the fundamental building blocks out of which all measure-preserving actions can be constructed. Consider the action of the infinite cyclic group Z on the unit disk $|z| \leq 1$ in the complex plane defined by setting $(z)n = ze^{in\theta}$ where θ/π is irrational. This action preserves the area measure in the disk but is far from ergodic. Indeed, for $0 < a < 1$, the set of all z with $|z| \leq a$ is both invariant and measure-theoretically proper. On the other hand, the circles $|z| = r$ fiber the disk, and the area measures may be recovered from the arc measures in the fibers by an integral over $0 \leq r \leq 1$. Moreover, these fibers are invariant so that the given action is in an obvious sense a "direct integral" of the actions in the fibers. Finally, by an argument already sketched, the fiber actions are all ergodic. Thus the given measure-preserving action decomposes as a "direct integral" of ergodic actions. Von Neumann's Theorem 2 asserts (with mild regularity assumptions) that a fibering (not necessarily with good topological properties) permitting such a decomposition always exists and is essentially unique. Theorem 5 makes a beginning on the difficult and still unsolved problem of classifying the possible "essentially different" ergodic actions of R by giving a complete classification in a special case. Let $t \to V_t$ be the unitary representation of the real line associated with the action by Koopman's construction. Let $V_t = e^{iHt}$ where H is self-adjoint. If H has a basis of eigenvectors (equivalently if V is a discrete direct sum of irreducibles), then the action is said to have a *pure point spectrum*. Given such an action, let $\lambda_1, \lambda_2,$ $\cdots$ be the eigenvalues of H. It is easy to see that the λ_j all occur with multiplicity one and constitute a *subgroup* of the additive group of the real line. Von Neumann showed that actions with a pure point spectrum are determined to within an obvious equivalence by the countable subgroup of eigenvalues of H, and that *every* countable subgroup occurs. The action is properly ergodic if and only if the subgroup is dense, and this happens if and only if the subgroup is not the set of all integer multiples of a fixed positive real number. Using group duality (which was discovered two years later), it is very easy to describe the properly ergodic action associated with a dense countable subgroup D of the real line. Think of D as a subgroup of $\hat{R}$ and let ψ be the natural isomorphism of D into $\hat{R}$. Then ψ^* (see section 17) is a dense imbedding of R in $\hat{D}$. The action of R on $\hat{D}$ defined by setting $(\chi)t$ $= \psi^*(t)\chi$ preserves Haar measure in $\hat{D}$ and is the required properly ergodic action. Theorem 6 is concerned with the properties of an interesting class of ergodic actions of R constructed out of ergodic actions of Z and real-valued functions. A main conclusion is that not only do there exist ergodic actions with point spectrum $\{0\}$, but that these are the rule rather than the exception — at least in the class considered by von Neumann. Von Neumann's paper stimulated many others and ergodic theory developed rapidly. By 1937 E. Hopf (1902—), one of the more active workers, was able to publish a short book on the subject. For a while there was hope that an ergodic action would be completely determined by its spectrum (as in the pure point spectrum case), but that hope turned out to be illusory. Even in the case of ergodic actions of the integers there is no immediate hope of obtaining either a complete classification or a characterization of the spectra that can arise.

Although ergodic theory originated in statistical mechanics, it had little impact on that subject until rather recently. In the first few decades of its existence, its most important applications by far were to the theory of stationary stochastic processes. Consider first the discrete parameter case and let $\cdots, f_{-1}, f_0, f_1, \cdots$ be the random variables of our stationary process. There is no essential loss in generality in assuming that the functions f_j separate the points of the universe Ω in that $\omega \to \cdots, f_{-1}(\omega), f_0(\omega), f_1(\omega), \cdots$ is a one-to-one map of Ω into the product space of countably many replicas of R. If μ' is the image of μ in the product space, the complement of the image of Ω will be of measure zero. Hence there is no loss in generality in identifying Ω with the space R^z of *all* real-valued functions on the integers. When this is done, $f_n(\omega)$ is just $\omega(n)$, and translation provides a natural action of Z on Ω such that $f_n(\omega) = f_0([\omega]n)$. Moreover, the definition of stationarity translates into the statement that μ is invariant under the Z action. In other words, a discrete parameter stationary stochastic process is uniquely determined by the system consisting of a measure-preserving action of Z on a probability measure space Ω, μ and a single measurable function f on Ω. The random variables of the process are of course the functions $\omega \to f([\omega]n)$ for $n = \cdots, -1, 0, 1, 2, \cdots$. Similarly, although the argument is less straightforward, a continuous parameter stationary stochastic process is defined by the system consisting of a measure-preserving action of R on a measure space Ω, μ and a measurable function f on Ω, the random variable f_t being $\omega \to f([\omega]t)$. While the measure-preserving action of R or Z canonically associated with a stationary stochastic process need not be ergodic, the underlying universe Ω can always be measure-theoretically fibered into ergodic parts by the theorem of von Neumann stated above. Moreover, any actual event $\omega \,\epsilon\, \Omega$ will belong to some ergodic part, and for most applications of the theory one can forget the rest of Ω. For these reasons there is little loss in generality in considering only stationary stochastic processes in which the action is ergodic. Actually, it is quite easy to show that when the random variables are independent (which can happen only in the discrete parameter case) then the underlying action is ergodic!

Suppose then that our stationary stochastic process is defined by an ergodic action of R or Z on a probability measure space Ω, μ and a real-valued measurable function f. If the random variables have expectations, so that $\int (f(\omega))d\mu(\omega) < \infty$, we may apply the pointwise ergodic theorem of G. D. Birkhoff. There is a discrete version as well as a continuous one, and it asserts that for almost all ω, $\displaystyle\lim_{n \to \infty} \frac{f(\omega) + f(\omega \cdot 1) + f(\omega \cdot 2) + \cdots + f(\omega \cdot n)}{n + 1}$ exists and equals $\int f(\omega)d\mu(\omega)$. But $f((\omega)n) = f_n(\omega)$, and $\int f(\omega)d\mu(\omega)$ is the common expectation of the random variables f_n. Thus the discrete version of the pointwise ergodic theorem is nothing more or less than the strong law of large numbers for arbitrary (ergodic) stationary stochastic processes. As such it is a considerable generalization of the Borel-Cantelli theorem, which states the strong law for independent identically-distributed random variables. It is curious that this interpretation of the pointwise ergodic theorem was not immediately recognized. While G. D. Birkhoff's proof was announced and sketched in late 1931, and Kolmogorov's book on measure theoretic foundations appeared in 1933, the book does not mention the ergodic theorem. The connection between the ergodic theorem and the strong law of large numbers was not mentioned in print until 1934. It was pointed out in three independent papers published in that year by Doob, E. Hopf, and Khintchine respectively.

Suppose now that the random variables have finite variances so that f is in $\mathscr{L}^2(\Omega, \mu)$. Let $G = Z$ or R depending upon whether we are in the discrete case or the continuous case. Let $x \to V_x$ be the unitary representation of G defined by the Koopman construction. Then $(V_x(f) \cdot f)$ is a positive definite function canonically associated with the process, and by the Bochner (or Herglotz) theorem $V_x(f) \cdot f = \int \chi(x) d\nu(\chi)$ where ν is a finite measure on $\hat{G}$, i.e., on the circle or the line. This measure is called the *spectrum* of the process and plays an important role in its theory. Another interesting consequence of the pointwise ergodic theorem is that the function $x \to (V_x(f) \cdot f)$, and hence the spectrum ν can be determined from the "complete past" of almost any sample function. For definiteness let us look at the continuous case. Then $V_x(f)(\omega) = f([\omega]x)$, and since f and $V_x(f)$ are both square integrable, their product $(V_x(f))f = g$ is integrable. Hence, by the ergodic theorem,

$$\lim_{T \to \infty} \frac{1}{T} \int_0^T g([\omega](t)) dt \text{ exists for almost all } \omega \text{ and equals } \int g(\omega) d\mu(\omega) = V_x(f) \cdot f.$$

Now $g([\omega](t)) = f([\omega](t))f([\omega] (x - t)) = f_\omega(t)f_\omega(x - t)$ where $t \to f_\omega(t) = f([\omega]t)$ is the sample function attached to the point ω of the universe Ω.

Thus, for almost all ω, $\lim_{T \to \infty} \frac{1}{T} \int_0^T f_\omega(-t)f_\omega(x - t) dt$ exists and equals $V_x(f) \cdot f$.

It can be computed for negative x when one knows $f_\omega(t)$ for $t \le 0$, and since it is an even function of x, this determines it for all x. The significance of this result lies in the fact that the theory of stationary stochastic processes is mainly applied to the statistical analysis of so-called "time series" — such as occur for example in economics and meteorology. One thinks of the sample functions $t \to f_\omega(t)$ as "possible" functions describing the temperature, say, or the price of wheat as a function of the time. One supposes that the variation of temperature or wheat prices actually observed is given by a function f_{ω_0} chosen "at random" from Ω according to the probability measure μ. Looking at how $f_{\omega_0}(t)$ has behaved for $t \le 0$ (i.e., in the past), one can compute the Fourier transform of the spectrum of the whole stochastic process. It is just the so-called "autocorrelation function" $\lim_{T \to \infty} \frac{1}{T}$

$\int_0^T f_{\omega_0}(-t)f_{\omega_0}(x - t) dt$ of the sample function f_{ω_0}.

Some appreciation of the significance of the spectrum of a process may be had by looking at its discrete components, if any. Suppose that the measure ν of which $V_x(f) \cdot f$ is the Fourier transform has an "atom," i.e., that $\nu(\{\lambda\}) \ne 0$ for some λ in $\hat{R}$. This happens precisely when there exists $\varphi \not\equiv 0$ in $\mathscr{L}^2(\Omega, \mu)$ such that $V_x(\varphi) \equiv e^{i\lambda x}\varphi$, and then φ is uniquely determined up to a multiplicative constant and f may be written uniquely in the form $f = c\varphi + f^\perp$ where φ and $f^\perp$ are orthogonal in $\mathscr{L}^2(\Omega, \mu)$. Now $f_\omega(t) = f([\omega]t) = c\varphi([\omega]t) + f^\perp([\omega]t) = ce^{i\lambda t}\varphi(\omega) + f^\perp([\omega]t)$. Thus every sample function has a canonical decomposition as the sum of a constant multiple of the periodic function $e^{i\lambda t}$ and a sample function of a new process defined by $f^\perp$. The spectrum of the new process is the same as that of the original except for the removal of the atom at λ. Other atoms can be removed similarly, and in fact one can write $f = \Sigma\ c_j\varphi_j + g$ where the φ_j and g are mutually orthogonal, $|\varphi_j(\omega)| \equiv 1$, $V_t(\varphi_j) = e^{i\lambda_j t}\varphi_j$, and g defines a process whose spectrum has no atoms. Correspondingly, the sample function in f_ω may be written $f_\omega(t) = \Sigma\ c_j\varphi_j(\omega)e^{i\lambda_j t} + g_\omega(t)$. The terms $c_j\varphi_j(\omega)e^{i\lambda_j t}$ are called the "hidden periodicities" of the sample function. Using the ergodic theorem just as before (but applied to $f\bar\varphi_j$), one finds that the coefficient $c_j\varphi_j(\omega)$ may be computed from the past of almost any sample function by the formula

$$c_j\varphi_j(\omega) = \lim_{T \to \infty}\ \frac{1}{T}\int_0^T f_\omega(-t)e^{-i\lambda_j t}dt.$$

This formula is essentially identical with that used by Bohr (see section 17) in computing the expansion coefficients of almost-periodic functions. Indeed, when the c_j do not go to zero too slowly, the difference $f_\omega(t) - g_\omega(t)$ is an almost-periodic function in Bohr's sense. In any case, it is a generalized almost-periodic function whose values for $t > 0$ are determined by its values for $t < 0$. There can be no true randomness in our stochastic process if the underlying ergodic action has a pure point spectrum.

The consequences of ergodic theory for the statistical study of time series just described were pointed out in the 1930s, the major publications being papers by Khintchine and Wold published in 1934 and 1938 respectively. They provided a justification for and a conceptual clarification of methods already in use by scientists and statisticians in studying specific time series. The determination of hidden periods goes back to work of the physicist Schuster (1851-1934) beginning in 1898. The use of the autocorrelation functions goes back to before 1920. A much-cited paper by Yule (1871-1951), published in 1927, makes use of autocorrelations in studying sunspot data.

In 1930 Norbert Wiener published a long memoir entitled "Generalized harmonic analysis," which though conceived in a different spirit was in effect a remarkable anticipation of the theory of stationary time series. He had been engaged in trying to help electrical communication engineers cope with some of the problems of circuit design, and these problems seemed to require applying harmonic analysis to functions which were neither square summable nor periodic and, moreover, were more general than the almost-periodic functions of Bohr. For reasons which I shall not attempt to describe, Wiener decided to study measurable functions for which the auto-correlation function exists and is continuous. As we have seen, it is a consequence of the ergodic theorem that the sample functions of stationary stochastic processes constitute a rich source of examples. While the ergodic theorem was not stated or proved until a year or so after Wiener's paper appeared, practically all the examples offered by Wiener were essentially sample functions and were defined using randomness. On the other hand, Wiener did not at the time think of himself as studying a whole statistical ensemble of functions at once as in the case of the f_ω, but as studying a single function. His chief concern was to define the spectrum of the function, which he had to do without using Bochner's theorem (Bochner's theorem was published in 1932).

In studying carefully the relationship between a function and its spectrum, Wiener found himself in need of a more powerful Tauberian theorem than any that existed (see section 14), and was thereby led to write his prize-winning paper "Tauberian theorems," which was published in 1932.

Later Wiener came to realize the merits of thinking of his functions as sample functions. This led him to the important insight that the coded messages with which the communication engineers had to deal were close analogues of time series. Modern communication engineering makes heavy use of his discovery.

From the point of view of this article the relationship between Fourier integrals, Fourier series, almost-periodic function expansions, and Wiener's generalized harmonic analysis is best viewed as follows: Let S, μ be a suitably restricted measure space and let the real line R act upon S to preserve μ. Harmonic analysis on R is concerned with decomposing the unitary representation V of R defined in $\mathscr{L}^2(S, \mu)$ by the Koopman construction. Now by von Neumann's decomposition theorem, the given action can be decomposed into ergodic actions, and correspondingly V decomposes as a direct integral of Koopman representations — one for each ergodic component. This part of the decomposition is basically geometry, and one may think of harmonic analysis proper as the decomposition of V when the action is ergodic. We now divide into four cases according to whether ergodic action is properly ergodic or essentially transitive, and also according to whether $\mu(S)$ is finite or infinite. When the action is essentially transitive, S may be taken to be R modulo a closed subgroup which is necessarily either $\{0\}$ or the subgroup of all integer multiples of a, and the action is then just translation on the quotient group. Depending upon whether the subgroup is $\{0\}$ or not, that is upon whether $\mu(S) = \infty$ or $\mu(S) < \infty$, one is reduced to the Fourier transform or to Fourier series. When the action is properly ergodic, one can no longer identify S with the real line or one of its quotient groups. However, the decomposition of the Koopman representation V may still be regarded as defining a decomposition of real-valued functions on the real line, namely the functions $t \to f([s]t)$ for each s in S and each f in $\mathscr{L}^2(S, \mu)$. Indeed, for each E and each s the function $t \to P_E^V(f)([s]t)$ may be regarded as the component of $t \to f([s]t)$ whose spectrum is in the set E. When $\mu(S) < \infty$, this analysis is essentially the generalized harmonic analysis of Wiener — carried somewhat further than Wiener carried it. When in addition V is a discrete direct sum of irreducible (and hence one-dimensional) representations, one recovers a slight generalization of Bohr's theory of almost-periodic functions (see section 17). The case in which $\mu(S) = \infty$ does not seem to have been investigated.

19. EARLY APPLICATIONS OF GROUP REPRESENTATIONS TO NUMBER THEORY—THE WORK OF ARTIN AND HECKE

In the first quarter-century of its existence, the theory of representations of finite groups had many applications to group theory itself. I have already mentioned the theorem on the structure of groups of order $p^\alpha q^\beta$. To my knowledge, however, there were no applications to other fields such as physics, number theory, or probability until the 1920s. The extensive applications to quantum mechanics, which began in 1927 (see section 16) have had the most publicity, but they were not the first. Appropriately, in view of Dedekind's role in inspiring Frobenius, the first application outside group theory seems to have been to Dedekind's own creation — the theory of algebraic number fields. It was made in a celebrated paper of Artin (1898-1962) entitled "Über eine neue Art von L-Reihen," published in 1923.

Before attempting to explain the nature and significance of what Artin did in this paper, I must devote a few paragraphs to sketching the origins of the theory of algebraic number fields and the course of its development between its beginnings in the 1870s and the publication of Artin's paper in 1923.

As observed by Gauss in 1830, the problem of finding the integer solutions of the equation $x^2 + y^2 = n$ may be usefully approached by factoring $x^2 + y^2$ as the product of $x + iy$ and $x - iy$. The set of all complex numbers of the form $x + iy$ where x and y are integers is a ring called the ring of *Gaussian integers*. Like the ordinary integers, the Gaussian integers admit "unique" factorization into "primes." One calls a Gaussian integer a *unit* if it has a multiplicative inverse, and a *prime* if it cannot be written as a product of Gaussian integers other than units and products of units with itself. It is then easy to prove that every Gaussian integer is a product of primes and that this factorization is unique up to reordering and multiplication by units. The units are $1, -1, i$, and $-i$. Now since $(a + ib)(a - ib) = a^2 + b^2$, it follows at once that every Gaussian prime is a factor of some ordinary integer and hence of some ordinary prime. To determine the Gaussian primes, it thus suffices to factor the ordinary primes, and it is obvious that an ordinary prime p which is not already a Gaussian prime must be of the form $x^2 + y^2$ and factor into $x + iy$ and $x - iy$. It is not difficult to show that $x + iy$ and $x - iy$ are always Gaussian primes and are equal mod units only when $p = 2$. In that case $(1 + i) = (1 - i)(i)$. Thus once one knows for which primes p one can solve $x^2 + y^2 = p$, one knows all Gaussian primes. To solve $x^2 + y^2 = n$, one then has only to factor n into Gaussian primes and then divide the factors into two classes in such a way that each class contains just one of each pair of conjugates.

The advantage of this approach is that it may be applied not only to quadratic Diophantine equations other than $x^2 + y^2 = n$, but to higher order equations as well. For example, when m is odd, one can study the equations $x^m + y^m = n$ by factoring $x^m + y^m$ as the product $(x + \omega y)(x + \omega^2 y) \cdots (x + \omega^{m-1} y)(x + y)$ where ω is a primitive mth root of 1. There is a difficulty, however, in that the ring generated by ω need not have unique factorization. If it did, one could obtain enough information about the solvability of $x^m + y^m = n$ to prove Fermat's famous conjecture about the nonexistence of integer solutions of $x^m + y^m = z^m$ when $m > 2$. Indeed, in the early 1840s, Kummer (1810-1893), overlooking the possible failure of unique factorization, thought he had a proof of the Fermat conjecture in the general case. This mistake led him to attack the problem of finding a substitute for the unique factorization law and to solve the problem for the particular case of the ring generated by the mth roots of 1. The "ideal numbers" he introduced as substitutes for primes sufficed to deal with various special cases of the Fermat problem, but the general case remains open to this day.

If one passes from $x^m + y^m$ to a general homogeneous form of the mth degree $a_m x^m + a_{m-1} x^{m-1} y + \cdots + a_0 y^m$, one can still approach the problem of solving Diophantine equations of the form $a_m x^m + a_{m-1} x^{m-1} y + \cdots + a_0 y^m = n$ by factoring the left-hand side. One can write it as $a_m y^m((x/y)^m + a_{m-1}(x/y)^{m-1} + \cdots + a_0) = a_m y^m((x/y) - \alpha_1)((x/y) - \alpha_2) \cdots ((x/y) - \alpha_m) = a_m(x - \alpha_1 y)(x - \alpha_2 y) \cdots (x - \alpha_m y)$, where the α_j are the roots of $a_m x^m + a_{m-1} x^{m-1} + \cdots + a_0 = 0$, and attempt to generalize Kummer's ideas to the ring generated by the α_j. It turned out to be not at all obvious how to do this. The problem remained unsolved until attacked in different ways by two younger mathematicians, Kronecker (1823-1891) and Dedekind (1831-1916). Although Kronecker was Kummer's pupil, his solution was published a decade after that of Dedekind, and proved the less popular. Dedekind's theory appeared in 1871 as supplement X to the second edition of Dirichlet's lectures on number theory. Various revised forms appeared in later editions.

Given an equation $a_m x^m + \cdots + a_0 = 0$ with integer coefficients, let $\mathscr{F}$ be the smallest set of complex numbers containing all the roots of the equation and closed under addition, multiplication, and division. In other words, let $\mathscr{F}$ be the so-called *algebraic number field* generated by the roots in question. Every member of $\mathscr{F}$ satisfies some polynomial equation with integer coefficients, and those that satisfy such an equation with leading coefficient 1 are said to be *algebraic integers*. The set R of all algebraic integers in $\mathscr{F}$ is a ring, and Dedekind concerned himself with factorization in this ring. For each non- zero algebraic integer x in R, one can form the set I_x of all xy

for y in R and prove that $I_{x_1} = I_{x_2}$ if and only if $x_1 = ux_2$ where u is a unit. Moreover, it is easy to see that for all x_1 and x_2, $I_{x_1 x_2} = I_{x_1} I_{x_2}$ where the latter is defined to be the set of all sums $y_1 z_1 + y_2 z_2 + \cdots y_r z_r$ where the y_j are in I_{x_1} and the z_j are in I_{x_2}. Thus factorization can be translated into properties of the sets I_x and then one doesn't have to be concerned about the arbitrariness produced by units. Dedekind's key observation is that in those rings for which unique factorization does not hold, one can augment the subrings I_x with other subrings in such a way that unique factorization is restored. Each I_x in addition to being a subring has the property that $z \in I_x$, and $y \in R$ implies $zy \in I_x$. Dedekind defined a subring to be an *ideal* whenever it has this property. When R has unique factorization, the ideals correspond one-to-one to the elements mod units. Otherwise there are *always* ideals which are not of the form I_x, and Dedekind thought of them as defining "virtual" or "ideal" elements — hence the word *ideal*. Defining an ideal I in R to be prime when it cannot be written in the form $I_1 I_2$ where I_1 and I_2 are (not necessarily distinct) ideals, Dedekind was able to prove that every ideal may be written in the form $I_1^{k_1} I_2^{k_2} \cdots I_r^{k_r}$ where the I_j are distinct prime ideals and are uniquely determined up to a rearrangement.

An ideal of the form I_x is said to be principal, and one can show that the factorization of principal ideals leads to all prime ideals — indeed that every prime ideal "lies over" one and only one ordinary prime p in the sense that it occurs in the factorization of I_p. For any ideal I other than $\{0\}$, one can introduce an equivalence relation in R by setting $x \equiv y \bmod I$ if $x - y$ is in I, and show that there are only a finite number of equivalence classes. The number of these is denoted by $N(I)$ and called the *norm* of the ideal. When R is the ring of ordinary integers $N(I_x)$ is just $|x|$. More generally, $N(I_1 I_2) = N(I_1)N(I_2)$, and when p is an ordinary prime $N(I_p) = p^m$ where m is the dimension of the field considered as a vector space over the rationals. Thus every prime ideal lying over p has a norm which is a power of p. Actually it turns out that with the exception of a finite number of so called "ramified" primes, all the prime ideals lying over p are distinct and have the same norm p^k. If there are ℓ of these, then $p^{k\ell} = p^m$, so k and ℓ are divisors of m. Knowing ℓ for each prime p and knowing how the ramified primes behave tells us $N(I)$ for all possible ideals I. This information is summed up in the Dedekind zeta function of the field, which is defined by the equation $\zeta_{\mathscr{F}}(s)$ $= \sum_I \dfrac{1}{N(I)^s}$ — the sum being over all non-zero ideals. Of course one has also $\zeta_{\mathscr{F}}(s) = \sum_{n=1}^{\infty} \dfrac{\varphi_{\mathscr{F}}(n)}{n^s}$ where $\varphi_{\mathscr{F}}(n)$ is the number of ideals of norm n.

In the special case in which the field $\mathscr{F}$ is a two-dimensional vector space over the rational numbers, a so-called quadratic extension, then $\zeta_{\mathscr{F}}(s)$ coincides with the function $\sum_{n=1}^{\infty} \dfrac{\varphi_{\mathscr{F}}(n)}{n^s}$ which Dirichlet attached to a binary quadratic form a third of a century earlier (see section 6) in his proof of the existence of an infinity of primes in an arithmetic progression. In fact, the theory of the binary quadratic forms of a fixed discriminant D is more or less equivalent to the ideal theory in the algebraic number field generated by $\sqrt{D}$, and the theory of algebraic number fields as worked out by Dedekind and his successors may be looked upon as a far-reaching generalization of the theory of binary quadratic forms of Gauss and Dirichlet. In this generalization, the algebraic number fields whose integers admit unique factorization correspond to quadratic forms of class number one. For more general algebraic number fields there is a finite commutative group that generalizes the group formed by the inequivalent classes of quadratic forms of a given discriminant under Gauss's composition law (see section 6). This is the group of ideal classes, which may be defined as follows: One declares the ideals I_1 and I_2 to be in the same class if there exist elements x and y such that $I_1 I_x = I_2 I_y$. This equivalence relation divides all ideals into a finite number of classes and the number h is called the *class number* of the field. It is obvious that the class of II' depends only on the classes to which I and I' belong, and that the resulting composition law makes the classes into a commutative group. In addition to the problem of determining $\varphi_{\mathscr{G}}(n)$ (the number of ideals of norm n), one has also the more delicate problem of determining for each ideal class c the number $\varphi_{\mathscr{G}}^c(n)$ of ideals of norm n in that class. This latter problem is the analogue of the problem of finding the number of representations of n by a particular quadratic form, and as with that problem $n \to \varphi_{\mathscr{G}}^c(n)$ is not a multiplicative function of n (see section 6), but a linear combination of such functions with one term for each character of the ideal class group.

Generally speaking, the theory of algebraic number fields parallels the theory of binary quadratic forms except for the extra complications produced by passing from a second to a higher degree equation. Of course, these added complications can be quite serious, and one has as complete a knowledge of the function $n \to \varphi_{\mathscr{G}}(n)$ as in the quadratic form case only when the Galois group (i.e., the group of automorphisms of $\mathscr{F}$) is a commutative group. In that case, one has a generalization of the quadratic reciprocity law in that whether or not I_p is a prime ideal — and more generally the numbers of prime ideals into which I_p decomposes — depends (for the unramified primes) only on the congruence class of p relative to some fixed modulus m. That this is so is by no means obvious. It is a consequence of Kummer's results on the field generated by the mth roots of unity and of a remarkable theorem conjectured by Kronecker in 1886 and first completely proved by Weber (1842-1913) in 1887. This theorem asserts that every $\mathscr{F}$ with a commutative Galois group is contained for some m in the field generated by the mth roots of unity. Very little is known about the dependence on

p of the prime decomposition law when the Galois group is not commutative.

Let $\mathscr{F}$ be as above with Galois group G, and for each subgroup H of G let $\mathscr{F}_H$ be the subfield of all x in $\mathscr{F}$ such that $\alpha(x) = x$ for all automorphisms α of H. In its modern form, one of the main results of the Galois theory of equations (see section 11) asserts that *every* subfield of $\mathscr{F}$ is an $\mathscr{F}_H$ and that $H \to \mathscr{F}_H$ sets up a one-to-one inclusion inverting correspondence between the subgroups of G and the subfields of $\mathscr{F}$. Now each $\mathscr{F}_H$ has a subring of integers R_H and one can consider generalizing the problem of factoring the ideals I_p to that of factoring the ideals in R generated by the prime ideals in R_H. In other words, one can study ideal factorization in $\mathscr{F}$ *relative* to an arbitrary subfield $\mathscr{F}_H$. Moreover, in view of the facts stated above about the case in which $H = G$ and $\mathscr{F}_H$ is the rational subfield Q, it is natural to hope for immediate results only when H is Abelian; in other words, to study first those extensions $\mathscr{F}|\mathscr{F}_H$ in which the *relative* Galois group is Abelian.

Extending the classical results of Gauss, Dirichlet, and Kummer to relatively Abelian extension fields turned out to be far from easy. Important preliminary results were obtained by Kronecker in 1882 and by Weber in 1891, 1897, and 1898. Hilbert is usually credited with having begun the systematic general theory, however, in a series of papers published between 1898 and 1902. Even the case of a relative quadratic extension proved to be difficult; it was the only one that Hilbert worked out in full detail. However, he outlined how the theory should look for a more general Abelian relative Galois group and made a number of conjectures which were established in the next two decades by his student Furtwängler (1864-1939) and by Takagi (1875-1966). Takagi, who brought the subject to a certain degree of completion in two important papers published in 1920 and 1922 respectively, not only proved Hilbert's conjectures, but made important conceptual advances as well. For a more complete account of the relationship between the work of Kronecker, Weber, Hilbert, Furtwängler, and Takagi the reader is referred to Hasse's article "History of class field theory," published in the proceedings of the 1965 Brighton Conference on algebraic number theory.

In generalizing the quadratic reciprocity law, Hilbert was led to an elegant new formulation in the classical case. It is based on a concept introduced by him and called the *norm residue symbol*. Let $\mathscr{F}$ be a quadratic extension field of the rational field Q. Then the norm residue symbol assigns a character $\chi_p^{\mathscr{F}}$ of the multiplicative group Q^* of Q to each prime p and to ∞. This assignment is such that $\chi_p^{\mathscr{F}}(r) = \pm 1$, and for each r and $\mathscr{F}$, $\chi_p^{\mathscr{F}}(r) = -1$ for only finitely many values of p. Thus $\prod_p \chi_p^{\mathscr{F}}(r)$ makes sense.

Hilbert showed that the classical quadratic reciprocity law (see section 6), together with its two supplements, is completely equivalent to the assertion that $\prod\limits_{p} \chi_p^{\mathscr{F}}(r) = 1$ for all non- zero rational numbers r (p of course ranges over all primes *and* ∞). Hilbert's definition of the character $\chi_p^{\mathscr{F}}$ depended on the factorization of I_p in the ring of integers of $\mathscr{F}$ in a way which it will be easier to describe in section 20 below.

For each prime, the characters $\chi_p^{\mathscr{F}}$ that arise as $\mathscr{F}$ varies over the quadratic number fields turn out to form a subgroup A_p of the group of all characters of order 2. Hilbert also showed that a system $p \to \chi_p$ of characters of order 2 arises as $p \to \chi_p^{\mathscr{F}}$ for some $\mathscr{F}$ if and only if the following conditions are satisfied: 1) $\chi_p \in A_p$ for all p including ∞; 2) for all but a finite number of values of p, $\chi_p(n) = 1$ whenever p does not divide n; 3) $\prod\limits_{p} \chi_p(r) = 1$ for all non-zero rationals r. In this sense, Hilbert described all possible quadratic extensions of Q in terms of the character groups A_p. In his theory of relative quadratic extensions, he found generalizations of both the quadratic reciprocity law and the theorem about the possible quadratic extensions of Q.

When one goes beyond the quadratic case to more general Abelian extensions, the system $p \to \chi_p$ of characters of order 2 must be replaced by a finite set of systems $p \to \chi_p$, where the χ_p are now only of finite order and where the finite set forms a group under pointwise multiplication. This finite group turns out to be isomorphic to the Galois group of the field $\mathscr{F}$. (Of course when the Galois group is of order two, it suffices to specify the unique system $p \to \chi_p$ which does not correspond to the identity.) In the generalization to relative fields, the prime ideals for the base field $\mathscr{F}_H$ replace the primes, and the symbol ∞ gets replaced by several such symbols — one for each possible dense imbedding of $\mathscr{F}_H$ in the real or complex number fields.

With this background it is possible to describe Artin's 1923 application of group representations to number theory. It is based on Artin's discovery that for each $\mathscr{F}$ there is a canonical way of assigning a Dirichlet series $s \to L(s, \chi, \mathscr{F}_H)$ to every pair $\mathscr{F}_H, \chi$ consisting of a subfield $\mathscr{F}_H$ of $\mathscr{F}$ and a character χ of the Galois group H of $\mathscr{F}$ relative to $\mathscr{F}_H$. Artin's definition of $L(s, \chi, \mathscr{F}_H)$ depends in turn upon certain facts about the action of H on the prime ideals in R which lie over a fixed prime ideal $\mathfrak{P}$ in R_H. It turns out that (with the exception of a finite number of "ramified" prime ideals $\mathfrak{P}$) H acts transitively on the prime ideals of R lying over $\mathfrak{P}$ and that the subgroup of H leaving a given one of these fixed is always cyclic and, moreover, has a canonical generator. Since the conjugacy class of this generator is evidently the same for all prime ideals on R lying over $\mathfrak{P}$, we have a natural map $\mathfrak{P} \to C_{\mathfrak{P}}$ of (unramified) prime ideals in R_H into the conjugacy classes in H. Now for each $\mathfrak{P}$ and each representation Γ of H, the determinant of $(I - \Gamma_x N(\mathfrak{P})^{-s})$ where I is the identity is easily seen to depend only on the conjugacy class to which x belongs — and of course on s and $\mathfrak{P}$. Choosing x to be any element in $C_{\mathfrak{P}}$ this determinant becomes a well-defined function $f_{\mathfrak{P}}(s)$ of s and $\mathfrak{P}$. The Artin "L function" $L(s, \chi, \mathscr{F}_H)$ is $\prod_{\mathfrak{P}} \dfrac{1}{f_{\mathfrak{P}}(s)}$ where Γ is any representation of character χ.

Although it is possible to define $f_{\mathfrak{P}}(s)$ for the ramified prime ideals as well, it suffices for many purposes to look only at the unramified ones and to identify two Dirichlet series when one can be transformed into the other by multiplying or dividing by a finite number of factors of the form $\dfrac{1}{Q(p^{-s})}$ where Q is a polynomial and p is an ordinary prime. Modulo such factors one can then verify the truth of the following relationships:

(1) When χ is the identity character, $L(s, \chi, \mathscr{F}_H)$ is the zeta function of $\mathscr{F}$ relative to $\mathscr{F}_H$.

(2) $L(s, \chi_1 + \chi_2, \mathscr{F}_H) = L(s, \chi_1, \mathscr{F}_H)L(s, \chi_2, \mathscr{F}_H)$ for all characters χ_1 and χ_2 of H.

(3) If χ^* is the character of H *induced* by a character χ of some subgroup H_1 of H (see section 15), then $L(s, \chi^*, \mathscr{F}_H) = L(s, \chi, \mathscr{F}_{H_1})$.

Applying (3) to the case in which H_1 contains only the identity and χ is one-dimensional, and using the fact that χ^* is then the sum of *all* irreducible characters $\chi_1, \chi_2, \cdots, \chi_\ell$ each occurring with multiplicity $\chi_j(e)$, one finds that $L(s, \Sigma \chi_j(e)\chi_j, \mathscr{F}_H) = L(s, 1, \mathscr{F})$. It now follows on applying (2) and (1) to the left- and right-hand sides respectively that $\prod_j L(s, \chi_j, \mathscr{F}_H)^{\chi_j(e)} = \zeta_{\mathscr{F}}(s)$. In other words, the zeta function of any $\mathscr{F}$ (modulo the equivalence relation mentioned above) can be factored as a product of the Artin L functions attached to the irreducible characters of any fixed subgroup H of the Galois group G of $\mathscr{F}$. This factorization is significant because when H is commutative, it follows from the work of Takagi that $\zeta_{\mathscr{F}}(s)$ factors as a product of generalized Dirichlet L functions and that Artin's L functions for the characters of H coincide (modulo equivalence) with the generalized Dirichlet L functions. Moreover, this coincidence of the two kinds of L functions is not an immediate consequence of the definitions, but is equivalent to one of the main theorems of class field theory. In fact, a few years later Artin was able to give a new proof of this theorem and reorganize the whole subject in a conceptually advantageous way by starting with a direct proof of the identity of the two kinds of L functions.

The identity of the two kinds of L functions in the commutative case provided for the first time a natural generalization of the Dirichlet L functions for fields with a non-commutative Galois group, and therefore a tool with which to attack "non-commutative class field theory." Using the relationship $L(s, \chi^*, \mathscr{F}_H) = L(s, \chi, \mathscr{F}_{H_1})$, one can express Artin L functions based on non-commutative characters in terms of generalized Dirichlet L functions to the extent that general characters on G can be expressed in terms of characters induced by one-dimensional characters of subgroups. Using this device, Artin was able to establish various important properties of his new L functions and to show that others would follow if one could prove that every character of an arbitrary finite group can be written as a linear combination with positive and negative integer coefficients of characters induced by one-dimensional characters of subgroups. A proof of this difficult theorem about finite groups was found by Brauer (1901-1977) and published in 1947. Brauer was a student of Schur and his most important immediate successor in developing the representation theory of finite groups.

Another early application of the theory of group representations to number theory was provided by Hecke (1887-1947) in 1928. It is perhaps more accurate to say that it was an application to the theory of modular forms, but the connection of the latter theory to questions in number theory — especially the theory of n-ary quadratic forms (see sections 10 and 22) — is so close that it seems appropriate to speak of an application to number theory. Let Γ_0 denote the subgroup of the group $SL(2, R)$ of all 2×2 real matrices of determinant 1 consisting of the matrices with integer coefficients. Let Γ_N be the normal subgroup of Γ_0 consisting of all $\left(\begin{smallmatrix} a & b \\ c & d \end{smallmatrix}\right)$ in Γ_0 for which $a - 1$, $d - 1$, b, and c are divisible by N. Then (see section 10) a modular form of weight k and level N is an entire function on the upper half-plane satisfying the identity $f\left(\frac{az+b}{cz+d}\right) = (cz + d)^{2k}f(z)$ for all $\left(\begin{smallmatrix} a & b \\ c & d \end{smallmatrix}\right) \epsilon \Gamma_N$ (and certain growth conditions as well). It was known that for each fixed k and N, the modular forms of weight k and level N form a finite-dimensional vector space. Hecke was concerned with the problem of finding an explicit basis for this space. For forms of level 1, such a basis was described in section 10. The basic idea of Hecke's paper was to break this problem down into subproblems as follows: For each modular form f of level N and weight k, consider the set of all functions $z \to f\left(\frac{az+b}{cz+d}\right)(cz + d)^{-2k}$ for $\left(\begin{smallmatrix} a & b \\ c & d \end{smallmatrix}\right) \epsilon \Gamma_0$. When $\left(\begin{smallmatrix} a & b \\ c & d \end{smallmatrix}\right) \epsilon \Gamma_N$, these all reduce to f itself. More generally, since Γ_N has finite index in Γ_0, these functions span a finite-dimensional vector space $\mathcal{M}_f$. For each $\left(\begin{smallmatrix} a & b \\ c & d \end{smallmatrix}\right) \epsilon \Gamma_0$, then $g \to g\left(\frac{az+b}{cz+d}\right)(cz + d)^{-2k}$ is a linear transformation $L_{\left(\begin{smallmatrix} a & b \\ c & d \end{smallmatrix}\right)}$ of $\mathcal{M}_f$ into itself, and the mapping $\left(\begin{smallmatrix} a & b \\ c & d \end{smallmatrix}\right) \to L_{\left(\begin{smallmatrix} a & b \\ c & d \end{smallmatrix}\right)}$ is a representation of Γ_0. Since $L_{\left(\begin{smallmatrix} a & b \\ c & d \end{smallmatrix}\right)}$ is the identity whenever $\left(\begin{smallmatrix} a & b \\ c & d \end{smallmatrix}\right)$ is in the normal subgroup Γ_N, this representation is in effect a representation of the finite quotient group Γ_0/Γ_N. Let $\mathcal{M}_f = \mathcal{M}_1 \oplus \mathcal{M}_2 \oplus \cdots \oplus \mathcal{M}_e$ be a decomposition of $\mathcal{M}_f$ as a direct sum of irreducible L-invariant subspaces, and let f_j be the component of f in $\mathcal{M}_j$. It is evident that each f_j is a modular form of weight k and level N but is special in being intrinsically associated with a particular irreducible representation of Γ_0/Γ_N. Hecke suggested the strategy of looking at the irreducible representations of Γ_0/Γ_N one at a time and for each representation W seeking a basis for those particular modular forms intrinsically associated with

W. It is not difficult to see that this problem is more or less equivalent to the following: For each irreducible representation W of Γ_0/Γ_N, let us define an entire function g from the upper half-plane to the vector space of W to be a modular W form of weight k if it satisfies the identity

$$g\left(\tfrac{az+b}{cz+d}\right) = (cz + d)^{2k} W_{\left(\begin{smallmatrix} a & b \\ c & d \end{smallmatrix}\right)} g(z) \text{ for all } \left(\begin{smallmatrix} a & b \\ c & d \end{smallmatrix}\right) \epsilon \Gamma_0$$

as well as an appropriate growth condition. The equivalent problem then is to find a basis for the modular W forms of weight k.

In order to carry out this program, one has first to determine the irreducible representations of Γ_0/Γ_N. When $N = N_1 N_2$ with N_1 and N_2 relatively prime, one has $\Gamma_0/\Gamma_N \simeq \Gamma_0/\Gamma_{N_1} \times \Gamma_0/\Gamma_{N_2}$. Hence it suffices to consider the case in which N is a prime power. In the case in which N is actually a prime p, Γ_0/Γ_N is isomorphic to the group of all 2×2 matrices of determinant one with coefficients in the field of p elements. Its representations are relatively easy to find and were described by Frobenius in 1896. For higher powers of p, the problem is more difficult and was not completely solved until very recently. Hecke confined himself to the case of prime N. The representations of Γ_0/Γ_{p^2} were determined in the 1933 thesis of Hecke's student Praetorius and independently by Rohrbach a year earlier.

20. IDÈLES, ADÈLES, AND APPLICATIONS OF PONTRJAGIN-VAN KAMPEN DUALITY TO NUMBER THEORY, CONNECTIONS WITH ALMOST-PERIODIC FUNCTIONS, AND THE WORK OF HARDY AND LITTLEWOOD

The representation theory of finite groups and compact Lie groups on the one hand, and the Pontrjagin-van Kampen duality theory on the other, constitute two rather different generalizations of the harmonic analysis of the nineteenth century. The first of these began to have applications to number theory and to physics in the middle 1920s; these applications have been discussed in sections 16 and 19. Applications of the Pontrjagin-van Kampen duality theorem began in 1936 with the introduction by Chevalley (1909—) of the concept of an *idèle* and the idèle group of an algebraic number field. The idèle concept is based in turn on the notion of a p-adic number introduced in 1901 by Hensel (1861-1941). If p is any prime, the p-*adic distance* $\varrho_p(r_1, r_2)$ between two rational numbers r_1 and r_2 is defined to be p^{-k} where k is determined by the relationship $r_1 - r_2 = \dfrac{m}{n} p^k$, m and n being integers not divisible by p. The p-adic numbers are then the elements of the field Q_p obtained by completing the rational field Q with respect to the p-adic distance just as the real field is obtained by completing Q with respect to the distance $\varrho_\infty(r_1, r_2) = |r_1 - r_2|$. It turns out that every p-adic number can be written uniquely in the form $p^n(a_0 + a_1 p + a_2 p^2 + \cdots)$, where each $a_j = 0, 1, 2, \cdots, p-1$, $a_0 \neq 0$, and that every sequence of such a_j's occurs. Those p-adic numbers for which $n \geq 0$ are called p-*adic integers*. They form a compact open subring of the field of all p-adic numbers which is accordingly locally compact. The additive group of Q_p and the multiplicative group $Q_p{}^*$ of all non-zero elements of Q_p are both totally-disconnected, locally-compact commutative groups. Moreover, the additive group is isomorphic to its own dual. In terms of $Q_p{}^*$ it is quite easy to complete the definition of the norm residue symbol of Hilbert mentioned in the last section. If the relevant quadratic extension field $\mathscr{F}$ of Q is generated by $\sqrt{D}$, it can be shown that the set of all $x^2 - y^2 D$ is a subgroup of $Q_p{}^*$ of index 2 and one defines $\chi_p^{\mathscr{F}}$ to be the restriction to Q^* of the unique character of $Q_p{}^*$ which is 1 on the subgroup and -1 otherwise. For fixed p, the $\chi_p^{\mathscr{F}}$ for different quadratic extension fields $\mathscr{F}$ are precisely those characters of Q^* of order two which are continuous in the p-adic topology. Those p-adic integers which are units in the ring of all p-adic integers, that is those of the form $a_0 + a_1 p + \cdots$ with $a_0 \neq 0$, are called the p-*adic units*. They form a compact open subgroup U_p of the group $Q_p{}^*$, and the quotient group is the infinite cyclic group.

For the special case of the rational number field Q, Chevalley's idèles are the members of a certain subgroup I of the infinite product $\left(\prod_p Q_p{}^*\right) \times R^*$. This subgroup consists of all members $\{x_p\}$, x of this product group such that $x_p \in U_p$ for all but a finite number of primes p. While this entire infinite product group is not itself a locally-compact group in any simple or natural way, the subgroup of idèles can be given a simple locally-compact topology. Quite generally, let $G_1, G_2, \cdots$ be any sequence of locally-compact groups, each member G_j of which admits a compact open subgroup K_j. Let G be the subgroup of $\prod_j G_j$ consisting of all sequences $x_1, x_2, \cdots$ with $x_j \in K_j$ for all but finitely many indices j. Then G contains the compact product group $K = \prod_j K_j$ as a subgroup and there are only countably many K cosets. Defining a subset O of G to be open whenever its intersection with each right K coset is open, one obtains a topology in G which converts it into a locally-compact topological group in which K is a compact open subgroup. The group G is called the restricted direct product of the G_j with respect to the K_j. The idèle group is the direct product of R^* and the restricted direct product of the $Q_p{}^*$ with respect to the U_p, and as such has a locally compact topology. The subgroup $\prod_p U_p$ is compact in this topology and $\prod_p U_p \times R^*$ is open.

Since Q^* has a natural dense imbedding in each $Q_p{}^*$ as well as in R^*, one has a natural imbedding of Q^* into the full product group $\prod Q_p{}^* \times R^*$. Moreover, it is easy to see that for each $r \in Q^*$, r as an element of $Q_p{}^*$ is in U_p for all but finitely many p. Hence the image of r in $\prod Q_p{}^* \times R^*$ is actually in the idele group I. In other words, one has a natural imbedding of Q^* as a subgroup I_0 of the idèle group I, and the members of this subgroup I_0 are called the *principal idèles*. It turns out to be possible to prove that the subgroup I_0 of all principal ideles is a closed subgroup so that the quotient group I/I_0 — the so-called idele class group — also has a natural locally-compact topology.

The significance of the idèle class group can be most easily appreciated by considering its characters of order two and confronting their determination with Hilbert's formulation of the quadratic reciprocity law in terms of his norm residue symbol. A character of the idèle class group is of course a character of the idèle group which is identically one on I_0. A character of the idèle group is uniquely determined by a system $\{\chi_p\}, \chi_\infty$ where χ_p is a character on Q_p^* and χ_∞ is a character on R^*. Not every system occurs, however. One sees easily that a system $\{\chi_p\}, \chi_\infty$ arises from some character of I if and only if for all but finitely many p, $\chi_p(u) = 1$ for all $u \in U_p$. Remembering that every character of Q_p^* defines a character on Q^* which determines it uniquely, we see that the characters on I correspond one-to-one to certain systems $\{\chi_p\}, \chi_\infty$ of characters on Q^*, and that the condition that such a system satisfies Hilbert's criteria (see section 19) for being associated with a quadratic extension field is precisely that it defines a character of order two on I which is identically one on I_0. In other words, Hilbert's description of all possible quadratic extension fields in terms of systems of characters of order 2 on Q^* may be reformulated as the statement that they correspond one-to-one in a natural way to the characters of order two on the idèle class group I/I_0. Of course, characters of order two correspond one-to-one to closed subgroups of index two, and more generally the Abelian extension fields of Q of finite degree correspond one-to-one to the closed subgroups of I/I_0 of finite index.

Now consider the dual $(\widehat{I/I_0})$ of I/I_0. Its subgroups of finite order correspond one-to-one to the closed subgroups of finite index of I/I_0 and in fact are the duals of the corresponding quotient groups. Moreover, it follows from the Artin reciprocity law that each quotient group is canonically isomorphic to the Galois group of the corresponding extension field. Thus the subgroups of finite order of $(\widehat{I/I_0})$ are the duals of the Galois groups of the finite Abelian extensions of Q. More generally, one can consider the infinite extension fields generated by countable sets of finite Abelian extensions of Q, including the maximal one consisting of all algebraic numbers. They correspond one-to-one to the infinite subgroups of the group $(\widehat{I/I_0})_f$ of all elements of finite order of $(\widehat{I/I_0})$. Their Galois groups may be identified with the duals of these infinite subgroups and so given a compact, totally disconnected topology. The group $(\widehat{I/I_0})_f$ itself is the dual of the quotient of I/I_0 by its connected component. Thus the quotient of I/I_0 by its connected component is a totally disconnected compact commutative group which can be identified with the Galois group of the maximal Abelian extension of Q

and whose closed subgroups correspond one-to-one to the finite and infinite Abelian extension fields of Q. As indicated by the title of his 1936 paper, *"Généralisations de la théorie du corps de classes pour les extensions infinies,"* Chevalley's original motivation in introducing idèle groups was to have a method of describing infinite Abelian extension fields analogous to that of Hilbert and Takagi for the finite ones. For this purpose, idèles were indispensible. It turned out, however, that the idèle group notion simplified the finite theory as well, and in 1940 Chevalley published a paper redoing the whole of class field theory in terms of idèles. In his thesis of 1933, Chevalley had simplified class field theory in other ways, and his 1940 paper blended the two kinds of simplification. In particular, he replaced many complicated arguments using Dirichlet series and complex analysis with simpler arguments involving the theory of topological groups. For the sake of simplicity we have defined the idèle group and the idèle class group only for the rational field Q. However, Chevalley dealt with the general case in which Q is replaced by an arbitrary algebraic number field $\mathscr{F}$ and the Q_p by the completions of $\mathscr{F}$ with respect to metrics defined by the prime ideals in the ring of integers of $\mathscr{F}$.

An additive analogue of the idèle group of an algebraic number field was introduced in 1945 by Artin and Whaples (1914—). It is defined as a restricted direct product group over the prime ideals and the "infinite primes" just as in the idele case. But now the additive groups of the completed fields are used in place of the multiplicative ones, and the compact open subgroups with respect to which the restricted product is taken are not the unit groups but the closures in the $\mathfrak{P}$-adic topologies of the integers of the field. The members of this additive infinite "product" group were originally called *valuation vectors*, but are now usually called adèles. Adèles can be multiplied together as well as added; they form a ring (the adèle ring of the field) under these two operations. The idèle group is precisely the group of units of the adèle ring, but its topology as a subset of the adèle ring is not the same as its topology as an idèle group.

Artin and Whaples introduced adèles as a tool in giving an axiomatic characterization of algebraic number fields. In the course of doing so, they went further than Chevalley had in demonstrating the utility of idèles in formulating and proving the facts of algebraic number theory. Actually, their characterization of algebraic number fields was included in a characterization of a parallel class of fields having prime characteristic. Let p be a prime and let Z_p denote the finite field of p elements. Then the field $Z_p(x)$ of all rational functions with coefficients in Z_p is a countable field, which is in an obvious sense the simplest infinite field of characteristic p. As such, it is a characteristic p analogue of the rational field Q, and one can develop a theory of the finite algebraic extensions of $Z_p(x)$ which is quite analogous to the theory of algebraic number fields. Artin laid the foundations for such a theory in his 1921 Ph.D. thesis. This thesis, published in 1924, dealt with the quadratic extensions of the fields $Z_p(x)$. Artin and Whaples showed that any field satisfying certain simple axioms was necessarily a finite algebraic extension of either Q or $Z_p(x)$.

In his 1950 Ph.D. thesis, Tate (1925—), carrying out a suggestion of Artin, showed how to use harmonic analysis in adèle groups to prove a vast generalization of the well-known functional equation for the Riemann zeta function. An abstract announcing similar results was published by Iwasawa (1917—) in the Proceedings of the 1950 International Mathematical Congress. Tate's thesis was not published until 1967, when it appeared in the Proceedings of the 1965 Brighton Conference on algebraic number theory. However, copies of it were privately circulated long before this and it also appeared in rewritten form as a chapter in a book by Lang. The fact that the zeta function of an *arbitrary* algebraic number field has an analytic continuation and satisfies a functional equation analogous to that satisfied by the Riemann zeta function was first proved by Hecke in a paper published in 1917. Various special cases had been treated earlier by other authors. In the same year Hecke published a second paper doing the same thing for general Dirichlet L functions. A bit later, he introduced a more general kind of L function determined by an algebraic number field and a so-called *Grössencharakter* for the field. He studied these L functions in papers published in 1918 and 1920, and proved that they too satisfy (rather complicated) Riemann-type functional equations. The method of Tate and Iwasawa made it possible to obtain all of these results of Hecke at one stroke by applying a generalization of the Poisson summation formula. The classical Poisson summation formula asserts that, for reasonably general complex-valued functions on the line, $\sum_{n=-\infty}^{\infty} f(n) = \sum_{n=-\infty}^{\infty} \hat{f}(n)$, where $\hat{f}(x) = \int_{-\infty}^{\infty} f(y)e^{2\pi ixy}dy$. More generally, if G is any separable locally-compact commutative group and Γ is any countable closed subgroup such that G/Γ is compact, then $\Gamma^{\perp}$, the subgroup of the character group $\hat{G}$ consisting of all characters χ with $\chi(\gamma) = 1$ for all $\gamma \epsilon \Gamma$, is also closed and countable, and when the Haar measure μ in G is suitably normalized, one has

$$\sum_{\gamma \epsilon \Gamma} f(\gamma) = \sum_{\chi \epsilon \Gamma^{\perp}} \hat{f}(\chi)$$

for all suitably restricted functions f on G. Here $\hat{f}(\chi) = \int_G f(x)\chi(x)d\mu(x)$. Tate and Iwasawa take the adèle group of the number field for G and the subgroup of principal adèles for Γ. Hecke's *Grössencharaktere*, whose original definition was rather complicated, can be defined much more simply using idèles. They are just the characters of the idèle class group I/I_0 that are not of finite order. Hecke's proof also hinged on the Poisson summation formula, which he used to prove a generalized form of Jacobi's inversion formula. The difference between Hecke and Tate is that Hecke does his computations "at infinity," i.e., over the Archimedean primes, whereas Tate works over all primes simultaneously.

There is another way of defining the adèle group of a number field which is based directly on group duality and makes no use of p-adic metrics. Let R be the ring of all algebraic integers in the algebraic number field $\mathscr{F}$ and let R^+ be the additive group of R. R^+ is countable, and we make it into a locally-compact commutative group by giving it the discrete topology. The dual $\hat{R}^+$ is then compact and in fact is isomorphic to the direct product of n replicas of the circle group T where n is the degree of $\mathscr{F}$ over the rationals. Let $\hat{R}^{+f}$ denote the subgroup of $\hat{R}^+$ consisting of all elements of finite order. Then $\hat{R}^{+f}$ is a dense countable subgroup of $\hat{R}^+$ and we may regard its natural imbedding as an injective homomorphism θ of the discrete group $\hat{R}^{+f}$ into $\hat{R}^+$. Its dual θ^* (see section 19) is then an injective homomorphism of $\hat{\hat{R}}^+ = R^+$ onto a dense subgroup of the compact totally disconnected dual $\hat{\hat{R}}^{+f}$ of $\hat{R}^{+f}$. One of the easy general theorems about group duality asserts that a compact group is infinitely divisible if and only if its dual has no elements of finite order, and that it has no elements of finite order if and only if its dual is infinitely divisible. Since $\hat{R}^{+f}$ is infinitely divisible, it follows at once that $\hat{\hat{R}}^{+f}$ has no elements of finite order. Since it has no elements of finite order, it has a unique minimal completely divisible extension in which it has countable index $(\hat{\hat{R}}^{+f})^{\sim}$. This extension may be made into a locally-compact group by giving each $\hat{\hat{R}}^{+f}$ coset the topology of $\hat{\hat{R}}^{+f}$ and declaring a subset of the divisible extension to be open if its intersection with each coset is open. This locally-compact group has the additive group $\mathscr{F}^+$ of $\mathscr{F}$ densely imbedded, and is the so-called non-Archimedean component of the adèle group of $\mathscr{F}$. The actual adèle group is the direct product of this locally-compact group with an n-dimensional vector space over the real numbers called the Archimedean component of the adèle group. The latter can be defined in a manner vaguely analogous to that used in defining the non-Archimedean component. Let $\overline{R}^+$ denote the vector space of all homomorphisms of R^+ into the multiplicative group of all positive real numbers. Then the Archimedean component of the adèle group is the vector space dual of $\overline{R}^+$. It is an n-dimensional real vector space containing R^+ as a lattice subgroup and $\mathscr{F}^+$ as a dense subgroup. The dense φ_1 and φ_2 imbeddings of $\mathscr{F}^+$ into $(\hat{\hat{R}}^{+f})^{\sim}$ and $\overline{R}^+$ may be combined to give an imbedding $f \rightarrow \varphi_1(f), \varphi_2(f)$ of $\mathscr{F}^+$ into the product group, i.e., into the adele group of $\mathscr{F}$. The range of this imbedding can be shown to be closed and is the group of all principal adeles.

It is interesting to examine the results of Hardy and Littlewood on Waring's problem (see section 14) in the light of the theory of almost-periodic functions (see section 19) and its connection with group duality. Choose fixed positive integers k and r. For each positive integer n, let $f(n)$ denote the number of integer solutions of $x_1^k + \cdots + x_r^k = n$. Then $f(1) + f(2) + \cdots + f(n)$ is the number of points with integer coordinates inside and on the hypersurface $x_1^k + x_2^k + \cdots x_r^k = n$. Elementary arguments show accordingly that $(f(1) + f(2) + \cdots + f(n))$ is asymptotic to a constant multiple of $n^{r/k}$. It follows easily from this that if $f_0(n) = \dfrac{f(n)}{n^{r/k-1}}$, then $f_0(1) + f_0(2) + \cdots f_0(n)$ is asymptotic to a constant multiple of n; that is, that

$$\frac{f_0(1) + \cdots + f_0(n)}{n}$$

has a limit as n tends to ∞. In other words, the function $n \to f_0(n)$ behaves like an almost-periodic function on the integers to the extent that it has a mean value. Of course, the properties of sample functions of stationary stochastic processes (see section 18) warn us that having a mean value is far from implying almost periodicity. On the other hand, one does not expect f_0 to be like a random function, and even if it were, one could still compute its hidden periods (if any). All of this suggests investigating the existence of the mean of $n \to f(n)e^{-in\lambda}$ for the various real values of λ, and in these terms the main results of Hardy and Littlewood on Waring's problem may be summed up as follows: Suppose that $r \geq 2^k(2k + 1)$. Then

(1) $c_\lambda = \displaystyle\lim_{n \to \infty}\ \frac{1}{n}\,(f(1)e^{-i\lambda} + f(2)e^{-2i\lambda} + \cdots + f(n)e^{-ni\lambda})$ exists for all real λ, and $c_\lambda = 0$ whenever λ/π is irrational.

(2) If $a_\lambda = c_{2\pi\lambda}$, then the series $\displaystyle\sum_\lambda a_\lambda e^{2\pi in\lambda}$ (where the sum is over all rational numbers λ) converges for all n to a function $S(n)$ and the convergence is uniform.

(3) $\displaystyle\lim_{n \to \infty} S(n) - f_0(n) = 0$.

(4) If $M(f_0) = \displaystyle\lim_{n \to \infty}\ \frac{f_0(1) + \cdots + f_0(n)}{n}$ and $\lambda = p/q$ where p and q are relatively prime, then $a_\lambda = a_{p/q} = \dfrac{1}{M(f_0)}\dfrac{1}{q^r}\left(\displaystyle\sum_{m=0}^{q-1} e^{2\pi i(m^k)(p/q)^r}\right)$.

One sees in particular that $f(n)$ is asymptotic to $\dfrac{n^{(r/k-1)}}{M(f_0)} S_0(n)$ where $S_0(n) = S(n)M(f_0)$, and is an almost-periodic function of n with explicitly known Bohr-Fourier coefficients. The function $S_0(n)$, or rather the Bohr-Fourier expansion of it, is what Hardy and Littlewood call the "singular series."

Every almost-periodic function on the integers Z (see section 17) may be extended to be continuous on a certain compactification of Z. This compactification is obtained by starting with a countable discrete subgroup Γ of $\hat{Z}$ and taking the dual θ^* of the isomorphic imbedding θ of Γ in $\hat{Z}$. θ^* imbeds Z as a dense subgroup of the compact dual $\hat{\Gamma}$ of Γ. In the case at hand, the fact that $c_\lambda = 0$ except when λ/π is rational implies that the subgroup Γ of $\hat{Z}$ is precisely the subgroup of all elements of finite order. Thus $\hat{\Gamma}$ is an open compact subgroup in the non-Archimedean component of the adèle group of the rational field. The discrete group Γ has a natural direct product decomposition over the primes. Indeed, for each prime p, the subset Γ_p of all elements γ in Γ with $\gamma^{p^k} = e$ for some k is a subgroup and every element is uniquely a product of members of a finite number of the Γ_p. Correspondingly, the compact dual $\hat{\Gamma}$ is the full direct product of the compact groups $\hat{\Gamma}_p$, and it turns out to be easy to verify that $S_0(n)$ regarded as a function on $\hat{\Gamma}$ factors as a product of functions on the various $\hat{\Gamma}_p$. As mentioned earlier, $\hat{\Gamma}_p$ is isomorphic to the group of all p-adic integers. Moreover, it is not hard to interpret these p-adic components of S_0 in terms of p-adic solutions of the equation $x_1^k + \cdots + x_r^k = n$. Similarly, $\dfrac{n^{(r/k - 1)}}{M(f_0)}$ has an interpretation in terms of real solutions. Quite apart from these interpretations, the product $\dfrac{n^{(r/k - 1)}}{M(f_0)} S_0(n)$ may be looked upon as a function defined on the whole adèle group which factors according to the natural factorization of the adèle group.

The fact that there is a connection between almost-periodic functions and the Hardy-Littlewood results was pointed out (in the special case $k = 2$) by Kac (1914—) in 1940.

21. THE DEVELOPMENT OF THE THEORY OF UNITARY GROUP
REPRESENTATIONS AFTER 1945 — A BRIEF SKETCH
WITH EMPHASIS ON THE FIRST DECADE

With the exceptions mentioned at the end of section 19, the theory of unitary group representations was until 1946 exclusively concerned with groups that were either compact or both locally-compact and commutative. A more general theory encompassing all locally-compact groups began rather suddenly with the publication in 1947 of four long papers and half a dozen or so short notes and announcements. (Two of the long papers were preceded by short announcements published in 1946.) Since then, there has been an enormous development, which cannot begin to be summarized within the compass of this article. I shall content myself instead with some brief indications and refer the reader at the appropriate time to some lengthy survey articles for a more adequate account.

In attempting to generalize from compact groups to locally-compact groups, one is confronted with two major problems. In the first place, one can no longer decompose representations as discrete direct sums except in very special cases; and in the second place, one has to deal with irreducible unitary representations which are infinite-dimensional. The latter circumstance brings with it a further difficulty in that the trace of an infinite-dimensional unitary operator is undefined. This means that the character of an infinite-dimensional irreducible representation must be defined in a roundabout way when it can be defined at all. When the group is commutative, the lack of compactness is partially compensated for by the fact that the irreducible unitary representations are not only finite-dimensional but *one*-dimensional. In this case, one can combine Pontrjagin-van Kampen duality with the ideas of spectral theory to obtain an entirely adequate substitute for the Peter-Weyl theorem. Just how this works has already been explained in section 17.

Thus far I have said relatively little about the problem of actually finding the possible irreducible representations of our groups. This is because the problem is relatively easy in the commutative case, was solved more or less completely for the important compact Lie groups by Weyl in the 1920s, and for finite groups is more a problem in algebra than analysis. However, for groups which are neither compact nor commutative, the problem of finding the possible (usually infinite-dimensional) unitary irreducible representations is one of the main problems of the theory. It involves a heavy use of analysis and is by no means readily solved. All four of the long papers published in 1947 dealt with important special cases of it and so did the majority of the short notes. Another case was the subject of Wigner's 1939 paper mentioned at the end of section 17. Before the results of any of these papers are described, it will be convenient to discuss a method used in most of them which was first discussed in an abstract setting in a paper that I published in 1949. This is a method for constructing unitary representations of locally-compact groups out of unitary representations of closed subgroups which generalizes the Frobenius construction $\chi \rightarrow \chi^*$ mapping characters of subgroups of finite groups into characters of the whole group (see section 15). Frobenius found his construction to be a very useful tool in producing irreducible representations and characters, and its generalization has turned out to be equally useful.

Let G be an arbitrary separable locally-compact group. (We may restrict ourselves to the separable case because most if not all of the important examples are separable, and because in so doing we avoid various distracting technical complications.) Let H be a closed subgroup of G and suppose for the time being that the right coset space G/H admits a measure μ which is invariant under the natural action $(Hx)y = Hxy$ of G on G/H. It is not difficult to see that μ, if it exists, is uniquely determined up to a multiplicative constant. Now let L, $x \rightarrow L_x$ be any unitary representation of H in a separable Hilbert space $\mathcal{H}(L)$. (We always suppose that $(L_x(\varphi) \cdot \psi)$ is continuous for all φ and ψ in $\mathcal{H}(L)$; this implies that $x \rightarrow L_x(\varphi)$ is continuous for all φ in $\mathcal{H}(L)$.) Consider the set $\mathcal{F}_L$ of all Borel functions $x \rightarrow f(x)$ from G to the Hilbert space $\mathcal{H}(L)$ which satisfy the identity $f(hx) = L_h(f(x))$ for all h in H and all x in G. Now for each f in $\mathcal{F}_L$, the function $x \rightarrow (f(x) \cdot f(x))$ is a non-negative Borel function on G. Moreover, since $(f(hx) \cdot f(hx)) = (L_h f(x) \cdot L_h f(x)) = (f(x) \cdot f(x))$, it follows that $x \to (f(x) \cdot f(x))$ is a constant on the right H cosets and so may be regarded as a function on G/H. Let $\mathscr{F}_L^0$ denote the subset of $\mathscr{F}_L$ consisting of all f in $\mathscr{F}_L$ for which $\int_{G/H}(f(x) \cdot f(x))d\mu(\bar{x}) < \infty$ where $\bar{x}$ is the image of x on G/H and $(f(x) \cdot f(x))$ is thought of as a function on G/H. It is not hard to show that $\mathscr{F}_L^0$ becomes a Hilbert space if we define $\|f\| = \sqrt{\int_{G/H}(f(x) \cdot f(x))d\mu(\bar{x})}$ and identify two members when they are almost everywhere equal. Moreover, for each x, the mapping $f \to f_x$, where $f_x(y) = f(yx)$, is easily seen to be a unitary operator in the Hilbert space $\mathscr{F}_L^0$. If we denote this unitary operator by U_x^L, we verify that $x \to U_x^L$ is a unitary representation of G. It is called the *representation of* G *induced by the representation* L *of* H. When G is finite and χ is the character of L, one verifies without difficulty that the character of U^L is precisely the induced character χ^* of G defined by Frobenius. For the case in which G is compact, the definition given here is essentially to be found in Weil's book, cited in section 17. When G/H fails to have an invariant measure, a slightly more complicated definition involving quasi-invariant measures has to be given. I shall not repeat it here, but I assure the reader that U^L is a well-defined unitary representation of G for all unitary representations L of all closed subgroups H of G.

In terms of this definition, it is possible to state a general theorem of which the results of Wigner's 1939 paper as well as those of one of the 1947 notes are both special cases. Let the separable locally-compact group G admit a closed commutative normal subgroup N, and suppose that there exists a second closed subgroup H such that $N \cap H$ contains only the identity and $NH = G$. Then every element of G can be written in one and only one way, as a product nh where n is in N and h is in H. One says that G is a *semi-direct product* of N and H. Each h in H defines an automorphism $n \to hnh^{-1} = \alpha_h$ of N, and the mapping $h \to \alpha_h$ is a homomorphism of H into the group of automorphisms of N. Evidently one can reconstruct G knowing only N, H, and the mapping $h \to \alpha_h$. The general theorem I propose to state reduces the problem of finding the irreducible unitary representations of G to that of finding the irreducible unitary representations of certain subgroups of H — at least when the "adjoint" action of H on the dual of N has a certain regularity property. To explain this property, notice that for each automorphism α_h of N, there is a well-defined adjoint automorphism α_h^* of $\hat{N}$. Indeed, if $\chi \in \hat{N}$ and $h \in H$, then $n \to \chi(\alpha_h(n))$ is also a member of $\hat{N}$ which may be denoted by $[\chi]\alpha_h^*$. It is obvious that $\chi \to [\chi]\alpha_h^*$ is an automorphism and $h \to \alpha_h^*$ is a homomorphism. Let us say that the semi-direct product is *regular* if there exists a Borel set C in $\hat{N}$ which meets each H orbit in $\hat{N}$ in one and only one point; that is, if for each χ in $\hat{N}$ there is one and only one χ^1 in C such that $[\chi]\alpha_h^* = \chi^1$ for some h in H.

Now let G be a regular semi-direct product and let C be a Borel set which meets each H orbit in C in one point. The general theorem alluded to above states that the equivalence classes of irreducible unitary representations of G may all be obtained as follows: Choose $\chi \epsilon C$. Let H_χ denote the closed subgroup of H consisting of all h in H for which $[\chi]\alpha_h^* = \chi$. Choose an irreducible unitary representation L of H_χ. Then $n, h \to \chi(n)L_h$ is an irreducible unitary representation χL of NH_χ. Form $U^{\chi L}$, the unitary representation of G induced by χL. It can be shown that $U^{\chi L}$ is irreducible, that $U^{\chi_1 L^1}$ and $U^{\chi_2 L^2}$ are equivalent if and only if $\chi_1 = \chi_2$ and $L^1 = L^2$ are equivalent representations of H_χ and that every irreducible unitary representation of G is equivalent to some $U^{\chi L}$. When the semi-direct product is not regular, one can use the axiom of choice to find a subset C which meets each orbit just once, but C will not be a Borel set. In this case, the $U^{\chi L}$ can still be formed and proved to be irreducible and inequivalent as indicated. However, it is no longer true that every irreducible unitary representation of G is equivalent to one of the $U^{\chi L}$.

The inhomogeneous Lorentz group considered by Wigner in 1939 is a regular semi-direct product of a four-dimensional real vector group and the Lorentz group, the latter being isomorphic to the quotient of $SL(2, C)$ by its two-element center. Actually, Wigner studied the irreducible unitary representations of the two-fold covering group one gets by using the whole of $SL(2, C)$. His results are a restatement of what one finds by applying the theorem above. It turns out that, depending on the position of the character χ with respect to the "light cone," there are four possibilities for the subgroup H_χ. It is conjugate either to a) the compact subgroup of all unitary matrices, b) a non-compact subgroup isomorphic to the group generated by the translations and rotations in the plane, c) the subgroup $SL(2, R)$ of all matrices in $SL(2, C)$ with real coefficients, or d) the whole of $SL(2, C)$. In case a, the irreducible representations of H_χ are known from the work of Schur and Weyl. In case b, H_χ is a semi-direct product of two commutative groups, and the general theorem just cited can be applied. In cases c and d, H_χ is a non-compact semi-simple Lie group. At the time Wigner's paper was written, nothing was known about their irreducible representations. Wigner, in fact, determined only those irreducible unitary representations of the inhomogeneous Lorentz group falling under cases a and b. However, he gave cogent arguments suggesting that the others were not relevant to the physical applications he had in mind.

A much simpler example of a regular semi-direct product was dealt with in one of the short notes published in 1947. In this note, Gelfand (1913—) and Naimark (1909—) determined all irreducible unitary representations of the group of all one-to-one transformations of the real line into itself of the form $x \to ax + b$ where $a > 0$. This group, often referred to as the "$ax + b$ group," is a semi-direct product of the additive group of all real numbers with the multiplicative group of all positive real numbers. Here N is the additive group of the real line and there are just three orbits. These are $\{0\}$ and the positive and negative real axes. H is the multiplicative group of all positive real numbers and H_x is respectively H, $\{1\}$, and $\{1\}$. It follows that the "$ax + b$ group" has (to within equivalence) just two irreducible unitary representations in addition to the obvious one-dimensional representations defined by the characters of H. They are the representations induced by the characters $b \to e^{ib}$ and $b \to e^{-ib}$ of N. Both are infinite-dimensional.

Three of the four long papers published in 1947 were written respectively by Gelfand and Naimark, Bargmann (1908—), and Harish-Chandra (1923—). All of them were concerned with the determination of the unitary representations of $SL(2, C)$ and hence of the Lorentz group. Bargmann and Gelfand and Naimark also treated $SL(2, R)$, but only Bargmann gave a detailed analysis in that case. As a matter of fact, the papers of Bargmann and Gelfand and Naimark are complementary, in that Bargmann gave details only for $SL(2, R)$ and Gelfand and Naimark only for $SL(2, C)$. Harish-Chandra contented himself with determining the representations of the Lie algebra of $SL(2, C)$, while Bargmann and Gelfand and Naimark found the integrated form of the representations and also discussed the decomposition of the regular representation. The facts about the irreducible unitary representations of $SL(2, C)$ are easily stated in terms of the general concept of an induced representation. Let T be the subgroup of $SL(2, C)$ consisting of all matrices of the form $\begin{pmatrix} \lambda & 0 \\ a & 1/\lambda \end{pmatrix}$. Notice that $\begin{pmatrix} \lambda & 0 \\ a & 1/\lambda \end{pmatrix} \to \begin{pmatrix} \lambda & 0 \\ 0 & 1/\lambda \end{pmatrix}$ is a homomorphism of T onto the subgroup D of all diagonal matrices. Thus each (one-dimensional) character χ of the commutative group D may be lifted to define a one-dimensional representation χ^1 of T. Gelfand and Naimark showed that the induced representations U^{χ^1} are all irreducible and that U^{χ_1} and U^{χ_2} are equivalent if and only if $\chi_1 = \chi_2$ or $\chi_1 = \chi_2^{-1}$. They showed also that the representations U^{χ^1} constitute "almost all" irreducible unitary representations in the sense that they suffice for the decomposition of the regular representation (see below). They constitute what is known as the *principal series* of representations of $SL(2, C)$. In addition to the principal series, they described a second infinite series of irreducible unitary representations called the *supplementary series*. It turns out that one can modify the inducing process[7] in such a manner that some non-unitary representations induce unitary representations of the whole group. This is true of certain non-unitary one-dimensional representations of T, and the members of the supplementary series can all be so described. Gelfand and Naimark were able to prove that every irreducible unitary representation (except of course the trivial representation) is equivalent either to a member of the principal series or to a member of the supplementary series.

In the study of $SL(2, R)$, an interesting new phenomenon arises. One has an obvious analogue of the subgroup T of the principal series and of the supplementary series, but the irreducible unitary representations of $SL(2, R)$ so defined do not exhaust all equivalence classes and do not even suffice to decompose the regular representation. They have to be supplemented by the members of a third series. It is called the *discrete series* because the representations which belong to it occur discretely in the decomposition of the regular representation. Let K be the compact commutative subgroup of $SL(2, R)$ consisting of all matrices of the form $\begin{pmatrix} \cos\theta \sin\theta \\ -\sin\theta \cos\theta \end{pmatrix}$, and for each integer k let χ_k denote the character $\begin{pmatrix} \cos\theta \sin\theta \\ -\sin\theta \cos\theta \end{pmatrix} \to e^{ik\theta}$. Then the induced representations U^{χ_k} are reducible but contain $|k|$ inequivalent discrete irreducible subrepresentations. Moreover, if $k > 0$, U^{χ_k} contains exactly one discrete component which is not contained in $U^{\chi_{k-1}}$ and if $k < 0$, U^{χ_k} contains exactly one discrete component which is not contained in $U^{\chi_{k+1}}$. In either case, let us denote this irreducible representation by W^k. Then the W^k are mutually inequivalent and constitute the discrete series of irreducible unitary representations of $SL(2, R)$. Notice that there is a natural one-to-one correspondence between the members of the discrete series and the non-trivial characters of K, and also a natural one-to-one correspondence between the members of the principal series and pairs χ, χ^{-1} of characters of the diagonal subgroup D (excluding the two cases in which $\chi = \chi^{-1}$). This, combined with the fact that D and K are both maximal Abelian subgroups, turns out to be highly significant. I shall say more about this significance below.

There is an alternative description of the discrete series more closely related to that originally given by Bargmann and important for applications to the theory of modular forms. Consider the action of $SL(2, R)$ on the upper half H^+ of the complex plane defined by $[z]\binom{ab}{cd} = \frac{az+b}{cz+d}$, and let $(V^k_{\binom{ab}{cd}} f)(z) = f(\frac{az+b}{cz+d})(cz + d)^{-k}$ where $k = \pm 1, \pm 2, \cdots$. Then there is a measure μ on the upper half-plane — unique up to multiplication by a positive constant such that the $V^k_{\binom{ab}{cd}}$ are unitary in $\mathscr{L}^2(H^+, \mu)$. The unitary representation of $SL(2, R)$ in $\mathscr{L}^2(H^+, \mu)$ so defined is easily seen to be equivalent to the induced representation U^{χ_k}. The subspace of $\mathscr{L}^2(H^+, \mu)$ on which U^{χ_k} restricts to W^k is the subspace of all analytic functions or the subspace of all conjugate analytic functions, according as $k > 0$ or $k < 0$.

The determination of the irreducible unitary representations of $SL(2, C)$ and $SL(2, R)$ completed Wigner's 1939 paper very nicely. However, it was only Bargmann whose work was inspired by that of Wigner. Neither Harish-Chandra nor Gelfand and Naimark cite Wigner and presumably were unaware of the relevance of his work to theirs. Instead they cite a paper of Dirac, published in 1945. In this paper, Dirac had pointed out that the Lorentz group (and hence $SL(2, C)$) had infinite-dimensional unitary representations and had suggested that they might have physical relevance. Harish-

Chandra was a student of Dirac and made his investigations at Dirac's suggestion.

In view of the role played by physics in inspiring this work on the unitary irreducible representations of $SL(2, R)$ and $SL(2, C)$, it is interesting to note that I proved the theorem on the unitary irreducible representations of regular semi-direct products as a corollary to a much more general theorem obtained as the end product of a three-stage generalization of the Stone-von Neumann theorem on the uniqueness of the solutions of the Heisenberg commutation relations (see section 16). Let G be a separable locally-compact commutative group and let $\hat{G}$ denote its dual. Let μ be Haar measure in G and let A be the regular representation of $G : A_x(f)(y) = f(yx)$ for all $f \in \mathscr{L}^2(G, \mu)$. For each χ in $\hat{G}$ let B_χ denote the unitary operator $f \to \chi f$. Then $\chi \to B_\chi$ is a unitary representation of $\hat{G}$ (equivalent in fact to the regular representation of $\hat{G}$). Moreover, an obvious calculation shows that A and B satisfy the simple commutation relation $A_x B_\chi = \chi(x)B_\chi A_x$. But when G is the additive group of an n-dimensional real vector space, then so is $\hat{G}$, and these commutation relations reduce precisely to the Heisenberg commutation relations in the "integrated" or Weyl form as described in section 16. It is natural to ask whether there is an analogous uniqueness theorem in the general case and to approach the question by applying the spectral theorem to the unitary representation B. The spectral theorem says that B is uniquely determined by a projection-valued measure P on $\hat{\hat{G}} = G$. A simple calculation shows that A and B satisfy the generalized Heisenberg commutation relations written down above if and only if P and A satisfy the commutation relations $A_x P_E = P_{[E]x^{-1}} A_x$ for all x in G and all Borel subsets E of G. These transformed commutation relations have the interesting property that they refer only to G and not to $\hat{G}$. Moreover, they make sense whether or not G is commutative. Indeed, if G is any separable locally-compact group and μ is a right invariant Haar measure, one obtains a system satisfying the transformed commutation relations by defining A_x^0 and P_E^0 in $\mathscr{L}^2(G, \mu)$ by the equations $A_x^0(f)(y) = f(yx)$ and $P_E^0(f)(y) = \varphi_E(y)f(y)$. Here $\varphi_E(y) = 1$ if $y \in E$ and $\varphi_E(y) = 0$ if $y \notin E$. The question then arises, whether every irreducible pair A, P consisting of a unitary representation A of G and a projection-valued measure P on G and which satisfies $A_x P_E = P_{[E]x^{-1}} A_x$ is necessarily equivalent to the pair A^0, P^0 defined above. It is, and I published a proof of this generalization of the Stone-von Neumann uniqueness theorem in 1949. Still a further generalization is possible, however. The commutation relation $A_x P_E = P_{[E]x^{-1}} A_x$ makes sense even when P is not defined on G. It is only necessary that P be defined on some Borel space on which G acts as a group of one-to-one Borel set-preserving transformations. In this generality, while uniqueness fails, one has a complete analysis of all possibilities — at least in the case in which S is a coset space G/H. The irreducible solutions of the commutation relations $A_x P_E =$

$P_{[E]x^{-1}}A_x$ (more precisely their equivalence classes) correspond one-to-one in a natural way to the equivalence classes of irreducible unitary representations of H. Of course when $H = \{e\}$ so that $S = G$, H has only one equivalence class of irreducible unitary representations and the solutions of the commutation relations are unique. The correspondence between solutions of the commutation relations and unitary representations of H is set up by the inducing construction and does not involve irreducibility. Given any unitary representation L of H, the Hilbert space $\mathcal{H}(U^L)$ consists of functions f in G which satisfy the identity $f(hx) = L_h f(x)$ for all h in H and all x in G. If E is any Borel subset of G/H, the function $P_E^L(f)$, which one obtains by reducing f to zero on all right cosets in E and leaving it unchanged outside of these cosets, is also clearly in $\mathcal{H}(U^L)$ and the operator $f \to P_E^L(f)$ is a projection operator. It is easy to check that $E \to P_E^L$ is a projection-valued measure on G/H and that U^L and P^L satisfy the commutation relations $U_x^L P_E^L = P_{[E]x^{-1}}^L U_x^L$ for all x in G and all Borel sets $E \subseteq G/H$. In a short note published in 1949, I sketched a proof of the converse theorem. Given any pair A, P satisfying the commutation relations in question, there exists a unitary representation L of H (uniquely determined up to equivalence) such that the pair A, P is equivalent to the pair U^L, P^L. Moreover, the algebra of all bounded operators which commute with all L_h is isomorphic to the algebra of all bounded operators which commute with all U_x^L and all P_E^L.

To gain some insight into why such a theorem might be true and also to understand why it is called "the imprimitivity theorem" it is useful to consider the special case in which H is an open subgroup of G. In that case, the projection-valued measure P defines a discrete direct sum decomposition of $\mathcal{H}(A)$ whose summands are parameterized by the points of $G/H = S$. These summands are not invariant under the A_x. This would be true if and only if $A_x P_E = P_E A_x$ for all x and E. On the other hand, the condition $A_x P_E = P_{[E]x^{-1}}A_x$, which does hold, is precisely equivalent to the assertion that the A_x permute the summands among themselves and in particular that A_x carries the summand whose parameter is s onto that whose parameter is $[s]x$. A direct sum decomposition of a Hilbert space having this property with respect to a unitary representation A taking place therein is called a *system of imprimitivity* for the representation. When the group acts transitively on the subspaces and H is the subgroup leaving a subspace $\mathcal{M}_0$ fixed, the original representation defines a representation L of H in $\mathcal{M}_0$ which evidently determines A. The discrete case of the imprimitivity theorem thus has a trivial proof. For finite groups it was known to Frobenius.

To see how the semi-direct product theorem might be a consequence of the imprimitivity theorem one has only to note a) that a unitary representation V of a semi-direct product NH is uniquely determined by its restrictions B and A to N and H respectively, and b) that the condition that $n, h \to B_n A_h$

be a representation of G, given that B and A are representations of N and H respectively, is easily computed to be the commutation relation $B_n A_h = A_h B_{\alpha_h(n)}$. When B is replaced by the projection-valued measure P on $\hat{N}$ which determines it, this commutation relation reduces to the statement that P is a system of imprimitivity for A. When V is irreducible and the semi-direct product is regular, one shows that P is supported by an H orbit in $\hat{N}$. The rest is calculation.

The work of Harish-Chandra, Gelfand and Naimark, and Bargmann on the irreducible unitary representations of $SL(2, C)$ and $SL(2, R)$ has implications beyond its possible relevance to physics and to completing the work of Wigner on the inhomogeneous Lorentz group. The group $SL(2, R)$ is the non-compact semi-simple Lie group of lowest possible dimension, and the group $SL(2, C)$ is the non-compact semi-simple Lie group of lowest possible dimension which also has the structure of a complex manifold. Moreover, while the semi-direct product theorem described above can be generalized to a theorem dealing with groups having non-commutative normal subgroups, this method of analysis fails completely in dealing with simple groups, that is, with groups having no closed normal subgroups. A different approach must be used and $SL(2, R)$ and $SL(2, C)$ are in different ways the most elementary examples. Modulo its two element center, each of these groups is simple.

With these two groups under control, it was natural to go on to more complicated cases, and Gelfand and Naimark began a study of $SL(n, C)$ almost immediately. In fact, the fourth long paper of 1947 and several of the short notes published in 1946 and 1947 are papers by them concerned with the irreducible unitary representations of $SL(n, C)$. The results are analogous to those for $SL(2, C)$ but are less complete. Let T be the subgroup of all matrices that are zero above the main diagonal, and let D be the subgroup of all diagonal matrices. Then $D \subseteq T$, and just as in the case of $SL(n, C)$ there is a natural homomorphism of T onto D so that every $\chi \in \hat{D}$ can be lifted to be a one-dimensional unitary representation χ' of T. The induced representations $U^{\chi'}$ are all irreducible and constitute what Gelfand and Naimark call the *principal series*. As in the case of $SL(2, C)$, the members of the principal series suffice to decompose the regular representation. The theory of $SL(n, C)$ differs from that of $SL(2, C)$ chiefly in that finding the remaining irreducible unitary representations is much more difficult. In fact, except when $n \leq 3$, the problem of determining all equivalence classes of irreducible unitary representations of $SL(n, C)$ is still an open one in which there is considerable current interest. When $n > 2$ there exist proper closed subgroups of $SL(n, C)$ which properly contain T. For example, when $n = 3$, one has the subgroup of all matrices of the form $\begin{pmatrix} a_{11} & a_{12} & 0 \\ a_{21} & a_{22} & a_{23} \\ a_{31} & a_{32} & a_{33} \end{pmatrix}$, and when $n = 4$, the subgroup of all matrices of the form $\begin{pmatrix} a_{11} & a_{12} & 0 & 0 \\ a_{22} & a_{23} & 0 & 0 \\ a_{31} & a_{32} & a_{33} & a_{34} \\ a_{41} & a_{42} & a_{43} & a_{44} \end{pmatrix}$.

The subgroups admit non-trivial one-dimensional characters which can be induced up to $SL(n, C)$ to define new infinite-dimensional irreducible unitary representations — the members of the so-called *degenerate series*. As with T in $SL(2, C)$, certain non-unitary characters of both T and the proper closed subgroups containing it can be "induced" to form irreducible unitary representations of $SL(n, C)$. The chief difficulty in determining *all* irreducible unitary representations of $SL(n, C)$ lies in deciding just which non-unitary characters lead to unitary representations.

Gelfand and Naimark found little difficulty in extending their analysis to the other classical complex groups — the complex orthogonal groups and the complex symplectic groups of all dimensions. All of these groups admit analogues of the groups T and D, and one defines the principal and other series in a strictly analogous manner. While it has only recently been proved that *all* members of the principal series for the complex groups are irreducible, Gelfand and Naimark could prove that almost all of them are, and these suffice to decompose the regular representation. Gelfand and Naimark published a book in 1950 giving a systematic account of their work on all the complex classical semi-simple Lie groups including $SL(n, C)$.

Once one has found the irreducible unitary representations of a group G, the problem arises whether or not they are adequate for harmonic analysis on the measure spaces on which G acts. Given a space S, with a measure μ invariant under the G action, one can form a unitary representation U of G whose space is $\mathscr{L}^2(S, \mu)$ by setting $U_x(f)(s) = f([s])x)$. One can then ask whether there is a sense in which this representation can be decomposed as a discrete or continuous direct sum of irreducibles and whether this decomposition is unique in any useful sense. When G is commutative as well as separable and locally compact, these questions are taken care of very nicely by the spectral theorem of Stone, Ambrose, Godement, and Naimark and the spectral multiplicity theory of Hahn, Hellinger, and Stone (see section 17). To go further, one needed a theory of direct integrals or continous direct sums of Hilbert spaces. Such a theory had been worked out by von Neumann in the 1930s for use in his work with Murray on algebras of operators (see the end of section 17), and a more or less complete typed paper on the subject was in von Neumann's possession in 1938. However, this paper did not get published until 1949 and apparently no one interested in the theory of unitary group representations knew its contents until 1947. In that year von Neumann made the typescript available to Mautner (1921—), who based his 1948 Ph.D. thesis on applying direct integral decompositions to unitary group representations. I used some of the ideas in von Neumann's typescript in my proof of the generalized Stone-von Neumann uniqueness theorem. Mautner's first main result (announced in a short note published in 1948) was in essence as follows: Let U be a (continuous) unitary representation of a separable locally-compact group G in a separable Hilbert space $\mathscr{H}(U)$. Let $R(U, U)$ denote the commuting algebra of U; that is, the algebra of all bounded linear operators in $\mathscr{H}(U)$ that commute with all U_x. Then corresponding to every *maximal* commutative subalgebra of $R(U, U)$ there exists an essentially unique decomposition of U as a direct integral of *irreducible* unitary representations of G. The meaning of direct integral decomposition can be most quickly explained in the case in which all of the component representations are infinite-dimensional and so have isomorphic Hilbert spaces. Let there be given a (suitably restricted) measure space S, μ, and for each $s \, \epsilon \, S$ a unitary irreducible representation L^s of G in a fixed Hilbert space $\mathscr{H}$. Suppose that $L_x^s(\varphi) \cdot \psi$ is measurable in $S \times G$ for each φ and ψ in $\mathscr{H}$. Form the Hilbert space $\mathscr{L}^2(S, \mu, \mathscr{H})$ of all square integrable measurable functions from S to $\mathscr{H}$. For each x in G let M_x denote the operator in $\mathscr{L}^2(S, \mu, \mathscr{H})$ which takes $s \rightarrow f(s)$ into $s \rightarrow L_x^s(f(s))$. Then each M_x is unitary and $x \rightarrow M_x$ is a (continuous) unitary representation of G. It is called the *direct integral* or *continuous* direct sum of the representations L^s with respect to the measure μ. For each measurable subset E of S, one can associate the operator $f \rightarrow \varphi_E f$ where $\varphi_E(s) = 1$ if $s \, \epsilon \, E$ and is zero otherwise. Denoting this operator by P_E, one verifies that $E \rightarrow P_E$ is a projection-valued measure. When all (or almost all) the L^s are irreducible, it turns out that the P_E constitute all the projection operators in a maximal commuting subalgebra of $R(M, M)$. Conversely, according to Mautner's theorem, every maximal commuting subalgebra of every $R(U, U)$ arises in this way from some direct integral of irreducibles that is equivalent to U. Since maximal commutative subalgebras always exist (by Zorn's lemma), so do direct integral decompositions into irreducibles. Mautner's theorem says nothing about the uniqueness of the decomposition, and in fact different choices of the maximal commuting subalgebra of $R(U, U)$ can lead in some cases to radically different decompositions into irreducibles. The situation was clarified in the early 1950s using ideas derived from the von Neumann-Murray theory of "factors." Further details will be given below.

As far as harmonic analysis on certain groups and homogeneous spaces is concerned, the results may be expressed without considering direct integral decompositions as such. This was done by Gelfand and Naimark for $SL(2, C)$ (and later for the classical groups) before they knew the results of von Neumann and Mautner. Bargmann also had results in this direction for $SL(2, R)$. Consider what the Peter-Weyl theorem tells us about expansions of square integrable functions on a separable compact group G. For each irreducible unitary representation L of G, the functions $x \rightarrow L_x(\varphi) \cdot \psi$ generate a finite-dimensional two-sided invariant vector space $\mathscr{M}^L$ of continuous functions on G which depends only on the equivalence class of L and is called the space of matrix coefficients for L. These spaces for the various possible L's are mutually orthogonal, and every square integrable f on G may be written as a sum $f = \sum_L f_L$ where $f_L \in \mathcal{M}^L$. Moreover, f_L may be computed from f and the character χ^L of L by the simple formula $f_L(x) = \chi^L(e) \int_G f(xy^{-1})\chi^L(y)d\mu(y)$, which reduces in the commutative case to the formula for computing Fourier coefficients. There is an obvious possible generalization of this formula to any separable unimodular locally-compact group having only finite-dimensional unitary irreducible representations — for example, a semi-direct product of a commutative group and a finite group. One could define $f_L(x)$ as $\int_G f(xy^{-1})\chi^L(y)d\mu(y)$ for all f in $\mathcal{L}^1(G, \mu)$ and hope to find a measure $\hat{\mu}$ in the space of all irreducible characters (depending of course on the choice of μ) such that $f(x) = \int f_L(x)d\hat{\mu}(L)$ for all f in some dense subspace of $\mathcal{L}^1(G, \mu)$. To do the same for groups with infinite-dimensional irreducible unitary representations seems impossible at first because of the fact that Trace L_x never exists when L is infinite-dimensional. But there is a way out. For each f in $\mathcal{L}^1(G, \mu)$ there exists a unique operator L_f such that $(L_f(\varphi) \cdot \psi) = \int f(x)L_x(\varphi) \cdot \psi)d\mu(x)$ for all φ and ψ in $\mathcal{H}(L)$, and it often turns out that L_f does have a trace. $f \to$ Trace (L_f) is then a linear functional, which may often be shown to be of the form $f \to \int \chi(x)f(x)d\mu(x)$ where $\chi(x)$ is a measurable complex-valued function and is uniquely determined by L (up to changes on sets of measure zero). When L is finite-dimensional, χ always exists and is just χ^L. Thus whenever the above conditions are satisfied it is natural to extend the definition and say that L has a character equal to χ. First for $SL(2, C)$ and later for $SL(n, C)$ and the classical complex groups, Gelfand and Naimark showed a) that characters in this generalized sense exist for all principal series members and b) that there is an expansion formula as indicated. They also found explicit formulae for the characters and the measure $\hat{\mu}$. It is important to notice that the expansion formula does not involve *all* irreducible unitary representations of G. Members outside of the principal series do not occur. On the other hand, one can put them in and simply assign measure zero to the set consisting of all of them. In order to have an expansion formula it suffices to know "almost all" irreducible unitary representations.

More generally, let G be an arbitrary separable locally-compact group whose left and right Haar measures are the same. One might hope to prove the existence of a family $\mathcal{F}$ of irreducible unitary representations L of G having characters χ^L in the above sense together with a measure $\hat{\mu}$ defined on suitable subsets of $\mathcal{F}$ such that $f(x) = \int_{\mathcal{F}} [\int_G f(xy^{-1})\chi^L(y)d\mu(y)]d\hat{\mu}(L)$ for all f in a dense subspace of $L^1(G)$. While such a theorem can indeed be proved for a large and important class of groups, it is true only in a modified form for another large class. Its failure to hold as stated in general is intimately related to the failure of uniqueness in direct integral decompositions — both failures are due to the existence of the new infinite-dimensional generalizations of full matrix algebras discovered in 1936 by Murray and von Neumann (see the end of section 17).

Let U be a unitary representation of G and suppose that U is discretely decomposable so that $U \simeq L^1 \oplus L^2 \oplus \cdots$ where the L^j are irreducible. Let $L^{'1}, L^{'2}, L^{'3}, \cdots$ be a subset of the L^j including one and only one member of each equivalence class that occurs. Let M^j be the direct sum of all L^k, with L^k equivalent to $L^{'j}$. Then $U \simeq M^1 \oplus M^2 \oplus \cdots$ and each M^j is a direct sum of mutually equivalent irreducible representations. The two decompositions, first into the M^j and then into irreducibles, are unique in different senses. The first decomposition is absolutely unique in that for each j, the invariant subspace on which U reduces to M^j is uniquely determined. The further reduction of M^j into irreducibles is unique only up to equivalence. There are very many quite different direct sum decompositions of $\mathcal{H}(M^j)$ into invariant irreducible subspaces. The non-uniqueness in the general case occurs only at the second stage and only in certain instances. Specifically, let G be an arbitrary separable locally-compact group and let U be an arbitrary unitary representation of G in a separable Hilbert space $\mathcal{H}(U)$. Let $CR(U, U)$ be the *center* of the commuting algebra of U. Then Mautner's argument associating a direct integral decomposition of U to every *maximal* commutative subalgebra of $R(U, U)$ associates a direct integral decomposition to $CR(U, U)$ whose components U^λ are not necessarily irreducible but are so-called *factor representations*. A factor representation is by definition a representation whose commuting algebra has a *one-dimensional center*, i.e., is a factor in the sense of Murray and von Neumann. In the discrete case considered above, this decomposition is the decomposition into the M^j. Now whenever a factor representation is discretely decomposable, it is easy to show that all components are equivalent and that the commuting algebra is isomorphic to the algebra of all bounded operators on a finite- or infinite-dimensional separable Hilbert space. Conversely, if M is a factor representation whose commuting algebra is isomorphic to the algebra of all bounded operators on a finite- or infinite-dimensional separable Hilbert space, then M is a direct sum of equivalent irreducibles whose number and equivalence class is uniquely determined. In this case, one speaks of factors and factor representations of type I. When all (or almost all) the factor representations U^λ of U are of type I, one says that U is of type I, and then the decomposition of U into irreducible representations is as unique as in the classical case. As shown by Murray and von Neumann in 1936, however, there exist factors that are not of type I. When these occur among the commuting algebras of the U^λ (on a set of λ of positive measure), the situation is quite different and a further analysis is required.

Murray and von Neumann classified factors according to the behavior of the lattice of projection operators. The projection operators in $R(U, U)$ correspond one-to-one to the subrepresentations of U, and in the present context it is perhaps more illuminating to present the results of Murray and von Neumann in the language of group representations. Let us define two representations U and V to be *disjoint* if no subrepresentation of U is equivalent to any subrepresentation of V. It is then easy to see that a projection operator P in $R(U, U)$ is such that the subrepresentations U^P and U^{1-P} are disjoint if and only if P is in the center $CR(U, U)$ of $R(U, U)$. Hence $R(U, U)$ is a factor if and only if it is impossible to write U as a direct sum of two disjoint subrepresentations. Let U and V both be factor representations. If they are both of type I, then it is trivial to prove that they fail to be disjoint if and only if each is equivalent to a multiple of the same irreducible representation, and then one is clearly equivalent to a subrepresentation of the other. Even if they are not of type I, it is possible to prove that whenever U and V are not disjoint, then either U is equivalent to a subrepresentation of V or V is equivalent to a subrepresentation of U. Let us define U and V to be *quasi-equivalent* if they are not disjoint. One can then prove that quasi-equivalence is an equivalence relation, that the direct sum of any finite or countably infinite family of quasi-equivalent factor representations is a factor representation in the same quasi-equivalence class and that every subrepresentation of a factor representation is a factor representation in the same quasi-equivalence class. The equivalence classes of factor representations in a fixed quasi-equivalence class thus form an ordered semi-group. When the factor representations in the quasi-equivalence class are of type I, this ordered semi-group is clearly isomorphic to that formed by the positive integers and ∞. One of the main results of Murray and von Neumann is equivalent to the assertion that there are just two other possibilities: either this ordered semi-group is isomorphic to that formed by the positive real numbers and ∞, or it consists of only one element. In the first case, the factor representation is said to be of type II, and in the second case to be of type III. Every factor is the commuting algebra of some representation of some group and has the same type. Let U be a factor representation of type II which is finite in the sense that $U \oplus U$ and U are not equivalent. It can be shown that there exists a subrepresentation V of U, unique to within equivalence such that $V \oplus V$ is equivalent to U. Let $\frac{1}{2}U = V$. Then $\frac{1}{2^n}U$ is defined for all n. Given any positive real number λ, let $\lambda = n + \Sigma\, 1/2^{n_j}$ where n and the n_j are non-negative integers and $n_1 < n_2 < n_3 < \cdots$. One may consistently define λU as $U \oplus \overset{n\ \text{times}}{\cdots} \oplus U + \Sigma\, 1/2^{n_j}\, U$ and ∞U as $U \oplus U \oplus U \oplus \cdots$ and show that every member of the quasi-equivalence class of U is equivalent to λU for one and only one value of λ. If we define λ to be the multiplicity of λU (relative to U), it is clear that quasi-equivalence classes of type II factor representations behave as though some "ideal" or "virtual" irreducible were being repeated with a continuum of (relative) multiplicities. While type II factor representations can be decomposed as direct integrals of irreducibles, the irreducibles that occur are far from being uniquely determined and the decompositions seem to be of little if any use. It seems best to stop the decomposition at the first level. When U is a type II factor representation but is not necessarily finite, the different projections in $R(U, U)$ define subrepresentations of U, and their multiplicities (relative to some finite subrepresentation of U) define a so-called *relative dimension function* $P \to d(P)$ on the set of all projections P in $R(U, U)$. If A is any bounded self-adjoint operator in $R(U, U)$ and $E \to P_E^A$ is the projection-valued measure on the line assigned to A by the spectral theorem, then the P_E^A are all in $R(U, U)$ and $E \to d(P_E^A)$ is a measure on the real line. When x is integrable with respect to this measure (and it always is when U is finite), the integral is called the *relative trace* of A. In 1937 Murray and von Neumann published a second paper whose chief purpose was to prove the surprisingly difficult theorem that the relative trace, when it exists, is a linear function of the operator. Since every bounded linear operator is uniquely of the form $A + iB$ where A and B are self-adjoint, this linear functional has a well-defined extension to all bounded linear operators whenever $d(P_i^A) < \infty$, and to a large subset in any case.

Using the relative trace concept in conjunction with the theory of direct integral decompositions, it is possible to prove an expansion theorem for reasonably general functions on any separable locally-compact group whose left and right Haar measures coincide. This was done by Segal (1918—) and Mautner in independent papers published in 1950. When U is a factor representation, the operator $U_f = \int f(x)U_x d\mu(x)$ will lie in the commutator of $R(U, U)$ and this is always a factor of the same type as $R(U, U)$. If $T^R(U_f)$ denotes the (suitably normalized) relative trace of U_f, then $f \to T^R(U_f)$ will be a linear functional, which one can hope to write in the form $f \to \int f(x)\chi(x)d\mu(x)$. When this is possible, one can think of χ as the character of U (with respect to the normalization chosen). Now consider the regular representation of the group G under consideration and its decomposition into factor representations defined by the center of the commuting algebra. When almost all of these factor representations are of type I *and* the irreducibles that generate them have characters as indicated above, the Segal-Mautner theorem implies the obvious generalization of the expansion theorem of Gelfand and Naimark. However, it goes further in two directions. First of all, it is not necessary that the linear functionals $f \to$ Trace L_f be of the form $f \to \int f(x)\chi(x)d\mu(x)$. One can think of the linear functions themselves as characters and replace the expression $\int_G f(xy^{-1})\chi^L(y)d\mu(y) = \int_G f(xy)\chi^L(y)d\mu(y)$ by Trace $\overline{L}_{f_x}$, where f_x is the left translate of f and $\overline{L}$ is the complex conjugate of L. Mautner and Segal do this and also include the case in which type II factor representations occur by replacing the trace of the irreducible generator with the Murray-von Neumann relative trace. It can be shown that type III representations do not occur in the decomposition of the regular representation of a group whose left and right Haar measures coincide. The final result is called the Plancherel theorem for the group in question, and the measure $\hat{\mu}$ in the appropriate set of irreducible and factor representations is called the *Plancherel measure*. When the group is commutative, the Plancherel measure is Haar measure in the dual group. It is important to notice that the Plancherel theorem of Mautner and Segal is an abstract existence theorem. In specific cases, the problem remains of finding the irreducible and factor representations that are needed and of specifying the particular measure on this set which serves as the Plancherel measure.

When non-type I factor representations exist for a group G, its representation theory is considerably more complicated in several ways, and it no longer suffices to know the irreducible unitary representations in order to know all unitary representations. One must also know the factor representations of type II and type III. Since the latter are always difficult (if not impossible) to find in their totality, it is helpful that many of the most interesting groups can be shown to have no non-type I factor representations at all. Such groups are called type I groups and obviously include the compact groups and the locally-compact commutative groups. Mautner, who was the first to note the effect on uniqueness in decomposition theory of the existence of non-type I factor representations, was also the first to attempt to determine which groups were type I groups and which were not. In a paper published in 1950, he showed that both $SL(2, R)$ and $SL(2, C)$ are type I groups and conjectured that all semi-simple Lie groups are type I. He also found a five-dimensional solvable Lie group that is not of type I and gave examples showing that discrete groups tend never to be type I groups unless they are very close to being finite or commutative. Over a decade later it was shown by Thoma that a countable discrete group is a type I group if and only if it has a commutative normal subgroup with a finite quotient. Irregular semi-direct products are a rich source of groups that are not of type I. Mautner's non-type I five-dimensional solvable Lie group is an example in which the normal subgroup is a four-dimensional vector group.

A Lie algebra version of Mautner's conjecture about the type I-ness of semi-simple Lie groups was proved by Harish-Chandra in the course of a long paper published in 1951. (The original conjecture was proved in a second long paper by the same author published two years later.) Harish-Chandra combined his attack on the type I-ness question with a powerful and original attack on the problem of finding the irreducible Banach space representations of general semi-simple Lie groups. A complete classification is difficult to derive, and Harish-Chandra soon concentrated his efforts on finding enough irreducible unitary representations to decompose the regular representation and on finding an explicit Plancherel formula. In 1952 he found the Plancherel formula for $SL(2, R)$, and in 1954 published a long paper on the general case, including complete results for the complex semi-simple Lie groups. His results in the complex case go beyond those found earlier by Gelfand and Naimark in that the five exceptional groups are included and in that all cases are treated at once in a uniform manner. The non-complex case turned out to be much more difficult; in fact, it demanded Harish-Chandra's best efforts for over a quarter of a century. The final details have been written down only recently. The fundamental simplification that takes place in the case of a complex semi-simple Lie group is that there is a single commutative subgroup the conjugates of whose elements form a set whose complement has Haar measure zero.

Already in the case of $SL(2, R)$ no such subgroup exists. Instead there are two commutative subgroups, the conjugates of whose elements constitute two sets intersecting in a two element set, whose union has a complement of Haar measure zero. In more complicated real semi-simple Lie groups, many such non-conjugate "Cartan subgroups" exist. By 1952 Harish-Chandra had already suggested that to find enough irreducible unitary representations for a Plancherel formula it would be necessary to associate a different family to each conjugacy class of Cartan subgroups. In $SL(2, R)$ the principal series and the discrete series correspond respectively to the diagonal subgroup and the compact subgroup of all $\begin{pmatrix} \cos\theta & \sin\theta \\ -\sin\theta & \cos\theta \end{pmatrix}$, and their members are parameterized by finite subsets of characters of these two commutative groups. Work of Gelfand and Graev published a year later showed that this was definitely the case for $SL(n, R)$. Here the number of Cartan subgroups is $\frac{n+2}{2}$ or $\frac{n+1}{2}$ according to whether n is even or odd. In 1954 Harish-Chandra described a method[8] for assigning a family of irreducible (or nearly irreducible) unitary representations to each *non-compact* Cartan subgroup A, its members being parameterized by certain finite subsets of the character group $\hat{A}$ of A. The method involved using the inducing process $L \rightarrow U^L$ and forming L out of irreducible unitary representations of lower-dimensional semi-simple Lie groups. The method failed when A was compact, and Harish-Chandra saw the central problem as that of finding some other method for constructing the family that he felt sure existed in this case. This is the celebrated problem of the "discrete series." Harish-Chandra announced a solution in 1963—at least as far as finding the *characters* of the representations was concerned.

I have already explained how some irreducible unitary representations may be said to have "characters" either in the sense of actual functions on the group or at least as linear functions $f \rightarrow \ell(f)$ defined for f in some reasonably large vector space of complex-valued functions on G. In the early 1950s Godement was particularly active in attempting to construct a general theory of characters along such lines. He published a long memoir on the subject in 1951 and another (in two installments) in 1954. As far as semi-simple Lie groups are concerned, however, the most useful theory is due to Harish-Chandra. Let $\mathscr{D}$ be the space of all infinitely differentiable functions with compact support on the semi-simple Lie group G. In 1954 Harish-Chandra showed that for f in $\mathscr{D}$, Trace L_f exists for all irreducible unitary L (and some Banach space representations as well) and that $f \to$ Trace L_f is a distribution in the sense of L. Schwartz (1915—). Two years later he showed that this distribution reduced to an analytic function on an open set with a complement of measure zero. Finally in 1965 he showed that his distribution characters are actually locally summable functions on the whole group.

My aim in this section has been to give some rough idea of the beginnings and principal concepts of the huge theory that has emerged from the successful attempt to extend harmonic analysis to locally-compact groups which are not necessarily compact nor commutative. My account is at most three-quarters complete as far as the first decade is concerned, and the finding of the general Plancherel formula for semi-simple Lie groups is only a fraction of what has been added to the theory in the past two decades. For further details about the period 1946-1961 the reader is referred to my colloquium lectures [*14*], and for a survey of developments between 1955 and 1975 to the second half of my 1976 book [*15*]. (The first half of the book is a reprint of a set of lecture notes for a course given at the University of Chicago in 1955.)

22. APPLICATIONS OF THE GENERAL THEORY

In the earlier sections of this article I have indicated three different origins of the method of harmonic analysis: in probability theory, in mathematical physics, and in number theory. In this section we shall see that the general point of view that began to evolve with Frobenius's introduction of group characters in 1896, and which reached a certain level of completeness and coherence in the 1950s, has significant applications to all three subjects.

The applications to probability theory are still relatively undeveloped and will be dealt with quite briefly. The basic idea is that families of random variables occur in many contexts (not just strung out in time) and that the parameter space may have symmetry properties. In other words, there is a natural generalization of a stationary stochastic process (see section 18) in which the ergodic action of the real line or the integers in a probability measure space Ω, μ is replaced by an ergodic action of some other group.

For example, in considering the statistical mechanics of a gas (considered for convenience as occupying all of space), the number of molecules in a finite subset V of space is a random variable, and the set of random variables obtained by varying V constitutes a generalization of a stochastic process. Assuming the gas to have properties invariant under translation and rotation considerations analogous to those given in section 18 leads to a natural measure-preserving action of the Euclidean group $\mathcal{E}$ on the underlying probability measure space Ω, μ. The Koopman construction then yields a unitary representation of $\mathcal{E}$, and the decomposition of this representation can be used as in section 18 to contribute to the analysis of the statistical mechanical problem. A preliminary study of this kind of application was made in a note published in 1960 by A. M. Yaglom.

The applications of the theory of unitary group representations to quantum physics are by now so extensive that anything like a complete summary would require a book-length article. I shall content myself here with a few remarks and references. In section 21 it was explained how one could be led to the imprimitivity theorem through three successive generalizations of the Stone-von Neumann theorem on the uniqueness of the solutions of the Heisenberg commutation relations. The final results seem to have nothing to do with the original physical problem. It is thus of some interest that it turns out a) to be possible to deduce the Schrödinger equation for a single free particle from the general principles of quantum mechanics and group theoretical invariance postulates, and b) that the imprimitivity theorem is the chief tool used in carrying out the derivation. Indeed, using the imprimitivity theorem one can show that the Heisenberg commutation rules are essentially consequences of Euclidean invariance — and that the existence and properties of spin come along as a bonus.

The argument proceeds as follows: Let S denote physical space and let $\mathcal{H}$ be the Hilbert space of states for our one-particle system (see section 16). Let $\mathcal{E}$ denote the group of all isometries of S so that $\mathcal{E}$ acts transitively on S and preserves the volume measure ν. For each Borel subset E of S, let P_E be the self-adjoint operator in $\mathcal{H}$ corresponding to the observable, which is one when the particle is observed to be in E and zero otherwise. Then P_E must be a projection operator, and it is not difficult to accept hypotheses implying that $E \to P_E$ must be a projection-valued measure on S. This projection-valued measure determines all position observables in the following manner: If g is any real coordinate, i.e., any real-valued (Borel) function on S, then $E \to P_{g^{-1}(E)}$ is a projection-valued measure on the line, and the self-adjoint operator associated with it by the spectral theorem is the self-adjoint operator in $\mathcal{H}$ associated with the coordinate observable g. Let $t \to V_t$ denote the unitary representation of the additive group of the real line, defining the "dynamics" or time evolution of our system. Once P and V have been given in some concrete fashion, our system is completely determined and all questions about what happens can be reduced to mathematical calculation. However, without further assumptions the pair P, V could be an arbitrary pair consisting of a projection-valued measure on S and a unitary representation V of the additive group of the line. We now introduce the assumption that the whole system is invariant under $\mathscr{E}$. This means in the first instance that each $\alpha \epsilon \mathscr{E}$ is intrinsically associated with an automorphism of our quantum model, i.e., with a unitary (or anti-unitary) operator U, and that $U_{\alpha\beta} = U_\alpha U_\beta \sigma(\alpha, \beta)$ where σ is some projective multiplier for the group $\mathscr{E}$. It means in addition that P_E and $P_{[E]\alpha}$ are transforms of one another by the operator U_α; that is, that $U_\alpha^{-1} P_E U_\alpha = P_{[E]\alpha}$ for all E and α. Now when every element in $\mathscr{E}$ is the square of another, all the operators U_α must be unitary rather than anti-unitary. Moreover, by replacing $\mathscr{E}$ by a covering $\tilde{\mathscr{E}}$ one can get rid of the factor σ. Supposing these things done, one recognizes that P is a system of imprimitivity for U based on S. Since $S = \tilde{\mathscr{E}}/K$ where K is the closed subgroup of $\tilde{\mathscr{E}}$ leaving some origin s_0 in S fixed, the imprimitivity theorem implies that the pair U, P is equivalent to the pair U^L, P^L where L is some unitary representation of K. When K is compact, as it is for the usual Euclidean model for space, there is only a discrete countable set of possibilities for L and hence for the system P, U. The simplest case is that in which L is the one-dimensional identity. In that case, we at once reach the conclusion that $\mathscr{H}$ is isomorphic to $\mathscr{L}^2(S, \mu)$ in such a way that $U_\alpha^L(f)(s) = f(s\alpha)$ and $P_E^L(f)(s) = \varphi_E(s)f(s)$ where $\varphi_E(s) = 1$ or 0 according as $s \epsilon E$ or $s \notin E$. We shall not pursue the analysis further here. Let it suffice to say that in most of the particles occurring in physics the representation L is an irreducible unitary representation of $SU(2)$ and that $\frac{1}{2}(\dim L - 1)$ is called the *spin* of the particle. The representation $t \to V_t$ is limited by the requirement that $V_t U_\alpha^L = U_\alpha^L V_t$, and the possibilities under this limitation can be analyzed using the theory of unitary group representations.

The connection between systems of imprimitivity and particle position observables seems to have been first noted by Wightman (1922—). Wigner, in collaboration with T. D. Newton, published a paper on position observables for relativistic particles in 1949 — the same year that my statement and abbreviated proof of the imprimitivity theorem were published. Not long thereafter, Wightman read both papers and noted that the contents of one were just what was needed to make the other rigorous. However, he did not publish his results until 1962. I worked out the axiomatics of a particle discussed above after hearing a vague account of what Wightman had done.

Starting with the group representational model for a single particle described above, one can discuss systems of interacting particles in a similar spirit and finally fit almost the whole of quantum physics into the framework of the theory of unitary group representations. Further details will be found on pp. 328–357 of my book [*15*].

In some ways the applications to number theory are the most interesting of all, in part perhaps because they differ more from applications of the compact and commutative theory, and in part because they are still very incompletely understood and are at the center of a rapidly developing area of mathematics. I shall describe them at somewhat greater length than I did the applications to probability theory and physics.

Let F be a function analytic in the upper half-plane and let N be a positive integer. Let Γ_N denote the subgroup of $SL(2, R)$ consisting of all $\left(\begin{smallmatrix} a & b \\ c & d \end{smallmatrix}\right)$ in which $a - 1$, b, c, and $d - 1$ are integer multiples of N. Let k be a positive integer. One defines an (unrestricted) modular form of level N and weight $k/2$ (or dimension $-k$) to be a complex-valued function F analytic in the upper half of the complex plane such that $F\left(\dfrac{az+b}{cz+d}\right) = (cz + d)^k F(z)$ for all z in the upper half-plane and all $\left(\begin{smallmatrix} a & b \\ c & d \end{smallmatrix}\right) \epsilon \ \Gamma_N$. Setting $a = 1, b = N, c = 0, d = 1$, one sees in particular that $F(z + N) = F(z)$, so that F has the Fourier expansion $F(z) = \displaystyle\sum_{j=-\infty}^{\infty} \varphi(j) e^{(2\pi i j z)/N}$. When $\varphi(j) = 0$ for $j < 0$ and (when $N > 1$) certain other growth conditions are satisfied, one says that F is a *modular form* of level N and weight $k/2$. As already indicated in sections 10 and 19, the theory of modular forms has close connections with number theory because of the interesting number-theoretical properties of the functions $n \rightarrow \varphi(n)$, which occur as coefficient sequences. In particular, for every positive definite quadratic form with an even number of variables, let $\varphi_Q(n)$ denote the number of integer points on the hypersurface $Q(x_1, x_2, \cdots, x_\ell) = n$. Then $n \rightarrow \varphi_Q(n)$ occurs as the coefficient sequence for a modular form of weight $\ell/4$ and some level depending on the form.

As explained in some detail in sections 10 and 11, a fairly complete theory of modular forms of level one was given by Hurwitz in his thesis (published in 1881). Extending the theory to forms of higher level turned out to be quite difficult. Although Fricke and Klein made a certain amount of progress (see section 11), the situation in 1925 was that complete results were available for only a few small values of k and N. Then, beginning with an announcement in 1925, Hecke published a series of papers in which several important new ideas were used to carry the theory a great deal further. Since Hecke's ideas and results are vital in the application of unitary group representations to number theory, it will be necessary to devote some space to describing them.

The material announced in 1925 was worked out in detail in a paper published in 1926. The main idea was to try to construct previously unknown modular forms using coefficient functions from other parts of number theory. Hecke found it possible to get new modular forms of weight $\ell/2$ and various levels from the coefficients of the Dirichlet series expansions for certain zeta functions associated with *real* quadratic number fields. Let $n \to \varphi(n)$ be a complex-valued function defined on the nonnegative integers which satisfies suitable growth conditions at ∞. Then $\sum_{n=0}^{\infty} \varphi(n)e^{2\pi inz}$ will converge in the upper half-plane and define an analytic function F_φ such that $F_\varphi(z + 1) \equiv F_\varphi(z)$. Since $\left(\begin{smallmatrix}1&1\\0&1\end{smallmatrix}\right)$ and $\left(\begin{smallmatrix}0&1\\-1&0\end{smallmatrix}\right)$ generate the modular group, and since $F_\varphi\left(\dfrac{0 \cdot z - 1}{1 \cdot z + 0}\right) = F_\varphi(-1/z)$ and $F_\varphi\left(\dfrac{1 \cdot z + 1}{0 \cdot z + 1}\right) = F_\varphi(z + 1)$, it follows that F_φ is a modular form of weight k and level ℓ if and only if $F_\varphi(-1/z) \equiv z^{2k}F_\varphi(z)$. On the other hand, consider the restriction $y \to F_\varphi(iy)$ of F_φ to the positive real axis. Since the positive real axis is a commutative group under multiplication, we may form the Fourier transform $\int_0^\infty (F_\varphi(iy) - \varphi(0))y \cdot \dfrac{dy}{y} = \int_0^\infty (F_\varphi(iy) - \varphi(0))y^{\sigma-1}dy$ and consider its extension to complex values $s = \sigma + i\tau \to \int_0^\infty (F_\varphi(iy) - \varphi(0))y^{s-1}dy$. (When so written, the Fourier transform is usually called the Mellin transform.) Writing $F_\varphi(iy) - \varphi(0) = \sum_{n=1}^{\infty} \varphi(n)e^{-2\pi ny}$ and integrating term by term, one obtains a series expansion $\sum_{n=1}^{\infty} \varphi(n) \int_0^\infty e^{-2\pi ny}y^{s-1}dy$ for this Mellin transform.

Making the change of variable $y \to \dfrac{t}{2\pi n}$, the nth term becomes $\varphi(n)\dfrac{1}{(2\pi n)^s}$ $\int_0^\infty e^{-t}t^{s-1}dt = \dfrac{\Gamma(s)}{(2\pi n)^s} \varphi(n)$, where $s \to \Gamma(s) = \int_0^\infty e^{-t}t^{s-1}dt$ is the classical gamma function. Thus the Mellin transform becomes $s \to \dfrac{\Gamma(s)}{(2\pi)^s} D_\varphi(s)$ where $D_\varphi(s)$ has the Dirichlet series expansion $D_\varphi(s) = \sum_{n=1}^{\infty} \dfrac{\varphi(n)}{n^s}$. A straightforward calculation shows that F_φ satisfies the identity $F_\varphi(-1/z) = z^{2k}F_\varphi(z)$ if and only if D_φ has an analytic continuation to the whole complex plane and satisfies the functional equation

$$\frac{\Gamma(2k - s)}{(2\pi)^{2k-s}}D_\varphi(2k - s) = \frac{\Gamma(s)}{(2\pi)^s}D_\varphi(s) .$$

(The argument is simplest if $\varphi(0) = 0$, for then D_φ is entire. Otherwise D_φ has a pole at $2k$ and the residue at this pole determines $\varphi(0)$.) Thus F_φ is a modular form of weight k if and only if D_φ satisfies the functional equation in question. In his 1926 paper, Hecke used a more complicated variant of the correspondence just described to construct modular forms of weight $1/2$ and higher level from Dirichlet series D satisfying functional equations of the form $\dfrac{\Gamma(1 - s)}{a^{1-s}} D(1 - s) = \dfrac{\Gamma(s)}{a^s} D(s)$. Here a is a positive constant which depends upon the level. Certain two-fold products of Dirichlet L functions and products of Dirichlet L functions with the Riemann zeta function have functional equations of this form.

In 1927 and 1928 Hecke published two further and rather different contributions to the problem of determining the modular forms of higher level. The 1928 contribution has already been described in the last part of section 19. In the 1927 paper Hecke showed how to extend the theory of Eisenstein series and cusp forms (see section 10) to forms of higher level. The Riemann surface whose points are the orbits in the action of Γ_N on the upper half-plane can be compactified by adding a finite number of points (the cusps). For each weight $k = 1/2, 1, 3/2, 2, \cdots$, one has a generalized Eisenstein series for each cusp which "takes the value" 1 at that cusp and "the value" 0 at all other cusps. Every modular form of level N and weight k is uniquely a sum of a form which vanishes at all the cusps (a cusp form) and a linear combination of the Eisenstein series attached to the cusps. The Fourier coefficients of the Eisenstein series have simple number-theoretical properties analogous to those in the level one case. The Fourier coefficients of the cusp forms are of a lower order of magnitude than the Fourier coefficients of the Eisenstein series. Using the facts just stated, Hecke was able to obtain the Hardy-Littlewood asymptotic formula (see section 20) for representations of integers as sums of squares (but not higher powers) as a corollary of his theory. The formulae are exact rather than asymptotic whenever there are no cusp forms.

The connection between modular forms and Dirichlet series satisfying a Riemann-type functional equation, which Hecke had exploited in a special case in the 1926 paper mentioned above, became the basis of a far-reaching theory presented by Hecke in four long papers published in 1936 and 1937. Hecke applied his results to the theory of n-ary quadratic forms in a very long paper published in 1940. Detailed summaries of part of this work were published in 1935 and 1936. Instead of continuing to find new modular forms by transforming known Dirichlet series, Hecke reversed things and studied the Dirichlet series obtained by applying the Mellin transform to an arbitrary modular form. This study led to two kinds of results. On the one hand, by using known facts about the general theory of modular forms and by proving precise theorems about the bijectivity of the Mellin transform between well-defined spaces of Dirichlet series and modular forms, Hecke was able to deduce theorems characterizing various zeta and L functions by their functional equation and certain additional properties. In doing so, he found an extensive generalization of a characterization of the Riemann zeta function given by Hamburger (1889-1956) in 1921 and 1922. On the other hand, he found a method for showing that the Dirichlet series one gets by taking the Mellin transforms of a modular form are like those occurring in algebraic number theory not only in that a) they satisfy a Riemann-type functional equation, but also in that b) they can be written as finite linear combinations of Dirichlet series which both satisfy the functional equation *and have an "Euler product factorization"* with one factor for each prime.

Perhaps a word is in order here about the definition and significance of an Euler product factorization. Let $n \rightarrow \varphi(n)$ be a complex-valued function on the integers of such a character that $\sum_{n=1}^{\infty} \dfrac{\varphi(n)}{n^s}$ converges whenever the real part of s is sufficiently large, i.e., in some half-plane $\sigma > \sigma_0$ where $s = \sigma + i\tau$. Typically $\varphi(n)$ will be the unknown solution to some number-theoretical problem containing n as a parameter. If φ is multiplicative in the sense that $\varphi(nm) = \varphi(n)\varphi(m)$ for n and m relatively prime, then φ is completely known when it is known at the prime powers. Moreover, if φ is multiplicative, one verifies at once that $\sum_{n=1}^{\infty} \dfrac{\varphi(n)}{n^s} = \prod_{p} E_p(s)$ where $E_p(s) = \sum_{k=0}^{\infty} \dfrac{\varphi(p^k)}{p^{ks}}$. Conversely, if $\sum_{n=1}^{\infty} \dfrac{\varphi(n)}{n^s}$ factors in the indicated way, one says that the Dirichlet series $\sum_{n=1}^{\infty} \dfrac{\varphi(n)}{n^s}$ has an Euler product factorization and one can prove easily that φ is multiplicative. In the original case considered by Euler (see section 5), $\varphi(n) = 1$ and $E_p(s) = \sum_{k=0}^{\infty} \left(\dfrac{1}{p^s}\right)^k = \dfrac{1}{1 - p^{-s}}$. In the Euler products which occur in number theory it is usually (if not always) true that $E_p(s)$ is a *rational* function of p^{-s}, whose numerator and denominator have degrees that are bounded as a function of p. Thus, for each prime p, one need only know $\varphi(p^k)$ for a finite number of values of k to know $\varphi(n)$ for all n. Finally, suppose that one has a system $\varphi_1, \varphi_2, \cdots, \varphi_h$ of linearly independent functions of n which are not multiplicative, but that the vector space they span has a basis $\psi_1, \psi_2, \cdots, \psi_h$ where the ψ_j are multiplicative. Then, as one can easily check, it suffices to know $\varphi_i(p^k)$ for all i, p, and k to know $\varphi_i(n)$ for all i and n. Moreover, if the Euler products which occur in the factorization of the Dirichlet series have *rational* factors as indicated above, it suffices for each p and i to know $\varphi_i(p^k)$ for a finite number of values of k.

Hecke's theory is considerably simpler for modular forms of level 1 than for forms of higher level, and I shall attempt to describe only this simple case. As already mentioned in section 10, the Eisenstein series of level 1 and weight k has a constant multiple whose Fourier coefficients $n \rightarrow \varphi_k(n)$ are given by the simple formula $\varphi_k(n) = \sum_{d \mid n} d^{2k-1}$. It follows easily that $n \rightarrow \varphi_k(n)$ is multiplicative, and that an Euler product factorization for $\sum_{n=1}^{\infty} \dfrac{\varphi_k(n)}{n^s}$ not only exists but takes the special form

$$\prod_{p} \left(\frac{1}{1 - p^{-s}}\right)\left(\frac{1}{1 - p^{2k-1}p^{-s}}\right) = \prod_{p} \left(\frac{1}{1 - (1 + p^{2k-1})p^{-s} + p^{2k-1}p^{-2s}}\right).$$

Hecke's principal result was that for each $k = 1, 2, 3, \cdots$, there is a basis for the space of cusp forms of weight k such that if $n \to \psi_l^k(n)$ is the sequence of Fourier coefficients for the lth basis element, then the Dirichlet series $\sum_{n=1}^{\infty} \frac{\psi_l^k(n)}{n^s}$ has an Euler product expansion of the form $\sum_{n=1}^{\infty} \frac{\psi_l^k(n)}{n^s} = \prod_p \frac{1}{1 - \lambda_l^k(p)p^{-s} + p^{2k-1}p^{-2s}}$ where the coefficients $\lambda_l^k(p)$ remain unknown. This has exactly the same form as in the case of Eisenstein series where $\lambda_l^k(p)$ is known and equal to $1 + p^{2k-1}$. The lowest weight k for which cusp forms exist is $k = 6$, and in this case the space of cusp forms is one-dimensional. Let $\tau(n)$ denote the nth Fourier coefficient of the unique cusp form of weight 6 whose $e^{2\pi i z}$ coefficient is 1. Already in 1916 Ramanujan had conjectured that $\sum_{n=1}^{\infty} \frac{\tau(n)}{n^s}$ has an Euler product factorization of the form $\prod_p \frac{1}{1 - \tau(p)p^{-s} + p^{11}p^{-2s}}$, and in 1917 Mordell published a proof of this conjecture together with a proof of the fact that the Dirichlet series $D(s) = \sum_{n=1}^{\infty} \frac{\tau(n)}{n^s}$ satisfies the functional equation $\frac{\Gamma(s)D(s)}{(2\pi)^s} = \frac{\Gamma(12-s)D(12-s)}{(2\pi)^{12-s}}$.

Hecke's theory constituted a far-reaching generalization of these isolated results. Ramanujan also conjectured that $|\tau(p)| \leq 2p^{11/2}$, or equivalently that the polynomial $1 - \tau(p)p^{-s} + p^{11}p^{-2s}$ cannot have unequal real roots. However, neither he nor Mordell was able to prove this. Hecke's theory suggests a natural generalization which was formulated by Petersson (1902—) in 1939. The Ramanujan-Petersson conjecture states that $|\lambda_l^k(p)| \leq 2p^{(2k-1)/2}$. It was finally proved by Deligne (1944—) in 1974 as a corollary of a much more general theorem.

Hecke's discovery of a basis with multiplicative Fourier coefficients for spaces of modular forms was based on the consideration of certain linear operators T_n, which had been used by Hurwitz in the classical theory of the 1880s and 1890s. However, Hecke found new properties of these operators and used them in a new way, and as a result they are now called Hecke operators. Let F be a modular form of weight k and level 1, and for each n let $T_n(F)(z) = n^{2k-1}\sum_{a,b} F\left(\frac{az+b}{d}\right)d^{-2k}$ where $d = n/a$, a varies over all divisors of n and for each a, $b = 0, 1, 2, \cdots, d-1$. Then $T_n(F)$ is also a modular form of weight k and level 1, and $F \to T_n(F)$ is a linear operator. Hecke's basic observation was that these operators not only commute with one another but vary with n in a manner strictly analogous to the way in which the Fourier coefficients of Eisenstein series vary. Specifically, $T_n T_m = \sum_d d^{2k-1}(T_{nm/d^2})$ where d varies over all positive divisors of the highest common factor of n and m. This implies that $T_n T_m = T_{nm}$ whenever n and m are relatively prime, and that $I + \Sigma z^j T_{p^j} = \dfrac{1}{(I - T_p z + Ip^{2j-1}z^2)}$ where I is the identity operator. Whenever the T_n are all diagonalizable, it follows from their commutivity that they are *simultaneously* diagonalizable. One thus obtains the desired basis in the space of modular forms by choosing any (suitably normalized) basis which diagonalizes all of the T_n. Hecke was able to prove the diagonalizability only for certain values of k. In the 1939 paper mentioned above, however, Petersson was able to prove diagonalizability in general. He did so by introducing a natural inner product in the space of modular cusp forms of a given dimension with respect to which all the T_n are self adjoint. This inner product turned out to be useful in the general theory of automorphic forms and is now known as the Petersson inner product.

While Hecke and Petersson were developing the theory described above, C. L. Siegel (1896—) was applying analytical considerations to a different kind of extension of the theory of n-ary quadratic forms. Siegel generalized the problem of finding the number of representations of an integer by a quadratic form to that of finding the number of representations of one quadratic form by another. Instead of being content with asymptotic results as were Hardy and Littlewood, however, or attempting to get at the lower order terms via a generalization of Hecke's theory (which was just then being developed), Siegel replaced the study of individual forms by the study of certain averages. (As explained in section 6, the same device was used by Gauss to get exact results in the theory of binary quadratic forms.) Siegel's main result is a formula for the average solution number which specializes to the singular series of Hardy and Littlewood (see section 20) when one of the forms is a form in one variable. He showed how this result could be interpreted as a product over the primes (including ∞) of p-adic or real solution densities, and also that it was equivalent to an identity between two differently defined "generalized modular forms."

Siegel's generalized modular forms are analytic functions of several complex variables which are related to the n-dimensional symplectic group in the same way that the classical modular forms are related to the group $SL(2, R)$. Let V^{2n} be a $2n$-dimensional vector space equipped with a nondegenerate alternating bilinear form $[\ , \ ]$. The n-dimensional symplectic group $Sp(n)$ is the group of all non-singular linear transformations T of V^{2n} into V^{2n} such that $[T(\varphi), T(\psi)] = [\varphi, \psi]$ for all φ and ψ in V^{2n}. If K is a maximal compact subgroup of $Sp(n)$, then $Sp(n)/K$ has the structure of a complex manifold which can be identified with a space of complex matrices. When $n = 1$, $Sp(n)$ is isomorphic to $SL(2, R)$, and $Sp(n)/K$ becomes the upper half-plane.

Siegel's results for the special case of definite forms appeared in a long paper published in 1935. Their extensions to the indefinite case and to forms with coefficients in an algebraic number field were published in 1936 and 1937 respectively. Of course, it was now very much in order to extend the classical theory of modular forms to include Siegel's new modular forms on $Sp(n)/K$, and the foundations for such a theory were laid down by Siegel himself in two papers published in 1939 and 1943 respectively. Many mathematicians became interested and the new theory soon experienced a considerable development.

One of the more active workers in the development of Siegel's theory as well as other generalizations of modular form theory to several complex variables was H. Maass (1911—). Returning to the one variable case, Maass in 1949 published a long paper extending Hecke's theory in an unexpected direction. While Hecke's theory made it possible to give an abstract characterization of the zeta function of an imaginary quadratic extension of the rational field Q, it failed for real quadratic extensions because of the different form the functional equation takes in that case. Maass showed that it was possible to develop a modification of Hecke's theory which applied to the zeta functions of real quadratic fields if one replaced the analytic functions of the classical theory of modular forms by group invariant functions which are not analytic in the sense of complex analysis. Instead, they are eigenfunctions of the second order differential operator $y^2\left(\dfrac{\partial^2}{\partial x^2} + \dfrac{\partial^2}{\partial y^2}\right)$. This operator is (to within a multiplicative constant) the only second order differential operator that commutes with the action of $SL(2, R)$. In Maass's theory (which parallel's Hecke's in some respects but is quite different in others), the role of the weight of a modular form is played by the eigenvalue corresponding to an eigenfunction of $y^2\left(\dfrac{\partial^2}{\partial x^2} + \dfrac{\partial^2}{\partial y^2}\right)$.

That the unitary representation theory of the group $SL(2, R)$ might be closely connected with the classical theory of modular forms (and hence to number theory)· is immediately suggested if one compares the identity $f\left(\dfrac{az + b}{cz + d}\right)= (cz + d)^{2k}f(z)$ occurring in the definition of a modular form of weight k with one form of the definition of the discrete series of irreducible unitary representations of $SL(2, R)$ given in section 21. The member V^{2k} of the discrete series had as Hilbert space a space of analytic functions on the upper half-plane and was defined by the formula $V^{2k}(^{a\,b}_{c\,d})f(z) = f\left(\dfrac{az + b}{cz + d}\right)(cz + d)^{-2k}$. Ignoring growth conditions for the moment, one sees at once that being a modular form of weight k and level N is equivalent to being a member of the subspace of $\mathcal{H}(V^{2k})$ on which V^{2k} reduces to the identity when restricted to Γ_N. If the obvious analogue of the Frobenius reciprocity theorem (see section 15) were true in the present context, the dimension of this subspace would coincide with the multiplicity of occurrence of V^{2k} in the decomposition of the representation $U^{I^{\Gamma_N}}$ induced by the one-dimensional identity representation I_{Γ_N} of Γ_N. While neither of these two heuristic arguments can be made correct, the result they suggest is largely true. For $k = 2, 3, 4, \cdots$, the dimension of the space of all *cusp* forms of weight k and level N is precisely equal to the multiplicity with which V^{2k} occurs as a discrete direct summand of the induced representation $U^{I^{\Gamma_N}}$. The first hint that such a connection between modular forms and the discrete series might exist occurs in a paper published by Gelfand and Fomin in 1952. These authors were concerned not with modular forms or questions in number theory, but with using the theory of unitary group representations to prove the ergodicity of certain flows on *compact* homogeneous spaces of the form G/Γ where $G = SL(2, R)$ and Γ is a discrete subgroup. Because of the compactness of G/Γ, the groups Γ_N are excluded, but one has obvious analogues of modular forms in which Γ replaces Γ_N. These are the automorphic forms of Poincaré, and when G/Γ is compact all automorphic forms are cusp forms. For use in their study of ergodicity Gelfand and Fomin proved that U^Γ contains V^{2k} a number of times equal to the dimension of the space of Γ automorphic forms of weight k. The connection of the Gelfand-Fomin observation with modular forms and number theory seems not to have been noticed until after Selberg (1917—) published an extremely interesting and influential paper in 1956.

While Selberg made no mention of either the Maass non-analytic automorphic forms or of the unitary representations of semi-simple Lie groups, his paper can be most easily understood as a contribution to a generalization of the Maass theory in which a group representational interpretation of Maass's automorphic forms plays a key role. Let Γ be a discrete subgroup of $G = SL(2, R)$ such that G/Γ is compact, and consider the unitary representation $U^{I\Gamma}$ of G induced by the identity representation of Γ. It follows easily from the compactness of G/Γ that $U^{I\Gamma}$ decomposes as a discrete direct sum of irreducible unitary representations of G, each occurring with finite multiplicity. The cited theorem of Gelfand and Fomin states that the multiplicities with which half the members of the discrete series occur are equal to the dimensions of certain spaces of classical automorphic forms. What about the other irreducible unitary representations of G, especially the principal series? Maass's introduction of non-analytic automorphic forms received an extra vindication when it was pointed out by Gelfand and Pjateskii-Shapiro (1929—) in 1959 that for each principal series member L there is a real number λ with the following property: The multiplicity with which L occurs in the decomposition of $U^{I\Gamma}$ is equal to the dimension of the space of all Maass automorphic forms for the group Γ

and the eigenvalue λ. Moreover, λ may be computed from the character $\begin{pmatrix} a & 0 \\ \mu & 1/a \end{pmatrix} \to |a|^{i\sigma}$ which induces the principal series member L by means of the formula $\lambda = 1/4 - \sigma^2$. To understand this correspondence between eigenvalues and principal series members on a conceptual level, identify H^+, the upper-half plane, with the coset space G/K and consider the induced representation U^{I_K}. (Here K is of course the maximal compact subgroup of all $\begin{pmatrix} \cos\theta & \sin\theta \\ -\sin\theta & \cos\theta \end{pmatrix}$.) One shows that U^{I_K} has a commutative commuting algebra and hence is uniquely a direct integral of irreducibles, each of which occurs once. Since the operator $y^2 \left(\dfrac{\partial^2}{\partial x^2} + \dfrac{\partial^2}{\partial y^2} \right)$ commutes with all U^I_x, and since there are no multiplicities, the decomposition of U^{I_K} decomposes $y^2 \left(\dfrac{\partial^2}{\partial x^2} + \dfrac{\partial^2}{\partial y^2} \right)$ as a direct integral of constant operators. The irreducibles that occur in the decomposition of U^{I_K} are (modulo sets of measure zero) precisely the members of the principal series, and we have accordingly an assignment of an eigenvalue of $y^2 \left(\dfrac{\partial^2}{\partial x^2} + \dfrac{\partial^2}{\partial y^2} \right)$ to each principal series member. It is this correspondence between eigenvalues of $y^2 \left(\dfrac{\partial^2}{\partial x^2} + \dfrac{\partial^2}{\partial y^2} \right)$ and principal series members which makes it possible for Selberg to avoid talking about irreducible unitary representations of $SL(2, R)$. (An analogous correspondence involving a family of invariant differential operators permits Selberg to do likewise in his generalization to homogeneous spaces other than $SL(2, R)/K$.)

Selberg's paper centers about the existence and consequences of a formula (now known as the "Selberg trace formula") which he asserts can be considered as a generalization of the classical Poisson summation formula. In its most elementary form the latter asserts that $\sum\limits_{n=-\infty}^{\infty} f(n) = \sum\limits_{m=-\infty}^{\infty} \hat{f}(2\pi m)$ where $\hat{f}(y) = \int_{-\infty}^{\infty} f(x)\, e^{ixy} dx$ and f is a mildly restricted complex-valued function on the real line. Noting that $e^{iny} = 1$ for all integers n if and only if $y = 2\pi m$ for some integer m, one can rewrite this formula as $\sum\limits_{\gamma \in \Gamma} f(\gamma) = \sum\limits_{\chi \in \Gamma^\perp} \hat{f}(\chi)$, where Γ is the subgroup of the additive group R of the real line consisting of the integers, $\Gamma^\perp$ is the subgroup of the character group $\hat{R}$ consisting of all χ with $\chi(\gamma) = 1$ for all γ in Γ, and $\hat{f}(\chi) = \int_{-\infty}^{\infty} f(x)\chi(x)dx$. Once it is so written, there is an obvious generalization in which R is replaced by an arbitrary separable locally-compact commutative group G, and Γ is any discrete closed subgroup such that G/Γ is compact. (More generally still, Γ can be an arbitrary closed subgroup, and then the formula becomes $\int_\Gamma f(\gamma)d(\gamma) = \int_{\Gamma^\perp} \hat{f}(\chi)d\chi$ for suitably chosen Haar measures in Γ and $\Gamma^\perp$.) In order to be led naturally to a further generalization in which G can be non-commutative, one has only to look at the formula in the commutative case from the point of view of induced representations and their characters. Consider U^{I_Γ}, the representation of G induced by the trivial representation I_Γ of Γ. When Γ is discrete and G/Γ is compact, $\Gamma^\perp$ is also discrete, and it is easy to check that U^{I_Γ} is just the direct sum of the one-dimensional representations defined by the members of $\Gamma^\perp$. While Trace $(U_x^{I_\Gamma})$ does not exist, one can define the character of U^{I_Γ} as a linear functional by the device already discussed in section 20. One forms $U_f^{I_\Gamma}$ and defines the character to be $f \to$ Trace $(U_f^{I_\Gamma})$. Now an easy computation shows that Trace $(U_f^{I_\Gamma})$ = $\sum_{\gamma \epsilon \Gamma} f(\gamma)$. Moreover, for each $\chi \epsilon \Gamma^\perp$, the linear functional defining the character of the one-dimensional representation defined by χ is just $\int \chi(x)f(x)d\mu(x) = \hat{f}(\chi)$. In other words, the Poisson summation formula simply asserts the equality of the character of U^{I_Γ} (as a linear functional) to the sum of the characters of its irreducible constituents.

Now let G be any separable locally-compact group and let Γ be any closed discrete subgroup of G such that G/Γ is compact. Then $U^{I_\Gamma} = \sum n_j L^j$ where the L^j are distinct irreducible unitary representations of G. In this case one can hope to find a reasonably large family of functions f for which the two sides of the equation Trace $U_f^{I_\Gamma} = \sum n_j$ Trace L_f^j make sense and are equal. To the extent that one can do this, one will have a formula which is evidently a non-commutative generalization of the Poisson summation formula. For compact G/Γ, this formula is in essence Selberg's trace formula. However, Selberg introduced certain restrictions on G and f that made it possible to avoid various difficulties associated with not knowing all the L^j which might occur and with the (possible) non-existence of Trace (L_f^j). For Selberg, G was always a Lie group with a distinguished compact subgroup K such that G/K is a Riemannian manifold and U^{I_K} has a commutative commuting algebra. Moreover, Selberg only applied his trace formula to functions f which were constant on the $K : K$ double cosets. Under these circumstances, only those irreducible unitary representations L which contain the identity when restricted to K make a contribution to the right side of the trace formula. Moreover, each such L contains the identity of K just one time. Let φ be a unit vector such that $L_k(\varphi) = \varphi$ for all k in K. Then the function $x \to (L_x(\varphi) \cdot \varphi)$ is independent of the choice of φ and is called the *spherical function* associated with L. For an f which is constant on the $K : K$ double cosets, Trace L_f is equal to $\int f(x)(L_x(\varphi) \cdot \varphi)$ whenever the former exists, and the latter may be substituted for the former in general. For the Maass case $(G = SL(2, R)$, $K =$ all $\begin{pmatrix} \cos\theta \sin\theta \\ -\sin\theta \cos\theta \end{pmatrix})$), the irreducible unitary representations which contain the identity when restricted to K are just the trivial representation and the members of the principal series which are trivial on the center. For Selberg, the problem of finding which principal series members occur in the decomposition of $U^{I\Gamma}$ appeared as the problem of finding the eigenvalues of the operator $y^2\left(\dfrac{\partial^2}{\partial x^2} + \dfrac{\partial^2}{\partial y^2}\right)$ in the space of Γ orbits in the upper half-plane.

Given the importance of the classical Poisson summation formula in number theory (Dirichlet's method of evaluating Gauss sums and deducing quadratic reciprocity, the proof of the Jacobi inversion formula, functional equations for zeta and L functions, etc.), one can hope that a non-commutative generalization will have a host of interesting new number-theoretical consequences. Indeed, Selberg's paper was first found interesting not because it helped establish a connection between number theory and unitary group representations, but because of the immediate and prospective consequences of the trace formula for classical questions in number theory and the theory of modular forms. Among other things, Selberg found new relations among class numbers for binary quadratic forms, a new way of determining the dimensions of spaces of automorphic and modular forms, and perhaps most interesting of all, a way of computing the traces of the Hecke operators T_n which are so important in Hecke's theory of Dirichlet series with Euler products.

Actually, from the outset Selberg dealt with the generalizations of the trace formula described above, which one obtains by replacing the identity representation I_Γ with an arbitrary finite-dimensional unitary representation M of Γ. Moreover, in order to obtain his results on traces of Hecke operators, he generalized in another direction by replacing Trace U_f^M on the left-hand side by Trace $A U_f^M$ where A is a rather special member of the commuting algebra $R(U^M, U^M)$. When $M = I_\Gamma$, there is a possible A associated with each $\Gamma : \Gamma$ double coset containing only finitely many right and left Γ cosets. It would take us too far afield to give further details here.

For applications to number theory, Selberg had to modify his theory to allow for the possible non-compactness of G/Γ. Already when $G = SL(2, R)$, the most interesting Γ's for number theoretical purposes are the so-called principal congruence subgroups Γ_N, and for these G/Γ_N is never compact. Accordingly, the induced unitary representation $U^{I\Gamma}$ (more generally U^M) is in part a discrete direct sum and in part a direct integral. In the special case in which $G = SL(2, R)$ and Γ is a subgroup of Γ_1 of finite index so that there are only finitely many "cusps," it turns out that $U^{I\Gamma}$ is the direct sum of two parts. One summand is a discrete direct sum of irreducibles, as in the case in which G/Γ is compact. The other part has one contribution from each cusp, each of which is a simple known direct integral of members of the principal series. Selberg showed how to take these contributions into account in his non-compact trace formula by making use of Maass's analogue of Eisenstein series. In defining his Eisenstein series for non-analytic automorphic forms, Maass had found that these series converged only for inappropriate values of the eigenvalue parameter. But he could obtain what he needed by analytic continuation. This "analytic continuation of Eisenstein series" was important in Selberg's considerations as well. When one tries to deal with higher-dimensional groups G and spaces G/K for noncompact G/Γ, one can no longer compactify by adding a finite number of points. One has to add higher-dimensional spaces and the whole theory becomes much more complicated. Selberg gave important indications as to how to proceed but left many open problems.

It may appear at first sight that Selberg's trace formula can only give information about the non-analytic automorphic forms of Maass — at least in the special cases considered by Selberg. But let $G = SL(2, R) \times K^1$ where K^1 is the subgroup of $SL(2, R)$ consisting of all $\begin{pmatrix} \cos\theta\sin\theta \\ -\sin\theta\cos\theta \end{pmatrix}$. Let K be the subgroup of $K^1 \times K^1 \subseteq SL(2, R) \times K^1$ consisting of all k, k. Then if $\chi_n\begin{pmatrix} \cos\theta\sin\theta \\ -\sin\theta\cos\theta \end{pmatrix} = e^{in\theta}$, the irreducible unitary representation $L \times \chi_n$ contains the identity when restricted to K if and only if L restricted to K^1 contains $\overline{\chi_n}$. Thus every discrete series member W occurs in the decomposition of U^{I_K} as $W \times \chi_n$ for some χ_n, and only occurs once with any particular $\overline{\chi_n}$. This choice of G and K satisfies Selberg's axioms and he obtained his results on classical automorphic forms by applying his theory to this case.

Selberg's work was translated into the language of the theory of unitary group representations and integrated with the earlier work of Gelfand, Fomin, and Maass in seminar reports by Godement and in the 1959 paper of Gelfand and Pjateskii-Shapiro cited above. In describing the picture which emerges, it is convenient to make use of the concept of an intertwining operator. Given unitary representations V and W of the same group G, an *intertwining operator* for V and W is by definition a bounded linear operator T from $\mathscr{H}(V)$ to $\mathscr{H}(W)$ such that $TV_x = W_xT$ for all x in G. Evidently the set $I(V, W)$ of all intertwining operators for V and W is a vector space. Its dimension is defined to be the *intertwining number $i(V, W)$* of V and W. Of course, $I(V, V)$ coincides with $R(V, V)$, the commuting algebra of V. If $T \in I(V, W)$, let N_T denote the zero space of T and let $\overline{R_T}$ denote the closure of the range of T. It is obvious that N_T and $\overline{R_T}$ are closed invariant subspaces and easy to show that the subrepresentation of V defined by $N_T^\perp$ is equivalent to the subrepresentation of W defined by $\overline{R_T}$. Indeed, the most general intertwining operator for V and W is obtained by composing an equivalence between subrepresentations with members of $R(V, V)$ and $R(W, W)$ respectively. When V is irreducible, $i(V, V)$ is one-dimensional and it follows easily that for any W, $i(V, W)$ is the multiplicity with which V occurs as a discrete irreducible constituent of W. More generally, one can deduce information about the decomposition of unitary representations W

of unknown structure from information about the members of $I(V, \dot{W})$ where V has known structure.

In the special case of automorphic and modular forms defined in the upper half-plane, the principal facts connecting their theory with the theory of unitary group representations revolve around the structure of the induced representations U^{T_Γ} where Γ is a discrete subgroup of $G = SL(2, R)$. When G/Γ is compact or Γ has finite index in $SL(2, Z)$, then the multiplicity with which a member of the principal series of irreducible unitary representations of G is contained discretely in U^{T_Γ} is equal to the dimension of the space of all Maass cusp forms for the group Γ and a fixed eigenvalue λ. Similarly, the multiplicity with respect to which the discrete series member V^{2k} for $k = \pm 2, \pm 3, \cdots$ is contained discretely in U^{T_Γ} is equal to the dimension of the space of all ordinary cusp forms of weight k for the group Γ. Equivalently, one can say (in either case) that the dimension of a space of cusp forms is equal to the dimension of a space of intertwining operators. Moreover, it turns out that this statement can be strengthened to the statement that there is a canonical isomorphism of one space on another. Using this canonical isomorphism to identify cusp forms with intertwining operators, one can give group-theoretical definitions of the Petersson inner product and the Hecke operators. Let L be an irreducible unitary representation of $G = SL(2, R)$, and let T_1 and T_2 be members of $I(U^{T_\Gamma}, L)$. Then $T_1 T_2^*$ is a self-intertwining operator for L, and since L is irreducible this self-intertwining operator is a complex multiple $B(T_1, T_2)$ of the identity. The function $T_1, T_2 \to B(T_1, T_2)$ is the Petersson inner product. If A is any member of $I(U^{T_\Gamma}, U^{T_\Gamma})$, then for each T in $I(U^{T_\Gamma}, L)$, AT is in $I(U^{T_\Gamma}, L)$. Thus there is a natural ring homomorphism of $I(U^{T_\Gamma}, U^{T_\Gamma})$ into the space of linear operators in $I(U^{T_\Gamma}, L)$. As will be explained below, each $\Gamma : \Gamma$ double coset with suitable finiteness properties defines a member of $I(U^{T_\Gamma}, U^{T_\Gamma})$. The corresponding linear operators in $I(U^{T_\Gamma}, L)$ are the Hecke operators.

When G/Γ is not compact, there may be non-discrete components in U^{T_Γ}. However, at least when Γ is a subgroup of finite index of $SL(2, Z)$, these are all direct integrals with respect to Lebesgue measure of members of the principal series. They may be discovered by considering the intertwining operators of U^{T_N} with U^{T_Γ} where N is the nilpotent (actually commutative) subgroup of $SL(2, R)$ consisting of all $\left(\begin{smallmatrix} 1 & 0 \\ a & 1 \end{smallmatrix}\right)$. The structure of U^{T_N} is easily determined from two general theorems in the theory of induced representations. Let T denote the intermediate subgroup consisting of all $\begin{pmatrix} \lambda & 0 \\ a & 1/\lambda \end{pmatrix}$ and let W be the representation of T induced by I_N. By the theorem on inducing in stages, U^{T_N} is equivalent to U^W. On the other hand, since N is normal in T, W is the regular representation of T/N lifted to T. Since T/N is commutative, the regular representation of T/N is the direct integral of all characters with respect to Haar measure. That U^{I_N} is a corresponding direct integral of members of the principal series (each repeated twice) follows at once. To find intertwining operators between U^{I_N} and U^{I_Γ}, one exploits the fact that for finite groups one has a complete overview of *all* intertwining operators for two induced representations. When both inducing representations are the identity, these intertwining operators correspond one-to-one to the complex-valued functions on G which are constant on the double cosets for the two subgroups involved. Thus there is a basis for the space of intertwining operators consisting of those whose corresponding function is identically one on some double coset and zero on all others. An explicit formula can be written down for each of these "double coset intertwining operators" which makes *formal* sense even when the group is infinite. Thus one can seek intertwining operators for U^{I_Γ} and U^{I_N} by investigating the convergence properties of the formal double coset intertwining operators associated with the $N : \Gamma$ double cosets. The double coset intertwining operator for $\Gamma \times N$ involves an integration over the coset space $N/x^{-1}\Gamma x \cap N$, and the double cosets divide sharply into two categories according to whether or not this coset space has a finite invariant measure. Those for which a finite invariant measure exists are of course those for which the formal double coset intertwining operators exist as actual operators, and are also just those for which $x^{-1}\Gamma x \cap N \neq \{e\}$. At least one such intertwining operator will exist precisely when Γ contains elements conjugate to members of $N - \{e\}$, i.e., when Γ contains "parabolic" elements. In considering the $N : \Gamma$ double cosets, it is possible to lump together those which are in the same $T : \Gamma$ double coset. Since T normalizes N, the different double coset intertwining operators for a given $T : \Gamma$ double coset are obtainable from one another in a trivial way. Now the homogeneous space G/T is identifiable with the one point compactification of the real line, i.e., with the *boundary* of the upper half plane H^+. Correspondingly, since $T : \Gamma$ double cosets may be identified with "boundary points" of the space of Γ orbits in H^+, those $T : \Gamma$ double cosets whose $N : \Gamma$ double cosets lead to well-defined intertwining operators, as indicated above, may be identified in this way with the "cusps" of the space of Γ orbits. Thus one has one intertwining operator for each cusp, and (at least when Γ is a subgroup of finite index of $SL(2, Z)$) these collectively account for the entire continuous part of U^{I_Γ}.

These considerations show how intimately the theory of automorphic forms is related to the decomposition theory of unitary representations of the form U^{I_Γ}. The Selberg trace formula may be regarded as the tool that makes it possible to exploit this relationship to obtain information about automorphic forms from knowledge of the irreducible unitary representations of $SL(2, R)$.

In the spirit of Hecke's 1928 paper, described in section 19, it is useful (and possible) to extend the theory just described by replacing I_Γ by a more general finite-dimensional irreducible unitary representation of Γ. In particular, when Γ is a subgroup of finite index of $\Gamma_1 = SL(2, Z)$, then U^{I_Γ} is a finite direct sum of induced representations of the form U^M where M is an irreducible constituent of the (finite-dimensional) representation of $\Gamma_1 = SL(2, Z)$ induced by I_Γ. Moreover, the work of Siegel and others on modular forms in several complex variables makes it desirable to develop a corresponding theory in which $SL(2, R)$ is replaced by $Sp(n)$. Finally, once things have been properly formulated in conceptual terms, there is no reason why one should not go on from $Sp(n)$ to general semi-simple Lie groups. Indeed, one might even hope for interesting new number theoretical applications of such an extended theory. This is an immense program on which much progress has been made, but which cannot be described in further detail here. Instead I shall confine myself to a few remarks. In a short note published in 1959, Harish-Chandra formulated a definition of automorphic forms which applied to quite general systems consisting of a semi-simple Lie group G and a discrete subgroup Γ. He also indicated a proof of the finite dimensionality of the space of all generalized automorphic forms of given "type." As a corollary he was able to show under rather general conditions that U^{I_Γ} contains each discretely contained component with finite multiplicity. In the same year, Gelfand and Graev published a long paper describing what they called the "horospherical method" or "the method of integral geometry" for decomposing unitary representations of semi-simple Lie groups of the form U^{I_K} where K is a maximal compact subgroup of a semi-simple Lie group G. It is not difficult to see that this method is essentially that of examining double coset intertwining operators as indicated above for U^{I_K} and U^{I_N} where N is a maximal nilpotent subgroup of G. Soon thereafter, Gelfand and Graev applied their method to U^{I_Γ} where Γ is a discrete subgroup of a semi-simple Lie group G. In 1962 they announced the following result: Let $\mathcal{M}$ be the orthogonal complement of the linear span of all double coset intertwining operators in $I(U^{I_N}, U^{I_\Gamma})$ for all maximal nilpotent subgroups N of G. Then U^{I_Γ} restricted to $\mathcal{M}$ is a discrete direct sum of irreducibles.

The connection between unitary representations and automorphic forms as described so far is unsatisfactory in one important respect. Although it "explains" the Hecke operators in group-representational terms, it does little to throw light on the Euler product decompositions which exist for the eigenfunctions of these Hecke operators. This gap was filled in in the 1960s with the aid of an extension of the idèle-adèle notion to non-commutative groups, inaugurated by Ono (1928—) and Tamagawa (1925—) in the late 1950s simultaneously with the translation of Selberg's ideas into the language of unitary group representations. For each prime p let G_p be the group $SL(2, Q_p)$ of all 2×2 matrices with determinant one and coefficients in the field Q_p of all p-adic numbers (see section 20), and let $G_\infty = SL(2, R)$. Let K_p be the subgroup of G_p consisting of all members of $SL(2, Q_p)$ whose coefficients are in the ring of all p-adic integers. Equivalently, K_p may be defined as the closure in $SL(2, Q_p)$ of the subgroup $SL(2, Z)$. Then K_p is a compact open subgroup of $SL(2, Q_p)$, and one can consider the subgroup of the direct product $\prod_p SL(2, Q_p)$ consisting of all $\{x_p\}$ with $x_p \epsilon K_p$ for all but finitely many K_p. We denote this subgroup by $\prod_p' SL(2, Q_p)$. It will be a separable locally-compact topological group if the subgroup $K = \prod_p K_p$ has the usual product topology and if $\prod_p' SL(2, Q_p)$ has the unique topology such that $K = \prod_p K_p$ is an open subgroup. The adèle group G_A for $SL(2, Q)$ is then defined to be the product group $\prod_p' SL(2, Q_p)$ $\times SL(2, R)$. Let θ_p be the canonical imbedding of $SL(2, Q)$ as a dense subgroup of $SL(2, Q_p)$ and let θ_∞ be the canonical dense imbedding of $SL(2, Q)$ in $SL(2, R)$. Then $\gamma \to \{\theta_p(\gamma)\}, \theta_\infty(\gamma)$ is an injective homomorphism of $SL(2, Q)$ in the adèle group G_A, and the range of this homomorphism can be shown to be closed. Thus $SL(2, Q)$ appears in a natural way as a discrete subgroup of G_A. It is called the group of *principal adèles*. More generally, one can define such locally compact subinfinite product groups for every so-called "algebraic" subgroup of $GL(n, C)$ which is "defined over" an algebraic extension k of the rationals. This means that the group consists of all invertible $n \times n$ matrices which satisfy a certain finite set of polynomial equations with coefficients in k.

The theory of such *algebraic groups* was begun by Maurer (1859-1927) in 1894 and then apparently forgotten for a half-century. Chevalley and Tuan (1914—) returned to the subject in 1945, and in 1951 Chevalley published an extensive account as volume II of his treatise on Lie groups. Chevalley's treatment made heavy use of Lie algebra techniques, but in 1956 A. Borel (1923—) transformed the subject with a long paper showing how (with the help of certain ideas of Kolchin [1916—]) to analyze the structure of algebraic groups by means of global arguments involving algebraic geometry.

When Ono introduced non-commutative adèle groups in 1957, he was influenced not only by the revival of the theory of algebraic groups and Borel's paper in particular, but also by some ideas advanced by Eichler (1912—) a few years earlier concerning a possible unification of algebraic number theory with the theory of n-ary quadratic forms (see the introduction to a paper of Ono published in 1959). Ono found that certain finiteness theorems in algebraic number theory (such as the finiteness of the number of ideal classes) can be translated into statements about the idèle group of the field. These statements not only make sense for the adèle groups of algebraic groups, but when specialized to the adèle groups of the orthogonal groups they reduce to known finiteness theorems for quadratic forms. Ono suggested that one try to find unified proofs for the theorems about quadratic forms and algebraic number fields, respectively, by finding proofs of the general statements for the adèle groups of sufficiently comprehensive classes of algebraic groups. In his 1957 paper he defined adèle groups in general, but proved finiteness theorems only for commutative algebraic groups. In a second paper published in 1959 (and mentioned above) he was able to take care of certain solvable groups as well. More generally (as suggested by Ono) one could now attempt to encompass much of classical number theory in a much more general theory centering around the properties of the adèle groups of algebraic groups. This program turned out to be attractive and fruitful and was soon in a rapid state of development. One of its earliest successes was the discovery by Tamagawa and M. Kneser (1928—) that some of the main results of Siegel on quadratic forms (discussed earlier in this section) are essentially equivalent to the assertion that G_A/G_Q has measure 2 with respect to a canonically defined Haar measure in G_A. Here G_A is the adèle group associated with the orthogonal group of a non-degenerate rational quadratic form in at least three variables, and G_Q is the subgroup of principal adèles. This raised the question of generalizing Siegel's results by computing this measure (now called the Tamagawa number) for the adèle groups attached to other semi-simple algebraic groups. A detailed discussion, together with several Tamagawa number computations, appears in some 1961 lecture notes of Weil. Other aspects of Ono's program were worked out by Borel (with the assistance of some joint work with Harish-Chandra). For further details, the reader may consult the published version of Borel's 1962 address to the International Mathematical Congress in Stockholm. In the Proceedings of this same Congress, the reader will also find the texts of addresses by Gelfand and Selberg respectively. These describe the state of development of the theory of automorphic forms in general semi-simple Lie groups as seen in 1962 from two quite different perspectives.

Two years after the Stockholm Congress, the ideas described in Borel's address were combined with those described in the addresses of Gelfand and Selberg by considering extensions of automorphic forms from semi-simple Lie groups to the corresponding adèle groups and by replacing the study of unitary representations induced from discrete subgroups of semi-simple Lie groups to the study of unitary representations of adèle groups induced by the identity representation of the subgroup of principal adèles. The basic early publications include a short note published in 1964 by Gelfand, Graev, and Pjateskii-Shapiro and detailed papers published in 1964 by Weil and in 1965 by C. Moore (1937—). Moore on the one hand, and Gelfand, Graev, and Pjateskii-Shapiro on the other independently laid the foundations for studying the unitary representation theory of infinite product groups of the adèle type but applied their results in different ways. Moore was concerned with nilpotent Lie groups and the structure of U^{I_Γ} where Γ is a discrete subgroup with a compact quotient. The other three authors were concerned with automorphic forms (although they did not say so in 1964) and their principal example was the adèle group associated with $SL(2, R)$. They made their intentions clear in 1966 when they published a book (vol. 6 of Gelfand's series *Generalized Functions*) entitled (in English translation) *Theory of Representations and Automorphic Forms.*

On the surface, Weil's paper seems almost unrelated to the considerations in the book of Gelfand et al., and it has yet to be integrated with those aspects of the unitary representation theory of adèle groups that tie in with the Hecke theory, which are described below. On the other hand, it relates unitary representations and adèle groups to number-theoretical problems in a most interesting way. Unfortunately, the results in it do not lend themselves to brief description; the reader is referred instead to my rather lengthy review of the paper in volume 29 of *Mathematical Reviews*. Let it suffice to state here that a key role in it is played by a certain natural projective representation of a generalization of the symplectic group, that the existence of this representation is implied by the generalization to locally-compact commutative groups of the Stone-von Neumann uniqueness theorem (see section 20), and that one of the main results may be looked upon as another generalization of the Poisson summation formula. The paper was written to prepare the way for a second paper, published in 1965, which contains a group-representational proof of Siegel's main results on quadratic forms imbedded in the generalization which specifies the Tamagawa number of most semi-simple algebraic groups. Roughly speaking, one may describe Weil's proof of Siegel's theorems as the result of using the connection between group representations and automorphic forms to replace automorphic forms by group representations in Siegel's proof.

To return to the adèle group G_A associated with $SL(2, R)$, let G_Q denote the group of principal adèles and consider the representation $U^{I_{G_Q}}$ of G_A induced by the one-dimensional identity representation of G_Q. A good way to understand the relevance of the study of $U^{I_{G_Q}}$ to the theory of modular forms in the upper half-plane is to consider the restriction of $U^{I_{G_Q}}$ to the factor group $e \times SL(2, R)$, where e is the identity of $\prod_p' SL(2, Q_p)$. If one first restricts $U^{I_{G_Q}}$ to the intermediate subgroup $\prod_p K_p \times SL(2, R)$, where K_p is the compact open closure in $SL(2, Q_p)$ of $SL(2, Z)$, and uses certain general theorems about induced representations (cf. pages 305-308 of [*15*] for further details), one finds that this restriction is a discrete direct sum $\Sigma \dim(L)U^L$. In this sum, L varies over a certain set of finite-dimensional irreducible unitary representations of $\Gamma = SL(2, Z)$. The representations L which occur are precisely those which are continuous in the $SL(2, Q_p)$ topology for all p and include all those which reduce to the identity on the principal congruence subgroups Γ_N. Thus in analyzing $U^{J_{G_Q}}$, one is simultaneously analyzing a great many induced unitary representations of $SL(2, R)$ including all those of the form $U^{J_{\Gamma_N}}$.

To see the connection with Hecke's theory of Euler products, one has to bring together the following three components:

(a) The connection described above between modular forms and the decomposition of the representations U^L;

(b) the easily established fact that any decomposition of $U^{J_{G_Q}}$ as a direct sum or direct integral restricts down to a decomposition of each U^L and indeed to a decomposition of each factor component of each U^L; and

(c) the fact established by Gelfand et al. and Moore that each irreducible unitary representation of G_A defines and is defined by a sequence $\{M^p\}$, M^∞ where M^p is an irreducible unitary representation of $SL(2, Q_p)$ and M^∞ is an irreducible unitary representation of $SL(2, R)$.

The details are too complicated to give here, but the decompositions of the factor components of the U^L brought about by (b) are by (a) reflected in corresponding decompositions of spaces of modular forms with values in $\mathcal{H}(L)$. These decompositions are those defined by the Hecke operators, and the Euler product decomposition of an irreducible component is a reflection of the factorization of the corresponding irreducible representation of G_A as $\Pi \, M^p \times M^\infty$. In particular, the coefficients in the Euler factors are determined by the particular M^p's that occur.

Once the relationship between Euler products and the adèle group representation $U^{J_{G_Q}}$ is understood, one sees how to approach the question of extending the Hecke-Euler product theory to modular forms associated with more general semi-simple Lie groups. A strong motivation for undertaking such a program was provided in 1967 by independent work of Weil and Langlands (1936—). Weil showed how a relatively mild generalization of Hecke's theory might allow one to identify the zeta functions of elliptic curves with the Dirichlet series of modular forms, and Langlands indicated a path toward identifying the Artin L functions (see section 19) with the Dirichlet series arising in a generalization of Hecke theory to the general linear group of degree n. In Langland's case, n is the dimension of the irreducible representation of the Galois group parameterizing the Artin L functions. Langlands began his career by making extensive contributions to the development of the Selberg-Gelfand program, and for the past decade he has been the leader in developing the new program suggested by the above considerations. For further details about the very complex theory which has emerged, the reader is referred to Borel's 1974-1975 Bourbaki seminar report [*1*].

Naturally, a program like Langland's cannot progress very far without knowledge of the irreducible unitary representations of the *p*-adic analogues of the semi-simple Lie groups. Mautner began the study of the unitary representations of these *p*-adic groups in 1958 with an attempt to settle the type I-ness question and a determination of some irreducible unitary representations. He was soon joined by Bruhat (1929—). The road was blocked for a while by insufficient understanding of the structure of the groups, but there is now a rather large literature on both structure and representations which I cannot even sketch here. A summary account will be found on pages 316-327 of [*15*]. As with physics, a few decades earlier, the needs of number theory have greatly stimulated the development of the theory of infinite-dimensional group representations.

23. SUMMARY AND CONCLUSION

With the preceding sketch of the nature of modern applications of the theory of unitary group representations to probability, physics, and number theory, we come to the end of our story. My central theme has been the power and scope of what I have called "the method of harmonic analysis." In addition there have been several unannounced subthemes. One of these is that physics is not so mysterious as many mathematicians seem to consider it. It is rather that physicists have different values and a different viewpoint, and this leads them to explain things in a manner uncongenial to mathematicians. If one works at it, it is possible to translate practically all of physics into well-defined mathematics. Moreover, when one does so, one finds a beautifully coherent scheme, which can be rather briefly summarized.

Another subtheme is that a rather large part of modern mathematics has developed in a natural way out of attempts to understand the solutions of quite simple equations. On the one hand, much of modern algebra and number theory has arisen out of attempts to understand the solutions in integers of equations of the form $Ax^2 + Bxy + Cy^2 = D$ where A, B, C, and D are known integers (and certain straightforward generalizations). On the other, much of modern analysis originated in attempts to deal with the partial differential equations that were encountered in physics as Newton's ideas were applied to continuous matter and as electricity, magnetism, and light began to be understood more quantitatively. The fact that the method of harmonic analysis is a key tool in dealing with both types of equations helps make it possible to see much more unity in mathematics and in mathematics and physics together than usually meets the eye.

In order to develop these themes, I have presented some of the main ideas and concepts of physics and number theory in more or less chronological sequence with emphasis on the impact of harmonic analysis. Before 1800 (except for Laplace's use of generating functions in probability) there was no systematic harmonic analysis, and both number theory and mathematical physics remained in a relatively primitive state. I have tried to emphasize the considerable progress made possible in both subjects by the introduction of Fourier analysis and its (unrecognized) analogue for finite commutative groups. At the end of the nineteenth century Frobenius invented group representations—under the indirect inspiration of the needs of number theory. Thirty years later Hermann Weyl recognized the essentially group-theoretical nature of Fourier analysis and observed that the theory of Frobenius was just the finite special case of an extension of Fourier analysis from commutative to non-commutative groups. The ensuing decades saw this new and enlarged concept of harmonic analysis produce advances in physics and in number theory comparable with those made over a century earlier by the original commutative version. These advances are still being made—especially in number theory. Indeed, it is possible to hope that startling progress will be made in classical problems once the intricate interaction between unitary group representation, automorphic forms, and number theory is better understood than it is at present.

The preceding account may seem to neglect such powerful tools as the theory of functions of a complex variable. As I have shown in the text, however, this latter theory can be regarded as an integral part of harmonic analysis, as can certain other techniques which are superficially rather different.

Although probability theory has been mentioned both as one of the fields in which harmonic analysis originated and as one of the fields to which the modern theory of group representations may be applied, the connection has developed in rather a different manner than has been the case for number theory and physics. Probability *theory* did not advance much during the nineteenth century (although many new applications were found). It rejoined the mainstream of modern mathematics when it was integrated with measure theory in the 1930s, and the main applications of harmonic analysis to it have been via the developments in the classical commutative theory made possible by the introduction of measure theory. While ergodic theory and the ergodic theorem are not usually thought of as being a part of harmonic analysis, I have taken pains in the text to show that in fact they are. It is this relatively new and undeveloped branch of harmonic analysis which has the most far-reaching connections with probability theory at present.

NOTES

1. Shortly after the typescript of this paper had been sent to the editors, I received a pre-print of an article by K. I. Gross entitled "On the Evolution of Noncommutative Harmonic Analysis," scheduled to appear in the *American Mathematical Monthly*. The central theme of Gross's article is the same as that of this one. His execution is different in being more elementary, less than one fifth as long, and much less detailed. The reader may find it helpful to read Gross's treatment as an introduction to this one.

Sections 14 through 22 (about three-quarters of the paper) were written while I was a member of the Institute for Advanced Study in Princeton. I wish to express my gratitude to the Institute as well as to the following individuals who read all or part of the typescript and made helpful comments and corrections: Allan Adler, Armand Borel, Philip Green, Harish-Chandra, Howard Jacobowitz, Ian MacDonald, B. Simon, Robert Stanton, M. Taylor, David Vogan, and A. Weil.

2. To call the prime number theorem a conjecture of Riemann's is perhaps too loose a statement. Other mathematicians such as Gauss had suggested the truth of such a result rather earlier. Riemann's contribution was to suggest a method of proof.

3. Rayleigh himself suggested an *ad hoc* modification of his law which made it more reasonable at low temperatures and short wave lengths.

4. Since writing these paragraphs about Planck's contribution, I have examined the original papers and decided that the disclaimer "This of course is not how Planck proceeded" is an inadequate indication of the amount of "poetic" license I have taken in trying to make clear the essential point in Planck's discovery. This point is of course a lot clearer now than it was then. In actual fact Planck published two papers, a few months apart. In the first he found his now famous radiation law by making a (physically unmotivated) mathematical adjustment in Wien's derivation of Wien's law. He was not satisfied with his reasoning and published the law only because its predictions agreed with those of Wien's law at one end of the spectrum and with recent experimental results at the other. In the second paper he showed that the law could be derived in a more satisfying way by making his famous discreteness assumption. He does not talk about passing to the limit. It should also be mentioned that Planck did not refer to Rayleigh's law as such, but only to the experimental results confirming its validity at high temperatures. For a full account of the complex history of the old quantum theory, the reader is referred to Hermann [*12*] and Kuhn [*13*].

5. With this recognition of the group theoretical character of harmonic analysis, the further development of the subject proceeded along two semi-independent paths. The followers of one path concerned themselves with obtaining ever deeper and more refined results about harmonic analysis on the line and the circle (and later with extensions to n dimensions). The followers of the other path concerned themselves with extending the easier theorems to more general groups and to the new applications which this made possible. This paper does not pretend to be a complete history of harmonic analysis, but is concerned rather with harmonic analysis as a method for exploiting symmetry. Accordingly, I shall concern myself almost exclusively with the work of the followers of the second path, and with all due respect will say nothing about the work of such important mathematicians as Zygmund, Beurling, and Salem. For similar reasons, the reader will find no discussion of the generalization of harmonic analysis associated with the Gelfand map in the theory of Banach algebras.

6. I am informed by Professor Weil that his conversation with von Neumann took place before he had seen Koopman's paper and that he independently had the idea of introducing the representation V.

7. For details see pages 657-658 of [*14*].

8. For details see page 270 of [*15*].

REMARKS

(1) The description of Lagrange's work on number theory given in §5 is seriously incomplete in that it ignores his work on continued fractions and their application to the explicit determination of all integer and rational solutions of all equations of the form $Ax^2 + Bxy + Cy^2 + Dx + Ey + F = 0$, where A, B, C, D, E, and F are integers. Euler had studied many special cases but had no general method. In particular, there were equations of the form $x^2 - By^2 = C$ for which Euler was unable to decide whether or not integer solutions existed. This work was begun in 1768 and the results appeared in 1769 and 1770 in Lagrange's first three papers on number theory.

The relevance of the group $\mathrm{SL}(2, Z)^{\pm}$ to problems about the number theory of binary quadratic forms is explained in §5. The possible relevance of continued fractions can be guessed once one is aware of the following facts connecting them with $\mathrm{SL}(2, Z)^{\pm}$: $(a)\,\mathrm{SL}(2, Z)^{\pm}$ is generated by the two elements $T = \begin{pmatrix} 0 & 1 \\ 1 & 0 \end{pmatrix}$ and $L = \begin{pmatrix} 1 & 1 \\ 0 & 1 \end{pmatrix}$. (b) If $T_q = L^q T$, then every element of $\mathrm{SL}(2, Z)^{\pm}$ can be written in the form $T_{q_1} T_{q_2} \cdots T_{q_n}$, where the q_j are integers. (c) If $q_1, q_2, \ldots$ are all positive integers and $T_{q_1} T_{q_2} \cdots T_{q_n} = \begin{pmatrix} A_n & B_n \\ C_n & D_n \end{pmatrix}$ then $\lim_{n \to \infty} \frac{A_n + B_n}{C_n + D_n} = \mu$ exists and $q_1, q_2, \ldots$ are the coefficients of the continued fraction expansion of μ. Of course, the group concept did not exist in Lagrange's day although he contributed to its ultimate formulation. See §11 and his use of $\mathrm{SL}(2, Z)$ in defining equivalence of binary quadratic forms.

(2) Our statement at the end of §6 crediting Dirichlet with having been the first to give a proof of the convergence of a Fourier series under reasonably general conditions should perhaps be qualified. Cauchy offered such a proof slightly earlier but Dirichlet found it inadequate and published his proof to set matters straight.

(3) The third paragraph from the end of §7 on pioneers in introducing quantitative methods into chemistry, electricity, and magnetism should certainly have included the name of Henry Cavendish (1731–1810), the first to measure the densities of gases and to investigate systematically the properties of hydrogen. He was a shy recluse, disinclined to publish, and it was later discovered that he made experiments on electricity in the early 1770s, which anticipated much later work. He also was the first to evaluate the fundamental constant in Newton's theory of gravitation.

(4) The reader is referred to [6] for a much more detailed treatment of most of the material in §18.

(5) The assertion in §19 that Furtwangler was a student of Hilbert seems to be a case of jumping to conclusions suggested by circumstantial evidence. I am indebted to Olga Tausky-Todd, herself a student of Furtwangler, for informing us that she doubts that Hilbert and Furtwangler had ever met.

GEORGE W. MACKEY   155

BIBLIOGRAPHY

The references given below do not begin to exhaust the sources I have drawn upon in writing this historical survey. In addition to consulting many original papers, I have read articles in biographical dictionaries such as [6], obituary notices, survey articles, etc. Since much of this was done unsystematically and over a period of years, compiling a complete list of sources would be an impossibly difficult undertaking. I have chosen instead to give only a list of books and articles specifically referred to in the text or used especially extensively. I have not listed any of the original memoirs in which the various discoveries mentioned in the text were reported. There are far too many. The interested reader should have no difficulty tracking down those that interest him, using the date given in the text and such aids as collected works or abstracting and reviewing journals such as mathematical reviews. Items *14* and *15* contain two halves of a lengthy bibliography for sections 21 and 22.

[1] BOREL, A., *Formes Automorphes et Series de Dirichlet*]d'apres R.P. Langlands], Séminaire Bourbaki 1974/75, #466.

[2] BELL, E. T., *Men of Mathematics* (New York: Simon and Schuster, 1937).

[3] ———, *The Development of Mathematics,* 2nd ed. (New York and London: McGraw Hill, 1942).

[4] DAVENPORT, H., *Multiplicative Number Theory* (Chicago: Markham Publishing Co., 1967).

[5] FEIT, WALTER, "Theory of Finite Groups in the Twentieth Century," to appear in *Graduate Studies, Texas Tech University.*

[6] GILLISPIE, C. C., ed., *Dictionary of Scientific Biography.*

[7] GRATTAN-GUINNESS, I., *Joseph Fourier, 1768-1830* (Cambridge, Mass.: M.I.T. Press, 1972).

[8] HAWKINS, THOMAS, *Lebesgue's Theory of Integration: Its Origins and Development* (Madison: University of Wisconsin Press, 1970).

[9] ———, "The Origins of the Theory of Group Characters," *Archive for History of Exact Sciences* 7 (1970-71): 142-170.

[10] ———, "Hypercomplex Numbers, Lie Groups and the Creation of Group Representation Theory," *Archive for History of Exact Sciences* 8 (1971-72): 243-287.

[11] ———, "New Light on Frobenius' Creation of the Theory of Group Characters," *Archive for History of Exact Sciences* 12 (1974): 217-243.

[*12*] HERMANN, ARMIN, *The Genesis of Quantum Theory (1899-1913)* (Cambridge, Mass., and London: M.I.T. Press, 1971).

[*13*] KUHN, T. S., *The Black-Body Problem and the Quantum Discontinuity, 1894-1912* (New York and Oxford: Oxford Univ. Press, 1978).

[*14*] MACKEY, G. W., "Infinite Dimensional Group Representations," *Bull. Amer. Math. Soc.* **69** (1963): 628-686.

[*15*] ———, *The Theory of Unitary Group Representations* (Chicago and London: University of Chicago Press, 1976).

[*16*] WHITTAKER, E., *A History of the Theories of Aether and Electricity,* Vol. I, *The Classical Theories* (New York: Philosophical Library, 1951).

[*17*] ———, *A History of the Theories of Aether and Electricity, Vol. II, The Modern Theories (1900-1926)* (New York: Philosophical Library, 1954).

HARVARD UNIVERSITY

HERMANN WEYL AND THE APPLICATION
OF GROUP THEORY TO QUANTUM MECHANICS

George W. Mackey

I was especially pleased to be invited to address this Congress because Hermann Weyl's work has had such an enormous influence on my own. A large part of the latter, both on the general theory of unitary group representations, and on its applications to quantum mechanics grew out of my study of a celebrated paper by M.H. Stone (who some years earlier had been my thesis advisor). This paper in turn seems to have been directly inspired by the 1927 paper in which Weyl first sketched his own ideas on the importance of group theory in quantum mechanics. Indeed the whole purpose of Stone's paper was to give exact formulations and announce rigorous proofs of two theorems suggested by Weyl's work. Moreover when I later embarked on a serious attempt to understand quantum mechanics my most important sources were von Neumann's "Mathematische Grundlagen der Quantenmechanik" and Chapter II of the English translation of Weyl's "Gruppentheorie und Quantenmechanik". Although we had very little personal contact I must consider Hermann Weyl as one of my most important teachers.

Weyl's work on the applications of group theory to quantum mechanics was immediately preceded by important contributions to the abstract theory of group representations including his far reaching observations on the essentially group theoretical character of Fourier analysis. To see all of this work in its proper perspective it will be useful to begin with a sketch of the historical background.

During roughly the first quarter of the twentieth century three exciting new developments were being pursued in mathematics and physics which seemed to have nothing to do with one another. One was in analysis, one in algebra and one in physics. They may be described briefly as follows:

(a) The work of Hilbert on integral equations and the invention of the Lebesgue integral leading to the Riesz-Fischer theorem, Hilbert's spectral theorem and functional analysis.

131

(b) The work of Frobenius, Burnside and Schur in inventing and developing the representation theory of finite groups.
(c) The development of the so called "Old quantum theory" by Planck, Einstein and Bohr.

After making some remarks about each of these topics in turn I shall sketch the remarkable events of the years 1925-1927 in which the many anomalies and contradictions of the old quantum theory were removed by the invention of quantum mechanics and in which it turned out that the mathematical developments listed in (a) and (b) above were just what was needed for the proper formulation and implementation of this new and subtle refinement of classical mechanics. Moreover (a) and (b) were not just brought together by their common application to (c). It was found that Fourier analysis, spectral theory and the theory of group representations could all be regarded as special cases of one far reaching unified theory. In my opinion the work of Hermann Weyl was the single most important factor in bringing about this startling unification of these apparently quite diverse topics.

Hilbert began his work on integral equations immediately after hearing about Fredholm's 1900 work on the same subject in the winter semester 1900-1901. The Lebesgue integral was introduced in Lebesgue's thesis of 1902 and by 1907 had led to the celebrated Riesz-Fischer theorem which put Fourier analysis in a much more satisfactory and elegant form. The high point of Hilbert's work on integral equations was his celebrated spectral theorem for bounded self adjoint operators in Hilbert space which was in turn suggested by his strategy of exploiting the analogy between integral operators and the matrices of linear algebra.

Because of the far reaching role it will play in what follows we pause to explain the nature of this theorem of Hilbert's and to introduce some technical terminology. In broad terms it is an infinite dimensional generalization of the classical theorem stating that every $n \times n$ matrix $\|a_{ij}\|$ of complex numbers, which is self adjoint in the sense that $\bar{a}_{ij} = a_{ji}$, is diagonizable. This means that there exists a matrix $\|u_{ij}\|$ which is "unitary" in the sense that $\|u_{ij}\| \cdot \|\bar{u}_{ji}\| = I$ where I is the identity matrix and such that $\|u_{ij}\|^{-1} \|a_{ij}\| \|u_{ij}\| = \|a'_{ij}\|$ is a "diagonal" matrix in the sense that $a'_{ij} = 0$ when $i \neq j$. From a more

132

geometrical point of view one thinks of $\|a_{ij}\|$ as defining a linear operator T_A in an n-dimensional complex vector space with an "inner product" and then diagonizability means that there is a basis of mutually orthogonal vectors each of which is an "eigenvector" in the sense that $T_A(\phi_j) = \lambda_j \phi_j$ where λ_j is a real number. This can be restated in the form: T_A is a direct sum of "constant" operators each acting in a one dimensional subspace. Of course the operator taking any ϕ into $\lambda_j \phi$ is the constant operator in the jth subspace.

When one replaces the finite dimensional space by a complete infinite dimensional one — a so called Hilbert space — the obvious generalization of this theorem is no longer true. Many important self adjoint operators have no eigenvectors except 0. To understand Hilbert's generalization it is useful to reformulate the classical theorem in what may seem like a perverse manner. For each subset E of the real line let $\lambda_{i_1}, \lambda_{i_2}, ..., \lambda_{i_w}$ be the eigenvalues of T_A which happen to lie in E and let P_E denote the unique operator such that $P_E(\phi_{j_k}) = \phi_{j_k}$ and $P_E(\phi_j) = 0$ when λ_j is not one of the λ_{i_s} i.e. when λ_j is not in E. Then each P_E is a self adjoint operator with the property that $P^2_E = P_E$. Such self adjoint idempotent operators are called *projections* and P_E is in fact the projection of the whole space on the subspace spanned by the eigenvectors $\phi_{\lambda_{i_1}}, \phi_{\lambda_{i_2}}, ..., \phi_{\lambda_{i_w}}$. This projection valued set function $E \to P_E$ is easily seen to have the following simple properties:

(1) $P_\phi = 0$ and $P_R = I$ where I is the identity operator, ϕ is the empty set and R the whole real line.

(2) $P_E P_F = P_F P_E = P_{F \cap E}$ for all subsets E and F of R.

(3) If $E_1, E_2 ....$ are mutually disjoint subsets of R then $P_{E_1 \cup E_2 \cup ...}$
   $= P_{E_1} + P_{E_2} + P_{E_3} + ... $.

Moreover given this set function $E \to P_E$ we can easily reconstruct all the matrix elements of T_A and hence T_A itself. Indeed for two arbitrary vectors ϕ and ψ one proves that $(T_A(\phi) \cdot \psi) = \sum_{\lambda \in R} \lambda (P_{\langle \lambda \rangle}(\phi) \cdot \psi)$ where $\langle \lambda \rangle$ denotes the set whose only element is λ. Of course $P_{\langle \lambda \rangle} = 0$ except when λ_i is one of the eigenvalues of T_A and the sum on the right hand side is actually finite and equal to $\sum_{i=1}^{n} \lambda_i (P_{\langle \lambda_i \rangle}(\phi) \cdot \psi)$ In our perverse and awkward looking reformulation the classical

133

diagonalization theorem says that for every self adjoint operator T_A in a finite dimensional Hilbert space there exists a unique projection valued set function $E \to P_E$ having properties (1), (2) and (3) listed above such that

$$(T_A(\phi) \cdot \psi) = \sum_{\lambda \in R} \lambda \, (P_{\langle \lambda \rangle}(\phi) \cdot \psi).$$

The great advantage of this reformulation is that with minor modifications it is also true for self adjoint operators in any separable infinite dimensional Hilbert space; and in somewhat different form this is what Hilbert proved for all bounded self adjoint operators. After making a few preliminary definitions we shall give a precise statement of this form of Hilbert's theorem. We define a subset of the real line to be a Borel set if it can be built up out of open intervals by repeated application of the process of countable union, countable intersection and the taking of complements. We then define a *projection valued measure* on the line to be a function $E \to P_E$ from the Borel subsets of R to the projection operators in some separable Hilbert space H which has properties (1), (2) and (3) listed above. Of course in (2) and (3) we must restrict $E, F, E_1, E_2, \ldots$ to be Borel sets. If there exists a finite interval $J =$ all x with $-u \leqslant x \leqslant u$ such that $P_J = I$ one says that the projection valued measure P has *bounded support*. Finally we notice that if P is any projection valued measure on R and ϕ is any vector in the Hilbert space H of P then $E \to (P_E(\phi) \cdot \phi)$ is an ordinary numerical measure on the Borel subsets of R and it makes sense to consider $\int f(x) \, d(P_x(\phi) \cdot \phi)$ for suitably restricted complex valued functions i.e. to integrate f with respect to the measure $E \to (P_E(\phi) \cdot \phi)$. More generally one can integrate f with respect to the complex valued measure $E \to (P_E(\phi) \cdot \psi)$. With these preliminaries we may state:

Hilbert's spectral theorem: Let T be any bounded self adjoint operator in the separable Hilbert space H. Then there exists a unique projection valued measure with bounded support, $E \to P_E$, defined on the Borel subsets of R and whose values are projection operators in H such that for all ϕ and ψ in H we have the identity

$$(T(\phi) \cdot \psi) = \int_R x \, d(P_x(\phi) \cdot \psi).$$

Conversely it is rather easy to show that every projection valued measure with bounded support is related in this way to a unique

134

bounded self adjoint operator T. One finds also that T has a basis of eigenvectors as in the classical finite dimensional case if and only if the corresponding P has countable support; that is if and only if there exists a countable set $E = \langle \lambda_1, \lambda_2, ... \rangle$ such that $P_E = I$. In that case the ranges of the projections $P_{\langle \lambda_i \rangle}$ constitute a direct sum decomposition of the Hilbert space and in each of these T is a constant times the identity. In the general case $P_{\langle \lambda \rangle}$ can well be zero for every real number λ and then one can think intuitively of the spectral theorem as stating that the self adjoint operator is a "direct integral" of constant operators rather than a direct sum.

Let us now turn to topic (b).

The theory of representations and their characters for finite groups was invented in 1896 by G. Frobenius. He did this in a more or less deliberate attempt to generalize a much simpler notion that had been used at least implicitly in number theory since its first appearance in Gauss' celebrated "Disquisitiones Arithmeticae" published in 1801. The word character was introduced by Gauss. Let G be any finite commutative group. Then by definition a character χ of G is a complex valued function on G such that $\chi(xy) = \chi(x)\,\chi(y)$ for all x and y in G. Gauss considered only those characters χ such that $\chi^2 \equiv 1$ and only those groups G which occurred in the number theory of binary quadratic forms. One of his fundamental contributions was to show that the set of all equivalence classes of forms with a given discriminant would be made into a finite commutative group in a certain way and his "characters" were designed to distinguish between inequivalent forms with the same discriminant. Dedekind generalized Gauss' notion in 1878 by removing the restriction that $\chi^2 \equiv 1$ and three years later Weber pointed out that it made sense for arbitrary commutative groups. A large part of its importance stems from the following simple theorem:

Theorem: Any complex valued function on the finite commutative group G may be written uniquely in the form $f(x) = \sum_{\chi \in \hat{G}} c_\chi \chi(x)$ where $\hat{G}$ denotes the set of *all* characters of G. Moreover the complex coefficients c_χ may be computed from f by the formula

$$c_\chi = \frac{1}{o(G)} \sum_{\chi \in G} f(x)\overline{\chi(x)}$$

where $o(G)$ is the number of elements in G.

135

While this theorem was not explicitly formulated until much later one can, with hindsight, recognize its use as a key element in a number of significant proofs in nineteenth century number theory.

The theory of higher reciprocity laws in number theory which was created by Gauss in 1828 and carried to a sort of conclusion by Kummer in the 1850's led more or less directly to Dedekind's theory of general algebraic number fields in 1870. The symmetry groups (Galois groups) of these fields — unlike those considered by Kummer — were not always commutative and problems that Dedekind could solve in the commutative case, using characters as a tool, remained baffling in the non commutative case. Dedekind appealed to Frobenius for help and Frobenius responded by showing how to generalize the theory of characters from commutative to non commutative finite groups. Of course the definition of character makes sense for non commutative groups but it does not go far enough. Every finite group G has a largest normal subgroup N such that G/N is commutative. The classical characters of G are all trivial on N and reduce essentially to characters of the commutative quotient group G/N.

Frobenius' solution is a little easier to explain in a second version which he found a year later. One simply replaces complex numbers by non singular $n \times n$ matrices and defines a matrix representation of the finite group G to be a matrix valued function $x \to A(x)$ defined on G such that $A(xy) = A(x)A(y)$. When n is one we recover the classical characters but now the non commutativity of matrix multiplication makes it possible for our notion to be significant for the non commutative part of G. If $x \to A(x) = \|a_{ij}(x)\|$ is a matrix representation of G one defines the *character* of this representation to be $\sum_{j=1}^{n} a_{jj}(x)$. For those representations with n=1 the characters in this sense are precisely the classical characters of Gauss as generalized by Dedekind and Weber. We shall refer to these henceforth as *one dimensional* characters.

Just as with spectral theory it is often illuminating to think in terms of linear transformations rather than matrices and to define a representation accordingly. Since the diagonal sum $\sum_{j=1}^{n} a_{jj}$ is the same for *all* matrix representations of the same linear transformation there is no difficulty about defining characters just as before. A representation $x \to A_x$ of G by linear transformations in some vector

136

space H(A) is said to be *irreducible* if there are no proper subspaces M of H(A) such that A_x(M) = M for all x. One proves that every representation is a direct sum of irreducible representations in the same sense that every self adjoint operator is a direct sum of constant operators. Defining an irreducible character to be the character of an irreducible representation one sees easily that every character is of the form $n_1\chi_1 + n_2\chi_2 + ...n_j\chi_j$ where χ_1, χ_2, ..., χ_j are distinct irreducible characters, j and the n_j are positive integers. This decomposition is unique. One proves also that, for each finite group G, the number of distinct irreducible characters is finite — and in fact equal to the number of conjugacy classes in the group.

Unlike the commutative case in which all irreducible characters are one-dimensional and quite easy to determine; finding the irreducible representations and corresponding characters of a finite non commutative group can be very difficult. In the period 1896 to 1924 it was an exciting new field of investigation. Even today there are serious and interesting unsolved problems.

Topic (c) grew out of late 19th century attempts to explain the spectrum of radiation from a so called "black body" by combining electromagnetic theory with statistical mechanics. These attempts were successful only when the temperature was large relative to the frequency of the radiation. It is significant that other predictions of statistical mechanics also failed at low temperatures. Then in 1900 Max Planck made the remarkable discovery that one could derive a formula valid for all temperatures from the bizarre assumption that the energy of a harmonic oscillator of frequency ν could not take on a continuum of values but only values of the form nh ν where h is a universal constant (now known as Planck's constant) and n = 0,1,2,3... . In 1905 and 1906 Einstein used similar ad hoc "quantization" hypotheses to explain the low temperature behaviour of specific heats and the so called photo-electric effect. A bit later in 1913 H. Bohr found a similar explanation of the wavelengths occurring in the spectrum of atomic hydrogen. There was success after success but no real understanding because the various quantization hypotheses could not be reconciled with the principles of classical mechanics. For a quarter of a century physicists had to live with a level of logical incoherence which was quite strange to them.

137

So much for the historical background. We are now ready to talk about the contributions of Hermann Weyl. These began in 1924 with his work on extending the representation theory of finite groups to a class of infinite continuous groups — the compact Lie groups. They culminated in 1927 with Weyl's work in: (a) Unifying group representation theory with Fourier analysis, (b) Helping to clarify the structure of the new quantum mechanics that emerged to replace the old quantum theory of 1900-1924 after the fundamental discoveries of Heisenberg and Schrödinger in late 1924 and early 1925, (c) Unifying spectral theory with the theory of group representations while applying both to the new quantum mechanics.

In 1924 I. Schur, a student of Frobenius and one of the leading early workers in the theory of group representations, published a paper indicating how one could generalize certain features of the theory from finite groups to the orthogonal groups. The key idea was to replace summing over the group elements by integrating over the (compact) group manifold — using a definition of integration introduced earlier in another connection by Hurwitz. Weyl immediately became interested and wrote a letter to Schur indicating how one might go much further. This was soon followed by a celebrated sequence of papers "Theorie der Darstellung kontinuierlicher halbeinfacher Gruppen durch lineare Transformationen, I, II, III" published in the Mathematische Zeitschrift in 1925 and 1926. In these Weyl gave a very complete and elegant account of the irreducible representations and their characters for all of the compact semisimple Lie groups. Further details will be found in Professor Freudenthal's talk at this Congress as well as below where we will indicate certain connections with Weyl's work on quantum mechanics.

A year later in 1927 Weyl published a further paper on the representation theory of compact Lie groups, this one written in collaboration with his student F. Peter. It was entitled "Die Vollständigkeit der primitiven Darstellungen einer geschlossenen kontinuierlichen Gruppe" and contains the now celebrated Peter-Weyl theorem. This theorem can be stated in several ways but can perhaps best be understood here as a new and unexpected generalization of the fact that quite general periodic functions on the real line may be expanded in Fourier series $f(x) = \sum_{n=-\infty}^{\infty} c_n e^{inx}$ where $c_n = \frac{1}{2\pi} \int_{-\pi}^{\pi} f(x) e^{-inx} \, dx$; that is, that the complex trigonometric

138

functions $e^{2\pi i n x}$ form a "complete" set of (orthogonal) functions amongst functions of period 2π. Observe now that the set of all real numbers is a continuous group under addition and that the "quotient group" T obtained by identifying numbers that differ by an integer multiple of 2π is a compact continuous group. Observe also that a complex valued function of period 2π may be regarded as a complex valued function on T and that in particular the functions $x \to e^{inx}$ $n = 0, \pm1, \pm2, \ldots$ are *characters* of T; indeed one can show that every continuous character of T is of this form. Thus one can restate the completeness of the complex trigonometric functions $x \to e^{inx}$ as the completeness of the continuous characters on a certain compact commutative Lie group. The idea of Peter and Weyl was that there should be an analogous result for any compact Lie group, commutative or not, with continuous irreducible representations replacing continuous characters. More precisely one replaces characters by matrix elements of irreducible representations. The proof of Peter and Weyl made use of the theory of integral equations and yielded a *new* proof in the classical Fourier case. It is interesting that their theorem can also be interpreted in purely group representational terms where it asserts that the so called regular representation of the group (which for non finite groups is infinite dimensional) may be decomposed as a direct sum of irreducible representations and every irreducible representation appears with a multiplicity equal to its dimension. The corresponding result for finite groups is much easier and was one of the earliest theorems of Frobenius.

From the point of view of the general structure of mathematics the Peter-Weyl theorem is especially interesting in that it unifies Fourier analysis with the theory of group representations and at the same time points out and underlines the essentially group theoretic nature of Fourier analysis. It is well known that Fourier analysis has been a well nigh indispensable tool in mathematical physics since its introduction by Fourier in 1807 — above all because of its power in solving linear partial differential equations with constant coefficients. Moreover as I have indicated in part (b) of my remarks on the historical background for Weyl's work, the characters of finite commutative groups have played an almost equally important role in the nineteenth century development of number theory - especially through the use of the formula $f(x) = \sum_{\chi \in \hat{G}} c_\chi \chi(x)$ where $c_\chi = \frac{1}{o(G)} \sum_{\chi \in \hat{G}} f(x)\overline{\chi(x)}$. If the reader will compare these formulae with

139

the formulae $f(x) = \sum\limits_{n=-\infty}^{\infty} c_n e^{inx}$ where $c_n = \frac{1}{2\pi} \int\limits_{-\pi}^{\pi} f(x)e^{-inx}dx$, he or she will have no difficulty in seeing that both pairs are special cases of one general assertion about compact commutative Lie groups. In other words the paper of Peter and Weyl can be regarded as showing that one of the most important methods of nineteenth century mathematical physics is in essence the same as one of the most important methods in nineteenth century number theory. If the reader knows of any earlier recognition of this important fact the author will be most interested in hearing about it.

Of course the unification just described is only part of the story. The Peter-Weyl theorem applies not only to commutative compact Lie groups but to compact Lie groups in general. We have before us not only a *unification* of the Fourier analysis of periodic functions on the line with its analogue for functions on finite commutative groups but a natural *generalization* in which the group T and the finite commutative groups are replaced by arbitrary compact Lie groups. (We are here extending the definition of Lie groups to include all topological groups in which the connected component of the unit element is a Lie group in the restricted sense. In particular all finite groups are included). This suggests the possibility that this extended non commutative theory might have applications, even more far reaching than the nineteenth century applications of the commutative theory, to both physics and number theory. In the ensuing half century this possibility has become an important reality the full extent of which is far from widely appreciated.

The applications to physics began in 1927, the very year in which the Peter-Weyl theorem was published. But before describing Weyl's remarkable contributions to this development we must go back to 1924 and pick up another thread, the development of quantum mechanics out of the "old quantum theory" which started in late 1924 and early 1925 with remarkable discoveries of W. Heisenberg and E. Schrödinger respectively. In what seemed to be quite different ways each had managed to derive the discrete energy levels of the hydrogen atom *without making a priori discreteness assumptions* as Bohr had done. The immediately ensuing developments took place so rapidly that it is difficult to know what happened. Much communication took place by word of mouth and private letters and one cannot trace the course of events through an orderly sequence of

140

publications. Let it suffice to say that many individuals were involved, including not only Heisenberg and Schrödinger, but also Born, Bohr, Pauli, Jordan and especially Dirac. By 1927 physicists had at their disposal a subtle refinement of classical mechanics which (a) was internally consistent, (b) automatically implied the mysterious "quantization rules" of the old quantum theory, (c) reduced to classical mechanics in the limiting case of large masses and energies and, (d) explained much that was beyond the reach of both classical mechanics and the old quantum theory. A key feature of this new mechanics was its abandonment of the idea that the future of a system of particles was determined by the values of the coordinates of the particles and their rates of change at some particular time t_o. Indeed it was decided that exact simultaneous values for all coordinates and velocities could not be determined, or rather did not in principle exist. It did make sense however to assign simultaneous probability distributions to all dynamical variables (observables) and the aim of the theory became that of studying how these probability distributions changed with time — and also which distributions could exist simultaneously.

The answers to these questions about changing probabilities, as formulated by the physicists, left something to be desired both in unity and in mathematical precision. These deficiencies were removed by J. von Neumann in an extremely influential paper published in 1927. Von Neumann observed that Hilbert's spectral theorem, suitably generalized to unbounded self adjoint operators, was just the tool that was needed. In his formulation the set of all observables in a quantum mechanical system are in a definite one-to-one correspondence with the (not necessarily bounded) self adjoint operators in a separable infinite dimensional Hilbert space H and the possible states (possible simultaneous probability distributions) similarly correspond one to one to the unit vectors in H, except that ϕ and $e^{i\lambda}\phi$ correspond to the *same* state whenever λ is real. The significance of this correspondence is as follows. If A is the self adjoint operator corresponding to some observable O and ϕ is a unit vector definig a state s then the probability that O will be found to have a value in the set E of real numbers when measured in the state s will be $(P_E^A(\phi) \cdot \phi)$. Here $E \to P_E^A$ is the projection valued measure canonically associated to A by the generalized spectral theorem. As noted above the function $E \to (P_E^A(\phi) \cdot \phi)$ is a measure

141

and when $\|\phi\| = 1$ it is a probability measure. Studying how these probability measures change with time reduces to studying how the state vectors ϕ change with time and this is accomplished by integrating a differential equation of the form $\frac{d\phi}{dt} = i\, H(\phi)$ where H is a suitable self adjoint operator.

At very nearly the same time von Neumann pointed out to Wigner that an analysis the latter had recently made in connection with the application of quantum mechanics to the understanding of atomic spectra could be clarified and extended by using the representation theory of the group S_n of all permutation of n objects. In the applications n is the number of electrons in the atom in question and Wigner had managed to deal with the case n = 3 by elementary methods. In taking up von Neumann's suggestion and developing his own method accordingly, Wigner became the pioneer in applying the representation theory of finite groups to the new quantum mechanics. Soon thereafter he discovered how to clarify the classification of spectral terms by the angular momenta of the atomic states by using the representation theory of the rotation group in three dimensions – a non commutative compact Lie group.

At this point we are ready to return to the work of Hermann Weyl. In that magic year 1927, Weyl published a paper entitled "Quantenmechanik und Gruppentheorie" in which he applied the theory of group representations to quantum mechanics in a rather different way than Wigner and at the same time contributed in a significant way to von Neumann's clarification of the conceptual foundations of this new mechanics. While so doing he indicated that Hilbert's spectral theorem could be regarded as a theorem about the unitary representation theory of a certain *non compact* connected Lie group – the additive group of the real line, thus pointing the way to encompassing the spectral theory of self adjoint operators as a special case of an enlarged theory of group representations.

In the introduction to his paper Weyl begins with the statement that one can distinguish sharply between two questions in (the foundations of) quantum mechanics:
(1) How does one arrive at the self adjoint operators which correspond to various concrete physical observables? (2) What is the physical significance of these operators; i.e. how does one deduce

142

physical statements? He goes on to say that question (2) has been satisfactorily answered by von Neumann (in the paper we have just discussed) but that von Neumann's treatment can be supplemented in certain ways. Coming back to question (1) he asserts that it is a deeper question which has not yet been satisfactorily treated and that he proposes to do so with the help of group theory. "Hier glaube ich mit Hilfe der Gruppentheorie zu einer tieferen Einsicht in den wahren Sachverhalt gelangt zu sein". This may be translated as "Here with the help of group theory I believe I have succeeded in arriving at a deeper insight into the true nature of things". In a footnote he cites the work of Wigner and says (Translation) "this connection with group theory lies in quite a different direction than the researches of Mr. Wigner who .....".

The part of Weyl's article following the introduction is divided into three parts of which part II will be our principal concern. In part I Weyl introduces the fundamental distinction between mixed and pure states. Von Neumann found this independently but did not publish it in the paper cited above. Weyl acknowledges the overlap in a footnote added in proof. The concept of mixed state, which is fundamental for quantum statistical mechanics, is usually mistakenly attributed to von Neumann alone. For example one often speaks of the "von Neumann density matrix". In part II he addresses the first of the two questions raised in the introduction. His particular concern is to find some a priori justification for the fact that the self adjoint operators Q_1, Q_2, ..., Q_n, P_1, P_2, ..., P_n which correspond to position coordinates and momentum components respectively should satisfy the now celebrated Heisenberg commutation relations $P_j Q_j - Q_j P_j = \frac{h}{i}$ (where h is Planck's constant) with all other pairs commuting. To discuss this problem he makes use of a fundamental connection between self adjoint operators and continuous unitary representations of R, the additive group of the real line. Indeed if A is a finite dimensional self adjoint operator one can make sense of e^{iAt} in several ways, in particular by diagonalizing A and replacing each eigenvalue λ_j by $e^{i\lambda_j t}$. One sees easily that $e^{i(t_1 + t_2)A} = e^{it_1 A} e^{it_2 A}$ so that e^{itA} is such a unitary representation and it is not hard to show conversely that for every continuous unitary representation $t \to U_t$ of R there is a unique self adjoint operator A such that $U_t = e^{iAt}$. In fact $A = \frac{1}{i} \frac{d}{dt} U_t \,]_{t=0}$. Weyl suggests that by using Hilbert's spectral theorem one can probably extend this correspondence to the infinite

143

dimensional case with unbounded self adjoint operators included. If so one can replace the Q_i and P_j by unitary representations U^i and V^j where $U_t^i = e^{iQ_i t}$ and $V_s^j = e^{iP_j s}$ and attempt to rephrase the question in terms of commutation relations for the U_t^i and V_s^j. This is easily done and the answer is that $PQ - QP = \frac{h}{i}$ if and only if $e^{isP} e^{itQ} = e^{\frac{ist}{h}} e^{itQ} e^{isP}$ for all real numbers s and t. It is of course more or less obvious that two self adjoint operators A and B will actually commute with one another if and only if e^{iAt} and e^{iBt} commute with one another for all t and s. Thus the Heisenberg commutation relations for a system with n particle coordinates may be rewritten in the form

$$e^{isP_k} e^{itQ_j} = e^{itQ_j} e^{isP_k} \qquad \text{when } j \neq k,$$
$$= e^{\frac{ist}{h}} e^{itQ_j} e^{isP_k} \quad \text{when } j = k,$$
$$e^{isP_k} e^{itP_j} - e^{itP_j} e^{isP_k} = e^{isQ_k} e^{itQ_j} - e^{itQ_j} e^{isQ_k} = 0$$

for all j and k.

Notice now that if we define

$$U_{t_1, t_2, \ldots, t_n} = e^{i(t_1 P_1 + t_2 P_2 + \ldots t_n P_n)} = e^{it_1 P_1} e^{it_2 P_2} \ldots e^{it_n P_n}$$

which we may do whenever the P_j commute with one another then

$$U_{(t_1, t_2, \ldots, t_n) + (t_1', t_2', \ldots, t_n')} = U_{t_1, t_2, \ldots, t_n} \, U_{t_1', t_2', \ldots, t_n'}$$

so that $t_1, t_2, \ldots, t_n \rightarrow U_{t_1, t_2, \ldots, t_n}$ is a unitary representation of the *commutative group* R^n of all n tuples of real numbers under addition. Similarly

$$s_1, s_2, \ldots, s_n \rightarrow V_{s_1, s_2, \ldots, s_n} = e^{i(s_1 Q_1 + s_2 Q_2 + \ldots + s_n Q_n)}$$

is also a unitary representation of the commutative group in question. Moreover assuming the truth of Weyl's conjecture about the general correspondence between self adjoint operators and unitary representations of R the additive group of the real line one shows easily that every (continuous) unitary representation of R^n may be written uniquely in the form

$$x_1, x_2, \ldots, x_n \rightarrow e^{i(x_1 A_1 + x_2 A_2 + \ldots + x_n A_n)}$$

144

where the A_j are mutually commuting self adjoint operators. In these terms as Weyl observed one may restate the Heisenberg commutation rules in the following terms. The Q_i and the P_j commute among themselves and the unitary representations U and V of R which this makes possible satisfy

$$U_{t_1,t_2,\ldots,t_n} V_{s_1,s_2,\ldots,s_n} =$$
$$V_{s_1,s_2,\ldots,s_n} U_{t_1,t_2,\ldots,t_n} \, e^{i(s_1 t_1 + s_2 t_2 + \ldots + s_n t_n)} \, .$$

These are the Heisenberg commutation relations in "integrated" or "Weyl" form.

Let us observe next (with Weyl) that if we define $W_{t_1,t_2,\ldots,t_n,s_1,s_2,\ldots,s_n}$ to be the operator $U_{t_1,t_2,\ldots,t_n} V_{s_1,s_2,\ldots,s_n}$ then $t_1,t_2,\ldots,t_n,s_1,s_2,\ldots,s_n \to W_{t_1,t_2,\ldots,t_n,s_1,s_2,\ldots,s_n}$ is *not* a unitary representation of R^{2n}. The factor $e^{\frac{i}{\hbar}(t_1 s_1 + \ldots + t_n s_n)}$ interferes. However it *is* a so called projective or ray representation. Quite generally if G is a group and $t \to L_t$ is a linear operator valued function on G one says that L is a projective (or ray) representation if $L_{st} = \sigma(s,t) L_s L_t$ where $\sigma(s,t)$ is a complex number depending on s and t. When $\sigma(s,t) \equiv 1$ the definition reduces to that of group representation as given earlier. For finite groups projective representations were studied in some depth by I. Schur in papers published in 1904 and 1907. Weyl points out that projective representations are especially relevant in quantum mechanics because whenever V^1 and V^2 are unitary operators in a Hilbert space H and $V^2 = e^{i\Theta} V^1$ where Θ is a real number then V^1 and V^2 define *exactly the same* permutation of the pure states of the system. It follows that in dealing with groups of symmetries each unitary operator defining a symmetry is determined only up to a "phase factor" $e^{i\Theta}$ and so the identity $L_{st} = L_s L_t$ must be relaxed to read $L_{st} = \sigma(s,t) L_s L_t$ where $\sigma(s,t)$ is a complex number of modulus one.

Weyl observed not only that the Heisenberg commutation relations are equivalent to the statement that W is a projective unitary representation of the *commutative* group R^{2n} with respect to the "multiplier" σ where $\sigma(t_1,t_2,\ldots,t_n,s_1,s_2,\ldots,s_n,t_1',t_2',\ldots,t_n',s_1',s_2',\ldots,s_n') = e^{-\frac{i}{\hbar}(s_1 t_1' + s_2 t_2' + \ldots + s_n t_n')}$ but that to within a certain natural equivalence σ is the only possible "non degenerate" multiplier for R^{2n}. Thus Weyl demonstrated that one is led naturally to self adjoint operators satisfying the Heisenberg commutation relations if one simply considers the most general projective unitary representation of the 2n parameter commutative Lie group R^{2n}. (One can show that no non degenerate multipliers σ exist for R^{2n+1}). He attached great significance to this fact and to the fact (of which he gave a heuristic proof) that to within unitary equivalence there is *only one* projective unitary representation of R^{2n} which is irreducible and has non degenerate multiplier σ. Correspondingly of course there is to within unitary equivalence only one irreducible 2n-tuple of self adjoint operators satisfying the Heisenberg commutation rules. As Weyl expressed it "The kinematical structure of a physical system is expressed by an irreducible group of unitary ray representations in system space".

In close connection with the above Weyl showed that the smallest and simplest commutative group with a non degenerate projective multiplier σ is the four element group $z_2 \times z_2$. Once again there is an essentially unique projective representation. It is two dimensional and the four two by two matrices concerned may be taken as $\begin{pmatrix} 1 & 0 \\ 0 & 1 \end{pmatrix}$, and the three matrices $\begin{pmatrix} 1 & 0 \\ 0 & -1 \end{pmatrix}\begin{pmatrix} 0 & 1 \\ 1 & 0 \end{pmatrix}\begin{pmatrix} 0 & i \\ -i & 0 \end{pmatrix}$ now familiar as the Pauli spin matrices.

These remarks constitute Weyl's contribution to the first of the two problems stated in the introduction to his paper. While more suggestive than persuasive or logically compelling, they are of importance as perhaps the first step in the program of deriving fundamental relationships in quantum mechanics from group theoretical symmetry principles in a program which I feel it appropriate to call Weyl's program and to distinguish fairly sharply from the related important program inaugurated and much developed by Wigner. (Later of course each made contributions to the other's program).

As suggested in the introduction to this account of Weyl's contributions and as will be explained below, Weyl's uniqueness theorem, as rigorized by Stone and von Neumann in 1930, admits a sequence of natural generalizations. The last of these may be used to give a much more logically compelling deduction of the Heisenberg commutation relations — and quite a bit more besides. These developments show the fundamental soundness of Weyl's intuition is expressed in his semi mystical answer to problem 1.

146

As far as the third and final section of Weyl's paper is concerned we shall mention only his emphasis on the point that when one integrates the Schrödinger equation $\frac{d\phi}{dt} = iH\phi$ one obtains $\phi_t = e^{iHt}\phi_0$ so that unitary representations of the real line enter once again: The change of a state with time is explicitly described by the action of such a representation.

Last but not least we come to Weyl's celebrated book "Gruppentheorie und Quantenmechanik" published in 1928 and based on a course of lectures given at the E.T.H. in Zurich during the winter semester 1927-28. Of the five chapters of this book I and III largely consist of mathematical preliminaries. The first contains an exposition of the theory of (mainly) finite dimensional Hilbert spaces and the third an exposition of the unitary representation theory of finite groups and compact Lie groups. The heart of the book lies in Chapters II, IV and V. Chapter II contains one of the earliest systematic coherent accounts of quantum mechanics as a whole. Perhaps only Dirac had as complete an overall view earlier, but his early accounts are less complete and well organized. The section of this chapter dealing with the notion of a particle in an electromagnetic field contains the following words (quoted from the English translation of the second edition) "The field equation for the potentials ψ and ϕ for the material and electromagnetic waves are invariant under the simultaneous replacement of ψ by $e^{i\lambda}\cdot\psi$ and ϕ_α by $\phi_\alpha - \frac{h}{e}\frac{\partial\lambda}{\partial x_\alpha}$. Here λ is an arbitrary function of the space-time coordinates. This "principle of gauge invariance" is quite analogous to that previously set up by the author on speculative grounds, in order to arrive at a unified theory of gravitation and electricity. But I now believe that this gauge invariance does not tie together electricity and gravitation but rather electricity and matter in the manner described above." This enunciation of the principle of gauge invariance is again a remarkable anticipation of future work. A quarter of a century later it was generalized by Yang and Mills in a now famous paper and in its generalized form has revolutionized elementary particle physics since the middle 1960's.

Chapter IV entitled "Application of the theory of groups to quantum mechanics" is divided into four parts. Part A, subtitled "the rotation group" is a complete and detailed exposition of how the unitary representation theory of the rotation group explains, or-

147

ganizes and illuminates the theory of atomic spectra. This is the application mentioned above which was discovered and worked out by Wigner and von Neumann. Part B, "the Lorentz group", is based on Dirac's celebrated paper of 1928 presenting a relativistically invariant quantum mechanical theory of the electron. However it is much more than a simple exposition of Dirac's work. Weyl presents the material in a different way and discusses its significance and implications in depth. Part C, "the permutation group", is concerned with the implications for the quantum mechanics of n interacting identical particles of the natural action of S_n the permutation group on n things on the tensor product $H \times H \times \ldots \times H$ of n copies of the single particle Hilbert space. For each permutation π in S_n there is a unique unitary operator W_π which maps the vector $\phi_1 \times \phi_2 \times \ldots \times \phi_n$ in $H \times H \times \ldots \times H$ into $\phi_{\pi(1)} \times \phi_{\pi(2)} \times \ldots \times \phi_{\pi(n)}$ and the mapping $\pi \to W_\pi$ is a unitary representation of S_n. When one decomposes this representation as a direct sum of multiples of the various irreducible representations of S_n one obtains a direct sum decomposition of $H \times H \times \ldots \times H$, two components which play a special role in quantum mechanics. These are the components defined by the two one-dimensional representations the identity representation I and the representation J which takes every "odd" permutation into –1 and every even representation into 1. One calls the subspace corresponding to I the symmetric subspace and that corresponding to J the anti-symmetric subspace. Moreover one refers to these subspaces as the symmetrized and antisymmetrized tensor products respectively. It is a fact of great importance that the Hilbert space of states for a system of n identical particles (with one particle space H) is not $H \times H \ldots \times H$ as one might be inclined to suppose but either the symmetric or the antisymmetric subspace. Since either case may occur one has a fundamental division of all particles into two categories. Nowadays one speaks of them as bosons and fermions respectively. Weyl discusses how this circumstance implies that interchanging two identical particles makes no physical difference whatever, how the fact that electrons are fermions implies the Pauli exclusion principle, how the latter together with the fact that electrons have spin $\frac{1}{2}$ explains the periodic table and how "quantizing a field" leads to particles (field quanta) which are bosons or fermions according as one uses the Heisenberg commutation relations or the anti commutation relations of Jordan and Wigner. The final part D "quantum kinematics" is an exposition of part II of the paper of Weyl discussed at length above.

148

The final chapter V is widely considered to be the most difficult part of the book. Its starting point is Wigner's observation about the utility of the representation theory of the permutation group in the analysis of atomic spectra. However Weyl carries the work much further and applies it to the structure of molecules and the elucidation of the chemical notion of valence. This program requires a considerable purely mathematical development and around seventy percent of the chapter can be read as pure mathematics. It will be easier to explain more fully after we have explained the notions of "induced representation" and "system of imprimitivity" which arise naturally when one generalizes the Stone-von Neumann rigorization of Weyl's theorem on the uniqueness of the irreducible solutions of the Heisenberg commutation relations.

Considering the comprehensiveness of Weyl's book, the early date at which it was written and the wealth of original ideas which it contains, one cannot fail to be tremendously impressed by Weyl's achievement or to understand why it is considered one of the great classics of mathematical physics.

We turn our attention now to an account of how later developments inspired directly or indirectly by Weyl's paper of 1927 "Quantenmechanik und Gruppentheorie" led ultimately to a considerable improvement of Weyl's answer to problem I. It is interesting (and instructive) to note that the intermediate stages as well as the final result seem far removed from physics and that the final result has other applications some of which are also important for quantum mechanics.

The first step came in 1930 with the publication of a celebrated short note by M.H. Stone entitled "Linear transformations in Hilbert space III. Operational methods and group theory". In this note Stone announces two mathematical theorems with a sketch of their proofs and states that an account of their significance for physics will be found in the above cited paper of Weyl. One of them is a theorem about arbitrary continuous unitary representations of the additive group of the real line R and bears the same relationship to the decomposability theorem for finite dimensional group representations as Hilbert's spectral theorem bears to the diagonizability theorem for finite dimensional self adjoint matrices. Just as in the spectral

149

theorem one associates to each such representation $t \rightarrow V_t$ a unique projection valued measure $E \rightarrow P_E$ defined on R. However here the relationship $(H(\varphi), \psi) = \int_{-\infty}^{\infty} \lambda (P_\lambda(\varphi), \psi)$ is replaced by $(V_t(\varphi), \psi) = \int_{-\infty}^{\infty} e^{it\lambda} d(P_\lambda(\varphi), \psi)$, the equation holding for all real t. The theorem goes on to state that the representation V and the projection valued measure P determine one another uniquely and that every P occurs. It will be convenient to refer to this theorem as the spectral theorem for unitary representations of R.

Consider now the two one to one correspondences set up by the spectral theorems for self adjoint operators and continuous unitary representations of R respectively. Having a common term (the projection valued measures on R) they define a one-to-one correspondence between self adjoint operators H and continuous unitary representations $t \rightarrow V_t$ of R. It is not difficult to check that H and V correspond in this way if and only if $V_t = e^{itH}$ for all t. The fact that every V is so related to some unique H is usually referred to as "Stone's theorem". The other theorem stated by Stone is simply the uniqueness theorem for operators satisfying the Heisenberg commutation relations in the form involving group representations given it three years earlier by Weyl. Stone did not publish his proof and the first published proof is due to von Neumann. One speaks of the Stone-von Neumann theorem.

The second step came in 1933 when A. Haar proved that every separable locally compact group admits a measure which is invariant under right (left) translation. That this measure is essentially unique was proved slightly later by von Neumann. The significance of this result is that it paved the way for extending the theory of unitary group representations from compact Lie groups to arbitrary compact topological groups and on to topological groups which are not even compact, provided that they are locally compact in the sense that every point has a compact neighbourhood. Haar himself observed that, using his existence theorem, one could extend the Peter-Weyl theorem to all compact topological groups.

The third step came in 1934 when L. Pontryagin and E.R. van Kampen developed their celebrated duality theorem for locally compact commutative groups. Let G be a locally compact commutative group. Let $\hat{G}$ denote the set of all continuous characters of G i.e. the group

150

of all continuous functions χ from G to the complex numbers of modulus one such that $\chi(xy) = \chi(x)\,\chi(y)$. The product of two continuous characters is evidently again such and under this operation $\hat{G}$ is again a commutative group. It is even a locally compact topological group with respect to a topology which may be loosely be described as the topology of uniform convergence on compact subsets. One calls it the character group of G or the dual group of G. Of course one may now form $\hat{\hat{G}}$ the dual of the dual and ask about its relationship to G. It is immediate that we may almost think of G as contained in $\hat{\hat{G}}$. Indeed for each $x \in G$ the function $\chi \to \chi(x)$ on $\hat{G}$ is in fact a continuous character on $\hat{G}$ and hence a member of $\hat{\hat{G}}$ which we may denote by f_x. The mapping $x \to f_x$ is clearly multiplication preserving and if it happens to be one-to-one we have an isomorphism of G onto a subgroup of $\hat{\hat{G}}$. The duality theorem, originated by Pontryagin and completed on various points by van Kampen, asserts that $x \to f_x$ is always one to one that the subgroup of $\hat{\hat{G}}$ onto which it maps G is the *whole* of $\hat{\hat{G}}$ and that this isomorphism of G with $\hat{\hat{G}}$ is an isomorphism of topological groups. In other words locally compact commutative groups occur in dual pairs — each member of any pair being the dual of the other. In some cases G and $\hat{G}$ may be isomorphic as topological groups so that G is self dual; for example this is so when G is finite and when G is the additive group of a finite dimensional real vector space. In general however this is no so. In particular $\hat{G}$ is compact whenever G is discrete and vice versa.

The fourth step came a decade later in 1944. Stone's formulation and proof of the spectral theorem for continuous unitary representations of R was, in effect, an extension of the theory of group representations to one particular non compact group. Of course R is locally compact as well as commutative and it is natural to wonder if one can go further and extend this theory to all locally compact commmutative groups. One can and this was realized independently at about the same time by Ambrose in the U.S.A., Godement in France, and Naimark in the Soviet Union. All of these mathematicians published their work in 1944. To see how to general-

151

ize the formula $\int e^{it\lambda} \, d(P_\lambda(\phi) \cdot \psi) = (V_t(\phi) \cdot \psi)$ one has only to realize that for each real λ the function $t \to e^{i\lambda t}$ is a character χ_λ of R and that $\lambda \to \chi_\lambda$ is an isomorphism of R with $\hat{R}$. Using this isomorphism, $E \to P_E$ can be thought of as a projection valued measure on $\hat{R}$ and one can rewrite Stone's formula as

$$(V_t(\phi) \cdot \psi) = \int_{\chi \in \hat{R}} \chi(t) \, d(P_\chi(\phi) \cdot \psi).$$

This formula of course makes equal sense when R is replaced by any locally compact commutative group G and $\hat{R}$ by $\hat{G}$, the Pontryagin dual of G. The theorem of Ambrose, Godement and Naimark now simply asserts the existence of a one to one correspondence between all continuous unitary representations V of G and all projection valued measures $E \to P_E$ on the dual $\hat{G}$ such that $(V_t(\phi) \cdot \psi) = \int \chi(x) \, d(P_\chi(\phi) \cdot \psi)$ for all $t \in G$ and all ϕ and ψ in H(V) the Hilbert space of V.

The fifth step came in 1949 when the present author published a paper in the Duke Mathematical Journal entitled "On a theorem of Stone and von Neumann". It was based on the following sequence of observations.

(1) The factor $e^{\frac{i}{h}(s_1 t_1 + \ldots + s_n t_n)}$ which occurs in the Weyl form of the Heisenberg commutation relations can be interpreted as $\chi(t)$ where $t = t_1, t_2, \ldots, t_n$ is an element of the group of all n tuples of real numbers under addition and χ is the character $t_1, t_2, \ldots, t_n \to e^{\frac{i}{h}(s_1 t_1 + \ldots + s_n t_n)}$.

(2) With this interpretation the Heisenberg commutation relations have an obvious generalization which may be written down for any pair U, V where U is a continuous unitary representation of an arbitrary locally compact commutative G and V is a continuous unitary representation of the dual group $\hat{G}$. This generaliziation reads:
$U_x V_\chi = \chi(x) \, V_\chi U_x$ for all $x \in G$ and all $\chi \in \hat{G}$.

(3) Using the generalized spectral theorem of Ambrose, Godement and Naimark one has a projection valued measure on $G = \hat{\hat{G}}$ canonically associated to V and an elementary argument shows

152

that V and U satisfy the commutation relation $U_x V_\chi = \chi(x) V_\chi U_x$
if and only if U and P satisfy the commutation relation

$$U_x P_E = P_{[E]x^{-1}} U_x$$

for all x and E where $[E]x^{-1}$ denotes the translate of E by x^{-1}.

(4) Because of (3) studying pairs U, V which satisfy the generalized commutation relation written down under (2) can be reduced to studying pairs, U, P which satisfy the commutation relation $U_x P_E = P_{[E]x^{-1}} U_x$ written down under (3). However both U and P are defined on the same group G, no reference whatever being given to the dual $\hat{G}$ of G. Moreover this new form of the commutation relation makes sense even when G is non commutative. This raises the question of proving a generalization of the Weyl-Stone-von Neumann uniqueness theorem that applies not only to dual pairs of locally compact commutative groups but to arbitrary locally compact groups – commutative or not. This can be done and the generalized theorem is the main result of the Duke Journal paper cited above. In this paper the group G is assumed to have a countable basis for the open sets but this restriction was removed by Loomis in 1952.

It is perhaps worth stating the theorem explicitly in such a way as to present a concrete example of a pair satisfying the commutation relation. Let G be a locally compact group and let μ be a right invariant Haar measure for G. Form the Hilbert space $\mathcal{L}^2(G,\mu)$ of all square summable complex valued functions on G. For each ψ in G let U_x^0 denote the unitary operator which takes f into g where $g(y) = f(yx)$. Then $x \to U_x^0$ is a continuous unitary representation of G known as the *regular representation*. Next for each Borel subset E of G let P_E^0 denote the projection operator which takes f into g where g is the function which agrees with f on E and is zero outside of E. One verifies that $E \to P_E^0$ is a projection valued measure on G and a straightforward computation shows that

$$U_x^0 P_E^0 = P_{[E]x^{-1}}^0 U_x^0 \qquad (*)$$

for all x and E. Finally one can prove that the pair U, P^0 is *irreducible* in the sense that no proper closed subspace of $\mathcal{L}^2(G,\mu)$ is invariant under all U_x^0 and all P_E^0. Now let U be *any* unitary re-

153

presentation of G and let P be *any* projection valued measure on G such that P_E and the U_x operate in the same Hilbert space H. Suppose that U and P satisfy (*) and are jointly irreducible. Our generalized uniqueness theorem then asserts that there exists a unitary operator W mapping H onto $\mathcal{L}^2(G,\mu)$ so that $WP_E W^{-1} = P_E^0$ and $W U_x W^{-1} = U_x^0$ for all x and E.

Shortly after this paper was submitted the author noticed that a still further generalization was conceivable. One can replace the projection valued measure $E \to P_E$ on G by a projection valued measure $E \to P_E$ defined on some space S on which G acts as a transformation group. The commutation relation $U_x P_E = P_{[E]x^{-1}} U_x$ still makes sense provided that we interpret [s]y as the transform of s in S by x in G. Moreover it reduces to the one discussed above when S = G and [s]x = sx is just group multiplication. The question that now presents itself is the following. Is the uniqueness theorem stated above still true at this new level of generality? The answer is no. However in an important special case one can analyze the non uniqueness completely and so produce a theorem that generalizes the uniqueness theorem stated above. More specifically one can find all possible solutions of the commutation relation in question. The special case in which this can be done is that in which S is the "homogeneous space" G/K defined by some closed subgroup K of G; that is the space whose elements are the right K cosets Kx and the transform of Kx by y is the right coset K(xy). It is evident that the action of G on G/K is "transitive" in the sense that given Kx and Ky in G/K there exists z in G such that [Kx]z = Ky. We need only choose $z = x^{-1}y$. Conversely every transitive G space S with suitable regularity properties is isomorphic to a coset space G/K. When S = G/K, what one finds instead of uniqueness, is that the possible solutions of the commutation relation above have equivalence classes that correspond one to one in a natural way to the equivalence classes of unitary representations of K. Moreover a solution of the commutation relations is irreducible if and only if the corresponding unitary representation of K is irreducible. Of course when K = ⟨e⟩ so that S = G there is only one irreducible representation of K. That is why there is uniqueness when S = G.

To gain some insight into the situation and a new point of view regarding the meaning of the commutation relation, it is useful to

154

consider the special case in which there is only a discrete countable infinity of right K cosets so that S is a countable discrete set of points. In that case the projection valued measure $E \to P_E$ is completely determined by the projections $P_{\langle s \rangle}$ assigned to the one point subsets $\langle s \rangle$ of S. Moreover the ranges of these projections constitute a direct sum decomposition of the underlying Hilbert space H. $H = \sum_{s \in S} H_s$ where H_s is the range of $P_{\langle s \rangle}$. A very simple computation now shows that the commutation relation $U_x P_E = P_{[E]x^{-1}} U_x$ holds if and only if each U_x carries the subspace H_s onto the subspace $H_{[s]x^{-1}}$.

Thus the operators U_x while they do not leave the subspaces H_s invariant in general, they do preserve their identity; merely permuting them among themselves. Now consider the special case in which $s = K$, the right coset s_0 containing the identity. Then for $x \in K$, $U_x(H_{s_0}) = H_{s_0}$ so that $k \to U_k$ defines a unitary representation L of the subgroup k by operators in the subspace H_{s_0}. A little reflection should convince the reader that once one knows K and the unitary representation L of K the pair U, P can be reconstructed and is uniquely determined by K and L. Thus finding all unitary equivalence classes of pairs U, P satisfying the commutation relations reduces, in this case at least, to finding all unitary equivalence classes of continuous unitary representations of the subgroup K.

Quite generally, if U is a unitary representation of a group G and $H(U) = H_1 + H_2 + \ldots$ is a direct sum decomposition of the underlying Hilbert space $H(U)$, one refers to this decomposition as a "system of imprimitivity" for the representation U provided that each U_x simply permutes the subspaces H_j among themselves. If in particular, for each pair i and j, there is an x such that $U_x(H_i) = H_j$ one refers to a *transitive* system of imprimitivity. This terminology is suggested by a closely related notion in the theory of permutation groups. The fact that a unitary representation of a finite group G together with a *transitive* system of imprimitivity for it is determined by a unitary representation L of a subgroup of G, was already known to Frobenius. We may now think of Frobenius's result as finding the most general solution of our last generalization of the Heisenberg commutation relation in the very special case of finite G and S. The fact that Frobenius's result has a more or less complete generalization to the case in which G is a general separable compact group and K is

155

an arbitrary closed subgroup is the content of a paper published by the present author in 1949 and entitled "Imprimitivity for representations of locally compact groups I". It will be useful to refer to the main result of this paper as the "imprimitivity theorem". In its replacement of a discrete system of imprimitivity by a projection valued measure it is strongly reminiscent of the passage from the classical diagonalization theorem for matrices to Hilbert's spectral theorem.

On the face of it the imprimitivity theorem seems to have left its original inspiration far behind and to have very little connection with physics. However, as we shall now indicate and as was promised earlier, it is just what is needed to give a more satisfying answer to Weyl's problem I. Let S denote the physical space and let $\mathcal{E}$ denote the group of all isometries of S. Then in the usual Euclidean model for space (and in certain other models as well) $\mathcal{E}$ acts on S as a *transitive* transformation group. Consider now the quantum mechanical model of a single free particle. Let H be the Hilbert space of pure states in the von Neumann formulation. The position coordinates and the velocity or momentum components will then be associated with certain self adjoint operators in H. Which operators? That is Weyl's problem I. To answer it we begin with the observation that the position coordinate observables may be discussed in a coordinate free way by replacing them with two valued observables which take on the value one or zero according as the particle is observed to be in a certain region E of space or not. Each such two valued observable will necessarily (according to the von Neumann scheme) be associated with a projection operator P_E and it is easy to argue that $E \to P_E$ must satisfy the conditions defining a projection valued measure on S. Once P is known it is easy to deduce the projection valued measure associated with *any* real valued coordinate i.e. any real valued Borel function f on S. It is just the projection valued measure $E \to P_{f^{-1}(E)}$ on the real line R. In these terms Weyl's problem I becomes (in part): What projection valued measure P on S has nature chosen (or must nature choose)? The answer is based on the hypothesis that nature's choice will reflect the symmetry of space as reflected in the action of the group $\mathcal{E}$ on S, that the laws of nature must be independent of position and orientation in space. It is not hard to argue that this principle implies the existence for each x in $\mathcal{E}$ of a certain unitary operator U_x which describes the transformation

156

of the states associated with a rigid motion of space and that the mapping $x \to U_x$ is a projective unitary representation associated with some multiplier σ. Of course U and P cannot be *independently* chosen; they must be so related that rigid motions change the position observables in the appropriate way. Analysis of this leads to the conclusion that U and P must be so related that $U_x P_E U_x^{-1}$ is just $P_{[E]x^{-1}}$. But this relation is equivalent to our generalized commutation relation $U_x P_E = P_{[E]x^{-1}} U_x$ and we may apply the imprimitivity theorem (which is valid also for projective representations). The conclusion is that, up to unitary equivalence there is just one possible pair P, U for each projective unitary representation L of the subgroup K of $\&$ leaving fixed an "origin" s_0 in space. K is of course just the compact group of all rotations about a fixed point and it is well known that it has precisely one irreducible unitary projective representation of every positive integer dimension. Thus one has an overall view of all possible pairs P, U and one sees in particular that there is only a discrete countable set of such. Given P, U one not only knows the operators corresponding to the coordinate observables but also the operators corresponding to the linear and angular momentum observables. The latter are derived from U by the general principle valid in both classical and quantum mechanics relating integrals of the motion to one parameter symmetry groups. In the special case in which the representation L of K is one dimensional one is led to the classical form for the operator corresponding to position and momentum observables for a particle *without* spin. More generally if L is the irreducible projective representation of dimension $j = 1,2,3,\ldots$ one is led to the classical form for the operators corresponding to the position and momentum observables for a particle of spin $\frac{j-1}{2}$. In particular one is led automatically to the Pauli matrices for particles of spin $\frac{1}{2}$. For further details including the extension to interacting particles and references to related work of Wigner and of Wightman the reader is referred to the middle sections of the author's book "Unitary group representations in physics, probability and number theory" W.A. Benjamin 1978.

We conclude with some very brief indications concerning connections of the above with the fifth chapter of Weyl's book. First of all given any unitary representation L of any closed subgroup K of any separable locally compact group G there always exists a pair P, U for the action of G on G/K such that the defining representation of K is L. The representation of U is uniquely determined by L and is known as the unitary representation of G *induced* by L. It is convenient to denote this induced representation by the symbol U^L. One finds that many interesting locally compact groups G have most if not all their irreducible representations of the form U^L where L is a lower dimensional representation of a proper subgroup. Moreover the imprimitivity theorem is a useful tool in proving such things. Indeed one can often detect a *transitive* system of imprimitivity for the representation in question and this implies that the representation is induced.

In his fifth chapter Hermann Weyl is concerned with the tensor product of n replicas of the same Hilbert space H. This Hilbert space is the space of states of a single particle moving in a potential field and is also the space $H(V)$ of a unitary representation V of a compact group K. As explained earlier in this paper there is a natural unitary representation W of the symmetric group S_n in H x H x ... x H. Now V x V x ... x V is a representation of the product group K x K x K x ... x K where in each case there are n factors and W and V x V x ... x V both act in the same product Hilbert space H x H x ... x H. These two representations combine to define a representation $V^n W$ of a certain "twisted" product of the two groups K x K x ... x K and S_n. Here we define the product of $x_1, x_2, ..., x_n$' π_1 and $y_1, y_2, ..., y_n$ π_2 to be $x_1 \pi_1(y_1), x_2 \pi_2(y_2), ..., x_n \pi_1(y_n)$. One is interested in determing the structure of the restriction of $V^n W$ to the subgroup of the twisted product consisting of all $x_1, x_2, ..., x_n, \pi$ with $x_1 = x_2 = ... = x_n$. This subgroup is of course isomorphic to $K \times S_n$. Consider now a decomposition of V into irreducibles $H(V) = H_1 + H_2 + ...$ where each H_j is an invariant subspace. Then each product space $H_{j_1} \times H_{j_2} \times ... \times H_{j_n}$ will be a subspace of H x H x ... x H and all of these subspaces together constitute a direct sum decomposition of H x H x ... x H. These subspaces are of course invariant under the representation V^n of K x K x ... x K but *not* under the representation $V^n W$ of the twisted product of K x K x ... x K with S_n. However, and here is the key point, they do constitute a discrete system of imprimitivity for $V^n W$. It is not a transitive system but one may decompose it into transitive pieces and thus decompose $V^n W$ as a discrete direct sum of induced representations. The general theory of induced representations tells one how to study the restrictions of these induced representations

158

to subgroups and in particular to $\widetilde{K}^n \times S_n$ where $\widetilde{K}^n$ is the "diagonal" subgroup of $K \times K \times \ldots \times K$ mentioned above.

Weyl does not explain what he is doing in these terms. However, if one does so and is familiar with the theory of induced representations, one finds oneself led automatically to many of the main arguments. The writer hopes some day to explain all this in detail in an article entitled "Weyl's fifth chapter revisited".*

Anschrift des Verfassers: Prof. Dr. George W. Mackey
Harvard University
Department of Mathematics
Science Center
One Oxford Street
Cambridge
Massachusetts 02138
USA

* This paper was written and typewritten while the author was a guest at the School of Theoretical Physics of the Dublin Institute for Advanced Studies. He wishes to thank the Institute for its hospitality and secretarial assistance.

159

THE SIGNIFICANCE OF INVARIANT MEASURES FOR HARMONIC ANALYSIS*
G. W. MACKEY

I.

In 1807 J. B. Fourier submitted a celebrated memoir to the French
Academy in which he introduced an important new method into mathematics: the method of harmonic analysis. In gross terms and in its original primitive form this method amounts to exploiting the fact that
any reasonably well behaved periodic function of period $2\pi\lambda > 0$ can
be expanded into a so called Fourier series; that is an infinite se-
ries of the form $\sum_{n=0}^{\infty} \left(a_n \sin \frac{nx}{\lambda} + b_n \cos \frac{nx}{\lambda}\right)$. Fourier originally applied his method to the study of the solutions of the partial dif-
ferential equation governing the flow of heat but it was soon recognized that it could be used effectively for all linear partial differential equations with constant coefficients. Since such partial
differential equations were playing a growing role in mathematical
physics, Fourier's discovery constituted a truly important step for-
ward.

One hundred years later in 1907 F. Riesz and Fischer independently used the recently discovered new integral of Lebesgue (1902)
to put the theory of Fourier series in a more complete and elegant
form by proving that every sequence $b_0, a_1, b_1, a_2, b_2 \ldots$ of real num-
bers such that $\sum_{n=1}^{\infty} a_n^2 + \sum_{n=0}^{\infty} b_n^2 < \infty$ actually occurs in the Fourier

---

* This paper is in final form; no version of it is, or will be submitted for publication elsewhere.

- 551 -

expansions of some measurable periodic function f such that
$\int_0^{2\pi\lambda} |f(x)|^2 \, dx < \infty$. For the first time one had a well defined and
quite general class of functions the Fourier coefficients of which
constituted an equally general and well defined class of infinite
sequences.

Two decades after that, in 1927, Hermann Weyl in collaboration
with F. Peter recognized the essentially group theoretical character
of harmonic analysis and set the stage for a vast generalization.
More precisely Peter and Weyl observed that expanding a square summable
(on a period) periodic function on the line into a Fourier series
could be interpreted as decomposing the "regular representation" of
a certain compact Lie group into irreducible components. In this way
they connected Fourier analysis with the relatively new branch of
algebra discovered by Frobenius in 1896 and known as the theory of
group representations.

At this point it will be useful to recall a few basic facts about
this latter theory. Let G be any finite group and let F(G) denote
the finite dimensional vector space of all complex valued functions
on G . For each x in G we may define a linear transformation
L_x on the vector space F(G) by defining $L_x(f)$ to be the right
translation of f by x ; that is the function g such that g(y) =
= f(yx) for all y in G . In other words L_x is defined by the
equation: $L_x(f)(y) = f(yx)$, holding for all y in G . It is evident both that L_x is a linear transformation of F(G) into F(G)
for all x and that the mapping $x \to L_x$ is a homomorphism of G
into the group of all non-singular linear transformations of F(G)
into F(G) . It is an example of a (linear) representation of G
where by definition a (linear) representation W of G is any homomorphism of G into the group of all non-singular linear transformations of some finite dimensional vector space H(W) into itself. One
defines a subspace M of the space H(W) of a representation W to

- 552 -

be *invariant* if $W_x(\phi) \in M$ for all $\phi \in M$ and $x \in G$. Evidently each invariant subspace which is more than $\{0\}$ defines a new representation whose space is M by simply replacing each W_x by its restriction to M . One calls this a subrepresentation and more precisely the subrepresentation defined by M . It is said to be *proper* if $M \neq H(W)$. The representation W is said to be *irreducible* if it has no proper subrepresentations. Two invariant subspaces M_1 and M_2 are said to be complementary if $M_1 \cap M_2 = \{0\}$ and every element $\phi \in H(W)$ may be written as $\phi_1 + \phi_2$ where $\phi_1 \in M_1$ and $\phi_2 \in M_2$. This decomposition is necessarily unique and W is in a certain obvious sense the *direct sum* of the two subrepresentations defined by M_1 and M_2 respectively. One shows that every proper invariant subspace admits a complementary invariant subspace - which however need not be unique. On the other hand the subrepresentations defined by two distinct complementary invariant subspaces can be shown to be *equivalent* in the sense of the following definition. Let W^1 and W^2 be any two representations of the same group G in complex vector spaces $H(W^1)$ and $H(W^2)$. If there exists an isomorphism T of $H(W^1)$ with $H(W^2)$ such that $TW_x^1 T^{-1} = W_x^2$ for all x in G one says that T is an equivalence between W^1 and W^2 and that W^1 and W^2 are equivalent representations of G . It follows from what has already been said that every representation can be decomposed into irreducibles and one proves further that these irreducibles are unique to within equivalence; that is if $W \simeq W^1 \oplus W^2 \ldots W^\ell$ and also to $V^1 \oplus V^2 \oplus \ldots \oplus V^{\ell^1}$ where $\oplus$ denotes direct sums and the W_i and V_j are irreducibles then $\ell = \ell^1$ and there exists a permutation π of $1,2,\ldots,\ell$ such that W_j and $V_{\pi(j)}$ are equivalent for all j . Of course some of the W_j may be equivalent to one another. It follows that to determine all possible representations of G (up to equivalence) it suffices to determine all equivalence classes of *irreducible* representations of G . It turns out that there are only a finite number of these for each G . Indeed, one

- 553 -

can show that every irreducible representation of G is equivalent
to a subrepresentation of the regular representation and that the
latter is a direct sum of irreducible representations each of which
occurs with a "multiplicity" equal to its dimension. Since the dimension of the regular representation is the number of elements $O(G)$
in G one sees at once that $d_1^2 + d_2^2 + \ldots + d_2^2 = O(G)$ where $d_1, d_2,$
$\ldots, d_2$ are the dimensions of the distinct equivalence classes of
irreducible representations of G .

If W is any representation of G then the complex valued
function $x \to \mathrm{Trace}(W_x)$ depends only on the equivalence class of
W (since $\mathrm{Trace}(TW_x T^{-1}) = \mathrm{Trace}(T^{-1}TW_x) = \mathrm{Trace}(W_x)$) and is called
the *character* χ_W of W . One sees easily that the character of a
direct sum is the sum of the characters so that every character is a
linear combination with non-negative integer coefficients of "irreducible characters". Here of course by an irreducible character one
means a character of an irreducible representation. One shows that
two representations are equivalent if and only if their characters
are identical. It is immediate from the definition that the character
of any representation of a finite group G is constant on the conjugacy classes of G . Conversely one shows that the irreducible
characters are linearly independent and form a basis for the vector
space of all complex valued functions on G which are constant on
the conjugacy classes. It follows that the number of distinct irreducible characters and (equivalently) the number of distinct equivalence classes of irreducible representations for a group G is
equal to the number of distinct conjugacy classes in G .

In the special case in which G is commutative one shows that
every irreducible representation is one dimensional and conversely
that if all irreducible representations are one dimensional then G
must be commutative. Any one dimensional representation W of a
group G may of course be written in the form $W_x = \chi(x)I$ where

I is the identity operator in the one dimensional vector space of complex numbers and χ is a complex valued function on G. Of course χ will actually define a representation of G if and only if it is never zero and satisfies the identity

$$\chi(xy) = \chi(x)\chi(y) \text{ for all } x \text{ and } y \text{ in } G.$$

Moreover $\text{Trace}(W_x) = \chi(x)$ so that χ is in fact the character of W. Such functions on finite commutative groups were called characters by Dedekind and Weber a quarter of a century before Frobenius discovered the theory of group representations in 1896. Indeed the work of Frobenius was inspired by Dedekind's need for a generalization of the character notion suitable for non-commutative groups. It is important to observe that for each $x \in G$ and each one dimensional character, χ, $x^m = e$ for some m so $(\chi(x))^m = 1$ so $\chi(x)$ is a root of unity and $|\chi(x)| = 1$. Note also that the reciprocal of a one dimensional character is also its complex conjugate and is itself a one dimensional character and that the product of two one dimensional characters is again such. It follows that the one dimensional characters form a group $\hat{G}$ under ordinary multiplication.

Let χ be any one dimensional character of the finite commutative group G. Then it is obvious that $\sum_{x \in G} \chi(x)\chi(y) = \sum_{x \in G} \chi(xy) = \sum_{x \in G} \chi(x)$ for all y in G. Hence $(\chi(y) - 1)\left(\sum_{x \in G} \chi(x)\right) = 0$ for all y in G and it follows that $\sum_{x \in G} \chi(x) = 0$ whenever $\chi(y) \neq 1$. Thus if χ_1 and χ_2 are two *distinct* one dimensional characters so that $\chi_1 \overline{\chi_2} \neq 1$ then $\sum_{x \in G} \chi_1(x)\overline{\chi_2(x)} = 0$. In other words the one dimensional characters of G are mutually "orthogonal" functions. It follows that they are linearly independent and from what has been said above that *every* function f on G may be written *uniquely* in the form

- 555 -

GEORGE W. MACKEY   193

$$(*) \qquad f(x) = \sum_{\chi \in \hat{G}} c_\chi \chi(x)$$

where the c_χ are complex coefficients. If one multiplies both sides by $\overline{\chi_1(x)}$ where χ_1 is any irreducible character and sums over all $x \in G$ one deduces from the orthogonality that $\sum_{x \in G} f(x)\overline{\chi_1}(x) = c_{\chi_1} O(G)$ so that one has the following formula for the coefficients c_χ in $(*)$:

$$(**) \qquad c_\chi = \frac{1}{O(G)} \sum_{x \in G} f(x)\bar{\chi}(x) \ .$$

Having derived formulas $(*)$ and $(**)$ let us return to Fourier analysis and recall that one obtains an equivalent theory if one replaces the functions $\sin\frac{nx}{\lambda}$ and $\cos\frac{nx}{\lambda}$ by the complex valued functions $\cos\frac{nx}{\lambda} + i\sin\frac{nx}{\lambda} = e^{\frac{inx}{\lambda}}$ where now $n = 0,\pm1,\pm2,\dots$. In this equivalent theory one finds that any reasonably well behaved complex valued function f on the real line of period $2\pi\lambda$ can be written in the form

$$(+) \qquad f(x) = \sum_{n=-\infty}^{\infty} c_n e^{\frac{inx}{\lambda}}$$

where

$$(++) \qquad c_n = \frac{1}{\lambda} \int_0^\lambda f(x) e^{-inx} \, dx \ .$$

The analogy between $(+)$ and $(++)$ on the one hand $(*)$ and $(**)$ on the other is striking. It becomes more so when one notices the following:

a) The functions $x \to e^{\frac{inx}{\lambda}}$ may be regarded as functions on the compact commutative Lie group, which one obtains from the additive group of all real numbers under addition by identifying numbers that differ by an integral multiple of λ . In other words they may be regarded as functions on the quotient group of the additive group of the real line by the closed subgroup of all integer multiples of λ .

   b) These functions satisfy the fundamental identity $e^{\frac{in(x+y)}{\lambda}} = e^{\frac{inx}{\lambda}} e^{\frac{iny}{\lambda}}$ defining one dimensional characters. In addition they are continuous.

   c) It can be shown that every continuous function on the group in question that satisfies the one dimensional character identity is equal to $e^{\frac{inx}{\lambda}}$ for some $n = 0,\pm 1,\pm 2,\ldots$.

In other words the basic theorem about Fourier series expansions of periodic functions of a real variable is a more or less complete analogue of a general theorem about functions on finite commutative groups; the finite commutative group being replaced by a compact one dimensional connected Lie group. Actually every possible compact connected commutative Lie group can be obtained from the additive group of all n *tuples* of real numbers for some $n = 1,2,\ldots$ by factoring out a closed discrete subgroup with n generators. Functions on this group may be identified with n tuply periodic functions of n real variables and the associated multiple Fourier expansions constitute an analogue of our character expansion theorem for the general compact, connected commutative Lie group. This is what I had in mind above when I said that Peter and Weyl had recognized the essentially group theoretical character of harmonic analysis.

The word character goes back to Gauss who used the notion in an essential way in his theory of "composition" for inequivalent quadratic forms of a fixed discriminant D in 1801. These forms constitute a finite commutative group under composition and Gauss' "characters" were in fact characters of order two for this group. From Gauss' time on characters and the finite analogue of the Fourier expansion theorem played an important and increasing role in number theory. It is interesting that these two versions of harmonic analysis so important for the development of nineteenth century mathematical physics and number theory respectively do not seem to have been recognized as related until the work of Peter and Weyl. Even today this connection is not as widely appreciated as one might hope and expect.

Now as mentioned briefly above Peter and Weyl did much more than bring to light an essential kinship between two important and superficially disparate tools. In indicating the group theoretical nature of this kinship they suggested a far reaching generalization. Why not let the group be non commutative? Perhaps such an extension would also prove to be an important tool in both number theory and mathematical physics.

Actually the paper of Peter and Weyl was an outgrowth of immediately preceding work of Schur and Weyl showing that the Frobenius theory could be extended from finite groups to compact connected Lie groups and more generally to those compact groups having a connected Lie group as an open and closed normal subgroup. In 1924 I. Schur, a student of Frobenius and one of the chief early developers of the theory of group representations, pointed out that one could carry over much of the theory to the orthogonal groups. The key idea was to replace summation over a finite group by an integration process over the group manifold introduced earlier by A. Hurwitz. Hermann Weyl became interested almost immediately and wrote a series of important papers between 1924 and 1927 extending Schur's ideas to more general compact Lie groups as indicated above and determining all irreducible unitary representations for the semisimple groups.

- 558 -

Once again let G be a finite commutative group and recall the
theorem which asserts that every complex valued function f on G
may be written uniquely in the form $f(x) = \sum_i c_i \chi(x)$ where χ varies over the one dimensional characters of G . Since $\chi(xy) =$
$= \chi(x)\chi(y)$ for all x and y in G it follows that the operators
R_y of the regular representation R of G carry each χ into a
constant multiple of itself and hence that each χ generates a one
dimensional invariant subspace. Thus the "Fourier" expansion theorem
for finite commutative groups is simply another formulation of the
theorem that the regular representation of such groups is a direct
sum of all one dimensional representations (or equivalently irreducible representations) each are occurring with multiplicity one.
In this spirit one can think of Frobenius' theorem about the decomposability of the regular representation of an arbitrary (non-commutative) finite group as a non-commutative generalization of the expansion theorem for finite commutative groups and so seek to generalize this theorem to the regular representation of compact Lie
groups. This is, in effect, what Peter and Weyl did.

II.

As far as applications are concerned, in 1923 just before Schur
and Weyl began to extend the Frobenius theory to compact Lie groups,
Artin found a very striking application of non-commutative characters
to number theory — specifically he showed that one could factor the
so called Dedekind zeta function of an algebraic number field F as
product of Dirichlet series of "lower degree" each defined by an irreducible character of the group G of all automorphisms of F . More
generally he showed how to associate one of these new Dirichlet series (now called Artin L functions) with every character (irreducible
or not) of every subgroup H of G and that this assignment has the
following two fundamental properties:

- 559 -

(a) $L(s,\chi_1+\chi_2) = L(s,\chi_1)L(s,\chi_2)$ whenever χ_1 and χ_2 are characters of H ;

(b) $L(s,\chi) = L(s,\tilde{\chi})$ whenever H_1 is a subgroup of H , χ is a character of H_1 and $\tilde{\chi}$ is a character of H uniquely determined by χ and known as the character of H *induced* by χ .

If $\tilde{\chi} = m_1\chi_1 + \ldots + m_r\chi_r$ where the χ_j are irreducible characters of H then it follows from (a) and (b) that $L(s,\chi) =$

$$= L(s,\chi_1)^{m_1}L(s,\chi_2)^{m_2}L(s,\chi_r)^{m_r} \ .$$ Thus any L function for H_1 may be factored as a product of L functions defined by the *irreducible* characters of H . Now it turns out that when χ is the identity character of a subgroup H of G then $L(s,\chi)$ is just the Dedekind zeta function for the subgroup F_H of all elements of F left fixed by H . Thus (b) implies a factorisation of the Dedekind zeta function of every subfield of F the details of which can be read off from the decomposition of the character of G induced by the trivial character of the corresponding subgroup H .

To understand the importance of such a factorisation one needs to be aware of several facts:

(1) The Dedekind zeta function of an algebraic number field F is by definition the function defined by the Dirichlet series $\sum_{n=1}^{\infty} \frac{\varphi_F(n)}{n^{\sigma}}$ where $\varphi_F(n)$ is the number of ideals I such that $\frac{R_F}{I}$ has n elements in the ring R_F of all "algebraic integers" in F .

(2) The functions $s \rightarrow \sum_{n=1}^{\infty} \frac{\phi_F(n)}{n^{\sigma}}$ and $n \rightarrow \phi_F(n)$ determine one another uniquely.

(3) The product $\left(\sum_{n=1}^{\infty} \frac{\phi(n)}{n^s}\right)\left(\sum_{n=1}^{\infty} \frac{\psi(n)}{n^s}\right)$ of two Dirichlet series is again a Dirichlet series $\sum_{n=1}^{\infty} \frac{\theta(n)}{n^s}$ where the function $n \rightarrow \theta(n)$ may be computed from φ and ψ by the simple explicit formula:

$\Theta(n) = (\varphi \star \psi)(n) = \sum_{d|n} \varphi(d)\psi(\tfrac{n}{d})$ where the sum is over all positive divisors of n .

(4) The functions $n \to \varphi_F(n)$ contain information about the ideal structure of R_F which plays a central role in studying the integer solutions of equations of the form

$$a_0 x^{\ell} + a_1 x^{\ell-1} y + \ldots + a_{\ell} y^{\ell} = m$$

where $a_0, a_1, \ldots, a_{\ell}, m$ are given integers and one wants to study the way in which the number of solutions depends upon m . Here F is the field generated by the roots of the polynomial $a_0 x^{\ell} + a_1 x^{\ell-1} + \ldots + a_{\ell}$. In the special case in which $a_0 = 1$, $a_1 = a_2 = \ldots = a_{\ell-1} = 0$ the ℓ -th power reciprocity laws may be regarded as statements about the explicit computation of the function φ_F .

(5) In earlier work, begun by Weber and Hilbert and brought to a certain level of completeness by Takagi in 1920 and 1922, it had been found that whenever H is commutative ζ_F could be factored as a product of ζ_{F_H} and certain Dirichlet series defined in quite a different way and known as generalized Dirichlet L functions. The older L functions were parametrized by the one dimensional characters of a commutative group H^1 defined by the intrinsic properties of the field F_H and through isomorphic to H was not related to it in any evident intrinsic fashion.

(6) Comparing his factorization (for commutative H) with these older ones Artin was led to conjecture that a certain mapping from prime ideals in the ring of integers in F_H to H in fact defined a canonical isomorphism of H^1 and H . He went on to show that if this could be proved it would establish the identity of the two kinds of L functions and of the two factorizations. While he could only prove this conjecture in special cases at first, he was able to give a complete proof four years later in 1927.

- 561 -

GEORGE W. MACKEY   199

In carrying out the program just described Artin made two fundamental contributions to the arithmetic of number fields:

(1) By finding a natural generalization of the older L functions he made an important first step in the direction of extending the theory of Hilbert, Takagi et al to extension fields with *non-commutative* Galois groups.

(2) He introduced an important conceptual revolution in the older theory by showing that the reciprocity laws were essentially corollaries of his isomorphism theory — one now speaks of this theorem as "The Artin reciprocity law". These contributions would hardly have been possible without the new theory of group representations; i.e. *non-commutative* harmonic analysis.

Applications to physics came soon afterwards. Indeed, they began in 1927, the year in which Artin proved his general reciprocity law, Weyl published the last of his three papers on the unitary representations of compact semi-simple Lie groups and Peter and Weyl recognized the group theoretical character of harmonic analysis. This was also the year in which the new quantum mechanics, inaugurated three years earlier by the discoveries of Heisenberg and Schrödinger achieved a coherent formulation and was put into a rigorous mathematical form by von Neumann. In that year Wigner and Weyl published papers applying the theory of group representation to the new quantum mechanics in significant but quite different ways. One of the fundamental problems solved by quantum mechanics consists in providing a well defined mathematical algorithm for predicting the different ways in which a given collection of atomic nuclei and electrons can clump together to form a more or less stable "composite particle". These composite particles are known as "atoms" when there is one nucleus and as "molecules" when there are several. Thus the algorithm in question, in principle at least, reduces chemistry and atomic spectroscopy respectively to mathematics. In practice the mathematical problems involved are extremely difficult and are far from being completely solved. They may be reduced to finding the eigenvalues of certain well defined second order differential operators in a large number of variables. Wigner's discovery is that the difficulty of the eigenvalue problem may be greatly reduced by using the theory of group representations to exploit certain obvious symmetries. Consider the case of an n electron atom and let H be the self-adjoint operator in the Hilbert space H whose eigenvalues are to be determined. If only it were true that the electrons ignored one another and interacted only with the nucleus one could reduce the problem to the one electron case which can be completely solved. In the case at hand one can exploit this fact as follows. One can write $H = H_0 + J$ where H_0 is what H would be if the electrons did not interact and $J = H - H_0$. Then one can define H_ε as $H_0 + \varepsilon J$ regard the eigenvalues of H_ε as functions of ε and attempt to determine the coefficients in a power series expansion about $\varepsilon = 0$. The coefficients of ε added to the known eigenvalues of H_0 would then be plausible approximations to the eigenvalues of $H = H_0 + J$. This procedure is known as first order perturbation theory which is usually carried out with H_0 replaced by a somewhat better approximation to H which has the same symmetry properties as H_0 and whose eigenvalues can be determined by solving a somewhat more difficult one electron problem.

When λ_0 is an eigenvalue of H_0 which occurs with multiplicity 1 and φ_0 is a corresponding eigenvector then a rather simple argument shows that the corresponding first order perturbation approximation is just $\lambda_0(J(\phi_0) \cdot \phi_0)$. However, when λ_0 is a multiple eigenvalue as it very often is, the situation is more complicated. The multiplicity is usually at least partially destroyed and λ_0 has accordingly a number of different first order perturbations. To determine these one must first determine the multidimensional subspace M_{λ_0} of all λ_0 eigenvalues and form the projection J^{λ_0} of J on this subspace; that is the operator $PJ^{\lambda_0}P$

restricted to M_{λ_0} where P is the projection operator with range
M_{λ_0} . The required perturbation approximations will then be the numbers $\lambda + \mu_k$ where $\mu_1, \mu_2, \ldots$ are the eigenvalues of J^λ .

It is perhaps ironic that the very symmetries which make the
eigenvalues of H_0 easy to determine often cause them to have rather
high multiplicities. On the other hand the symmetries that H shares
with H_0 can then be exploited to simplify the diagonalization of
J^{λ_0} and it is this that Wigner discovered. Here are some of the details.

Let K^n denote the n-fold direct product with itself of the
rotation group K and let S_n denote the group of all permutations
of n objects. There are natural unitary representations U and W
of these groups in the Hilbert space H and the symmetry of H_0
finds expression in the fact that H_0 commutes with all U_α and all
W_π . It is useful to note that U and W may be blended into a
single representation UW of a larger group containing isomorphic
replicas of K^n and S^n as subgroups. This larger group which we
denote by $K^n \ominus S_n$ consists of all pairs α, π where $\alpha \in K^n$ and
$\pi \in S_n$. To define the relevant multiplication of such pairs we note
that each π in S_n is canonically associated with an automorphism
of K^n ; this is the mapping $x_1 x_2 \ldots x_n \to x_{\pi(1)}\, x_{\pi(2)} \, \cdots \, x_{\pi(n)}$.
Writing $\pi(x, \ldots, x_n) = x_{\pi(1)}, \ldots, x_{\pi(n)} = \pi(\alpha)$ where $\alpha = x, \ldots, x_n$
the multiplication rule takes the form $(\alpha_1, \pi_1)(\alpha_2, \pi_2) =$
$= \alpha_1 \pi_1(\alpha_2), \pi_1 \pi_2$. UW is then the representation $\alpha, \pi \to U_\alpha W_\pi$. To
say that H_0 commutes with all U_α and all W_π is the same as say-
ing that H_0 commutes with all $(UW)_{\alpha, \pi}$.

Quite generally let H_0 be a self-adjoint operator in a Hilbert space H and let V be a unitary representation of a group
G such that $V_x H_0 = H_0 V_x$ for all x in G . An obvious argument
shows that every eigenspace M_λ of H_0 is an invariant subspace of

the group representation V . Hence if G is compact so that V is a direct sum of irreducible representations, this decomposition may be chosen so that each irreducible subspace lies in one of the eigenspaces M_λ . Thus for every irreducible subrepresentation V^j of V there must be at least one eigenspace M_λ whose dimension is greater than or equal to the dimension of V^j .

Now the most general irreducible unitary representation of $K^n \circledS S_n$ is determined by first choosing n irreducible unitary representations of $K^n, L^1, L^2, \ldots, L^n$ and forming the subgroup G of S_n consisting of all permutations which carry the product representation $L^1 \times L^2 \times \ldots \times L^n$ into itself. Next one chooses an irreducible unitary representation M of G and goes through the "inducing" process alluded to above to construct an irreducible representation of $K^n \circledS S_n$ from a representation of $K^n \circledS G$ obtained by combining $L^1 \times L^2 \times \ldots \times L^n$ with M . The dimension of the resulting irreducible representation of $K^n \circledS S_n$ will then be $\frac{n!}{O(G)} \times \dim U \times \dim L^1 \times \ldots \times \dim L^n$. In particular, when $L^1, L^2, \ldots, L^n$ are mutually inequivalent then $O(G) = 1$, $\dim M = 1$ and the dimension is $\dim L^1 \times \dim L^2 \times \ldots \times \dim L^n \times n!$. The number of electrons does not have to be very great before this dimension is large indeed.

Now these high dimensional eigenspaces M_{λ_0} in which the J^{λ_0} act and which are invariant subspaces for our representation UW of $K^n \circledS S_n$ are of course also invariant subspaces for the restriction of UW to any subgroup of $K^n \circledS S_n$. This is so in particular for the subgroup $\tilde{K} \times S_n$ where $\tilde{K}$ is the so called diagonal subgroup of K^n consisting of all $x_1, x_2, \ldots, x_n$ with $x_1 = x_2 = \ldots = x_n$. Of course $\tilde{K}$ is isomorphic to K and is elementwise invariant under the automorphisms of K^n defined by the elements of

S_n . The significance of this particular subgroup is that the operator H commutes with all operators in the restriction to it of UW .
From this it is easily shown that J^{λ_0} commutes with the operators
of this representation of $\tilde{K} \times S_n$. We must now explain how this fact
can be exploited to simplify the problem of finding the eigenvalues
of J^{λ_0} .

Quite generally let J^{λ_0} be any self-adjoint operator defined
on a finite dimensional Hilbert space H_{λ_0} and let V be a unitary
representation of some compact Lie group (not necessarily connected)
in this Hilbert space such that $V_x J^{\lambda_0} = J^{\lambda_0} V_x$ for all x in K .
For each irreducible character χ of K consider the operator P_χ
defined by the formula

$$P_\chi = \chi(e) \int_K \chi(x) V_x \, d\mu(x)$$

where the integration over K refers to the Hurwitz integration
process used three years earlier by Schur to extend the Frobenius
theory to the orthogonal groups. The P_χ then can be proved to be
projection operators whose ranges M_χ are mutually orthogonal *closed*
subspaces whose linear span is

$$H_{\lambda_0} = M_{\chi_1} \oplus M_{\chi_2} \oplus M_{\chi_r}$$

Here $\chi_1, \chi_2, \dots$ consist of all irreducible characters of K for
which $P_\chi \neq 0$. The subspaces M_{χ_j} are invariant both under J^{λ_0} and
under the V_x . In particular we have reduced the problem of diagonal-
izing J^{λ_0} into the easier problems of diagonalizing the restrictions
of J^{λ_0} to the subspaces M_{χ_j} . From another point of view, recalling
that diagonalizing an operator means decomposing it as a direct sum of

- 566 -

constant operators, producing the M_{χ_j} amounts to a partial diagonalization. This however is *not* the whole story. One can also show
that in each invariant subspace M_{χ_j}, V is a direct sum of replicas
of the unique irreducible representation L^{χ_j} whose character is
χ_j. Moreover the eigenspaces of J^{λ_0} in M_{χ_j} are invariant under
the V_χ. Hence each eigenspace has a dimension which is an integral
multiple of $\chi_j(e)$ the dimension of the irreducible representation
L^{χ_i}. Moreover determining these multiple eigenvalues can be reduced
to the problem of diagonalizing an operator in a space whose dimension is a factor of the dimension of M_{χ_j}; that is the multiplicity
with which L_{χ_j} occurs in the reduction of V into irreducibles.
If in particular V happens to be free of multiple occurrences of
irreducibles then there is *no* more work to do. The reduction of
J^{χ_0} to each M_{χ_j} is already a constant operator. In the general
case the problem of finding the eigenvalues of J^{λ_0} is reduced to
a number of eigenvalue problems in which the maximum vector space
dimension which occurs is the maximal multiplicity with which any
irreducible subrepresentation of V occurs in the decomposition of
V. Since $\dim H = \sum_j m_j \chi_j(e)$ where m_j is the multiplicity with respect to which L^{χ_j} occurs and $\chi_j(e)$ is the dimension of the space
of L^{χ_j} the m_j -s are in general much smaller than $\dim H$.

Wigner's original paper of 1927 dealt only with a special case; the
general case was worked out later, partly in collaboration with von Neumann. Moreover this application to the analysis of the possible composite particles that can be formed from a nucleus and a fixed finite
number of electrons and hence to the theory of atomic spectra is only

the beginning. The method can be adapted to apply to molecules, to
the theory of the formation of nuclei out of protons and neutrons
etc. and is also capable of extensive refinement.

It is interesting to note a rather close parallel between this
application of non-commutative harmonic analysis to quantum physics
and Fourier's pioneering use of commutative harmonic analysis in
classical physics over a century earlier. In essence Fourier used
the fact that the Laplace operator $\frac{\partial^2}{\partial x^2} + \frac{\partial^2}{\partial y^2} + \frac{\partial^2}{\partial z^2}$ commutes with
the regular representation of the translation group in space to
diagonalize the former and so trivialize the solution of the equation of heat conduction.

The applications of group representations to quantum mechanics
made by Hermann Weyl at the same time are of a rather different nature and are more concerned with the conceptual foundations of quantum mechanics than with the solution of specific problems. They are
discussed in full in a paper I presented at the end of June at a
conference in Kiel honoring the 100th birthday of Weyl. (Haar and
Weyl were both born in 1885.) That paper will be published in the
proceedings of that conference.

III.

The foregoing should convince the reader that the extension of
harmonic analysis made possible by Frobenius' discovery of group representations in 1896 and the work of Schur, Peter and Weyl between
1924 and 1927 has applications as deep and extensive as those of its
classical commutative version. However up to this point the underlying group has always been either a commutative Lie group or a compact
group whose connected component is a Lie group. Only in these cases
is the Hurwitz integration process or the classical Lebesgue integral
(in one or more dimensions) available to replace the "summing over

the group" so essential to the development of the Frobenius theory.
Thus it was a major advance in 1933 when Alfred Haar proved the following theorem. If G is any locally compact topological group which
satisfies the second axiom of countability then there exists a countably additive measure μ defined on a σ field of sets which includes all open and closed sets, which is finite on compact set and
positive on open sets and which has the fundamental property of invariance under right translation: $\mu(Ex) = \mu(E)$ for all sets E in
the σ field of definition and all x in G . (Of course the measure
$E \rightarrow \mu(E^{-1})$ will be a measure with similar properties but left invariant instead of right invariant. However it is only for the so called
unimodular groups that the same measure can be *both* right and left invariant.) The relatively easy uniqueness theorem: that any two Haar
measures for the same group must be constant multiples of one another
was found by von Neumann three years later.

Haar's remarkable discovery (its proof was complicated and far
from obvious) opened the door to extending Frobenius' theory to a much
wider class of groups unrestricted by requirements of either differen-
tiability or compactness and correspondingly extending the possibilities for applying the already powerful tool of harmonic analysis.
Haar himself pointed out that his theorem made possible an almost immediate extension of the results of Peter and Weyl from compact Lie
groups to arbitrary compact groups (having the second axiom of countability) and this last restriction was soon removed by the work of
André Weil and others. Moreover a year or so later Pontrjagin and van
Kampen showed how the old character theory for finite *commutative*
groups could be extended to their beautiful duality theory. Let G
be any locally compact commutative group and let $\hat{G}$ be the set of
all "continuous unitary characters" of G ; that is the set of all
continuous homomorphisms of G into the group of all complex numbers
of modules one. Then $\hat{G}$ itself is a group under pointwise multiplication and it becomes a topological group if one defines a subset O

- 569 -

GEORGE W. MACKEY   207

of $\hat{G}$ to be open if whenever $\chi_0 \in 0$ there exists an $\varepsilon > 0$ and a compact subset K of G such that $|\chi(x) - \chi_0(x)| < \varepsilon$ for all $x \in K$ implies that $\chi \in 0$. With this topology $\hat{G}$ may be shown to be locally compact. It is called the locally compact commutative group dual to G ; or simply the dual of G . Consider now $\hat{\hat{G}}$ the dual of $\hat{G}$. One sees at once that there is a natural homomorphism G into $\hat{\hat{G}}$. Indeed, if $x \in G$ then $\chi(x)$ for variable $\chi \in \hat{G}$ may be considered as a function $f_x(\chi)$ defined on $\hat{G}$: $f_x(\chi) = \chi(x)$. It is obvious that $x \to f_x$ is a homomorphism. One can prove further that it is one-to-one bicontinuous and has the whole of $\hat{\hat{G}}$ for its range. In other words it sets up a canonical isomorphism between G and $\hat{\hat{G}}$ and shows that the relationship between G and $\hat{G}$ is reciprocal; G is just as much the dual of $\hat{G}$ as $\hat{G}$ is the dual of G . From the point of view of the Frobenius theory $\hat{G}$ is precisely the set of all irreducible unitary representations of G and it follows from the Peter Weyl theorem that $\hat{G}$ must be discrete and countable whenever G is compact and has the second countability property. The converse is true and quite generally (without the second countability property) G is compact if and only if $\hat{G}$ is discrete.

The Pontrjagin—van Kampen extension of harmonic analysis to arbitrary locally compact commutative groups found an important application to number theory almost immediately. In 1936 Chevalley introduced substantial simplifications into the study of the finite abelian extensions of an algebraic number field F by lumping them all together into one infinite extension field $\tilde{F}$. The Galois group $G_{\tilde{F}}$ of all automorphisms of $\tilde{F}$ leaving the elements of F fixed is then in a natural way a projective limit of its finite quotient groups and as such a totally disconnected infinite compact commutative group. In the Artin reciprocity law described earlier the Galois groups of the individual extensions were canonically isomorphic to certain other (rather awkward to define) finite groups. Chevalley lumped these together too as the finite quotient groups of a new locally compact

- 570 -

group introduced by him and called the idèle class group. There is
an idèle class group canonically associated with every F. It is
defined without any reference to $\tilde{F}$ as a quotient I_0 of a subgroup I of the product of the multiplicative groups of the members
of an infinite family of "completions" of F. These are locally compact fields containing a replica of F as a dense subfield. There
is one of these for each prime ideal $\mathfrak{p}$ in the ring of integers of
F and a finite number of others. The former are finite extensions
of the field of p-adic numbers for various natural primes p and
the latter replicas of either the real or complex number field. The
replicas of the multiplicative group F^* of F in the various completions define an isomorphism of F^* with a closed subgroup of I
and this is the subgroup I_0. As reformulated by Chevalley the Artin
reciprocity law asserts a canonical isomorphism between G_F and the
quotient of I/I_0 by its connected component. Chevalley's approach
made it possible to replace a number of difficult analytical arguments in the work of Takagi and Artin by simple topological considerations.

This work of Chevalley showed in particular that harmonic analysis could have significant applications which involved groups that
were totally disconnected but not discrete. Later (in the 1950's)
Chevalley's ideas were extended to non-commutative algebraic groups
of Ono and Tamagawa and used for important applications to number
theory.

The systematic theory of group representations and harmonic
analysis for locally compact groups which are either compact or commutative made possible by the work of Peter, Weyl, Haar, Pontrjagin,
and van Kampen was expounded in two very influential books published
in 1938. These were "L'integration dans les groupes topologiques et
ses applications à l'analyse" by André Weil and "Topological groups"
by L. Pontrjagin. The latter was originally published in Russian but
an English translation appeared in 1939. Weil's book put particular

- 571 -

GEORGE W. MACKEY   209

emphasis on the connections with classical Fourier analysis and the fact that one has a "Fourier transform" and a "Plancherel formula" for the square summable functions (with respect to the measure of Haar) defined on any locally compact commutative group G and its dual $\hat{G}$. This Fourier transform $f \to \hat{f}$ (where $\hat{f}(\chi) = \int f(x)\chi(x)d\mu(x)$ for all f in a dense subset of $L^2(G)$) sets up a canonical unitary map of $L^2(G)$ on $L^2(\hat{G})$.

It was of course natural to attempt to extend the theory expounded in these two books to *all* locally compact groups including those which are neither compact nor commutative and to hope that such an expanded theory would have equally impressive applications to number theory and physics. Two serious difficulties stood in the way of any immediate implementation of this program. One is that many of the most interesting locally compact groups that are neither compact nor commutative have no finite dimension irreducible unitary representations except the one dimensional identity and most of the others have far "too few". Thus one must learn somehow to cope with an extension of the Frobenius theory in which one has infinite dimensional irreducible representations. The other difficulty is that practically all (if not all) non-compact locally compact groups admit unitary representations with *no* irreducible subrepresentations. A fortiori such representations cannot be decomposed as finite or infinite direct sums of irreducible representations and something new is needed if one is to reduce the study of general representations to the study of irreducible ones as is done in the theory of Frobenius — indeed if one is to carry out harmonic analysis at all where such groups are involved.

The key to the resolution of the second difficulty was found by Hilbert in a different context between 1904 and 1910. His celebrated spectral theorem for bounded self-adjoint operators in Hilbert space deals with a precisely analogous problem in which the difficulty is the failure of a self-adjoint operator to have eigen-

- 572 -

vectors in the general case. He resolved the difficulty by introducing measure theoretic consideration and, in effect replacing direct sums by "direct integrals". Actually the operator problem is completely equivalent to a very special case of the group representation problem because of the fact that for every bounded self-adjoint operator H , the mapping $t \to e^{itH}$ is a unitary representation of the additive group of the real line. In 1927 Hermann Weyl (in connection with his work on applications of group representations to quantum mechanics) conjectured that this correspondence could be extended to a one-to-one correspondence between *all* (continuous) unitary representations of the additive group of the real line and *all* self-adjoint operators by allowing the self-adjoint operators to be unbounded. This conjecture was proved by M. H. Stone in 1931 who made it corollary of an explicit theorem about the decomposition of unitary representations of the additive group of the real line — an analogue of the spectral theorem applying to one parameter groups of unitary operators rather than single self-adjoint operators. Once the Pontrjagin — van Kampen theory became available it became obvious how to formulate a generalization of Stone's theorem that applied to all locally compact commutative groups. One had only to have both things in mind at the same time. Nevertheless, no one formulated and proved such a generalization until 1944 when three people (Ambrose, Godement and Naimark) did so independently and more or less simultaneously. There was also an early (and at the time unrecognized) contribution to the problem of determining all the irreducible unitary representations of a locally compact group when some of them are infinite dimensional. In the 1927 paper of Weyl just alluded to there was another conjecture inspired by the needs of quantum mechanics. This stated the uniqueness of the solutions of the Heisenberg commutation relations in an integrated form given them by Weyl (which assumed the truth of the first conjecture). Stone's paper of 1931 actually announced proofs of both conjectures and a detailed proof of the second

- 573 -

was published soon after by von Neumann. It was later recognized
that this work implicitly contained a complete determination of all
unitary irreducible representations of a certain non-commutative,
non-compact nilpotent Lie group — now called the Heisenberg group.

While Hilbert's spectral theorem may be regarded as having provided the main clue for resolving the first difficulty problems arise
when one attempts to deal with the two difficulties simultaneously
and pass from "direct integrals" of the one dimensional irreducible
representation of commutative groups (where one can use integration
over $\hat{G}$) to direct integrals of infinite dimensional irreducible
representations. The necessary general theory of direct integral decompositions of Hilbert spaces was worked out by von Neumann and
existed in typescript form as early as 1938. However it was neither
published nor known to interested mathematicians until 1950. This
theory was developed in connection with the theory of "Rings of
operators" published by von Neumann and Murray between 1936 and 1944
in a celebrated series of four papers. It turned out to be highly
relevant to the decomposition theory of group representations. Indeed,
if $x \to U_x$ is any unitary group representation and $R(U,U)$ is the
set of all bounded operators which commute with all U_x then $R(U,U)$
is a "ring of operators" (von Neumann algebra in present day terminology) in the von Neumann—Murray sense. Moreover every von Neumann
algebra arises in this way. The discovery by von Neumann and Murray
of "non-type I factors" (von Neumann algebras with one dimensional
centers whcih are *not* isomorphic to the algebra of all bounded operators in some Hilbert space) turned out later to be equivalent to the
existence of a rather intractable anomaly in the decomposition theory
for the unitary representations of certain locally compact groups
(now called non-type I groups).

In addition to the work just described there were two quite
important direct contributions to the program of carrying the theory
of unitary group representations over to non-compact non-commutative

groups which were published before the end of World War II. The first
of these was a celebrated paper of Wigner published in 1939 and describing all "physically significant" irreducible unitary representations of the group generated by the Lorentz group and the four dimensional group of all translations in space time — the so called Poincaré group. The other was published in 1943 by Gelfand and Raikov and
contains a proof (based on positive definite functions and the theory
of convex sets) that every locally compact group has "sufficiently
many" irreducible unitary representations if one includes infinite
dimensional representations.

All this to the contrary notwithstanding it seems fair to say
that a systematic extension of harmonic analysis to groups which are
neither compact nor commutative started only in 1947 when four long
papers on the subject appeared. These were followed by an increasing
stream which has not yet shown serious signs of abatement. We shall
not attempt to follow the historical development beyond this point
and will conclude this paper with a section describing the general
nature of the harmonic analysis which is now possible; putting our
main emphasis on the important role played by measure theory in general and invariant measures in particular.

IV.

Let G be a group and let S be a so called G space; that
is a set on which G acts as a group of bijections or one to one
onto transformations. If we let [s]x denote the transform of
s ∈ S by x ∈ G then the G space structure of S is completely
described by the two properties

(i) [[s]x]y = [s](xy)

(ii) ([s]e) = s where e is the identity element of G .

In other words we may define a G space to be a set S together with a mapping $s,x \to [s]x$ of $S \times G$ into S such that (i) and (ii) hold for all $s \in S$ and all $x \in G$. Such pairs G, S consisting of a group G and a G space S are to be found everywhere in mathematics and physics — typically when S has a mathematical structure admitting non-trivial automorphisms (symmetries) and G is the group of all of these automorphisms or some significant subgroup.

Given a group G and a G space S let F denote the set of all complex valued functions on S. Then if we define $[f]x = g$ where $g(s) = f([s]x^{-1})$ F itself becomes a G space but a rather special kind in that it also has the structure of a vector space and in that the mappings $f \to [f]x$ are automorphisms of that vector space structure; i.e. non-singular linear transformations. It follows that if we define $U_x(f) = [f]x = g$ where $g(s) = f([s]x)$ then each U_x is a linear transformation and the mapping $x \to U_x$ is a (not necessarily unitary) representation of G. In most cases the space F of *all* functions on S is a little too large and unwieldy to be interesting and one replaces it by some subspace F^0 invariant under all the U_x. When S is countable a convenient and useful choice for F^0 is the Hilbert space of all complex valued functions f on S such that $\sum_{s \in S} |f(s)|^2 < \infty$. In that case the representation $x \to U_x$ is unitary (that is all operators U_x are unitary operators). This is important because unitary representations are susceptible to a much more complete analysis than non-unitary representations. When S is not countable there is often available a natural generalization of F^0; namely the Hilbert space $L^2(S,\mu)$ where μ is a measure defined on the Borel subsets of S with respect to some natural topological structure and μ is invariant under the action of G. Here again the representation $x \to U_x$ will be unitary and the countable case is of course just the special case in which the measure of every set is the number of points in it.

- 576 -

In the philosophy of the author "harmonic analysis" as a universal tool in mathematics is best considered not as the study of the most general possible group representation or even the most general possible unitary group representation but as the study of those particular unitary group representations which arise from G spaces S in the manner just indicated (and from certain generalizations to be described below). The most straightforward case is that in which $G = S$ and the action of G on S is right multiplication. In that case Haar's theorem of 1933 tells us that an essentially unique μ exists whenever G is locally compact and then U is the so called "regular representation" of G. As already stressed above, the classical theory of Fourier series is just the decomposition of the regular representation of G in the special case in which G is the group of rotations in the plane. In addition the classical theory of the Fourier transform can (less straightforwardly) be considered as the (direct integral) decomposition of the regular representation of the additive group of the real line. As a very early example in which $S \neq G$ consider the case in which G is the group of rotations about a fixed origin in three dimensional space and S is the surface of the sphere of radius one with center at the origin. In this example G has just one irreducible unitary representation of every odd dimension and U turns out to be the direct sum of all of these each occurring just once. Accordingly $L^2(S,\mu)$ (μ is of course the area measure on the sphere) has a decomposition as a direct sum $H_0 \oplus H_1 \oplus \oplus H_2 \ldots$ where H_j is $2j+1$ dimensional. The functions in these spaces H_j can be explicitly described in elementary terms; they are just the restrictions to the unit sphere of those polynomials P in three variables which are homogeneous of degree j and satisfy Laplace's equation $\dfrac{\partial^2 P}{\partial x^2} + \dfrac{\partial^2 P}{\partial y^2} + \dfrac{\partial^2 P}{\partial z^2} = 0$. They are called surface harmonics. The fact that a more or less general real valued function on S can be expanded in an infinite series of surface harmonics

- 577 -

GEORGE W. MACKEY   215

was known to Laplace and Legendre and used by them in the 1780's in their studies of the gravitational attraction of non-spherical masses. This happened over twenty years before Fourier's work.

The example of the rotation group acting on the sphere shares with the examples in which $S = G$ the property that the action of G on S is transitive; that is that for each pair s_1, s_2 of elements of S there exists an element x of G such that $[s_1]x = s_2$. There are many other examples of transitive G spaces in which $S \neq G$ and the transitive G spaces are, in a sense to be explained below, among the most important. Let us look at them systematically. Let G be any group and let S be any *transitive* G space. Choose an "origin" s_0 in S and let H_{s_0} denote the set of all x in G such that $[s_0]x = s_0$. Evidently H_{s_0} is a subgroup of G. Moreover given H_{s_0} we may reconstruct the G space S. Indeed the mapping $x \to s_0 x$ takes G into all of S and is such that x and y are mapped into the same element of S if and only if $s_0 x y^{-1} = s_0$; that is if and only *if* $xy^{-1} \in H_{s_0}$. In other words $x \to s_0 x$ defines a one-to-one mapping of the space G/H_{s_0} of all right H_{s_0} cosets $H_{s_0} x$ onto S.

In this mapping the given action of G in S is reflected in a very simple way. If $s \sim H_{s_0} y$ then $[s]x \sim H_{s_0} yx$. In fact, if H is *any* subgroup of G then the space G/H of all right H cosets Hx becomes a transitive G space if we define $(Hx)y = H(xy)$ and the subgroup leaving the coset H fixed is just H. The argument just given shows that every transitive G space S is "isomorphic" (in an obvious sense) to the coset space G/H for some subgroup H. As is easy to see G/H_1 and G/H_2 are isomorphic G spaces if and only if H_1 and H_2 are conjugate subgroups of G. In particular replacing s_0 by another "origin" s_1 simply replaces H_{s_0} by the conjugate subgroup H_{s_1} .

Having seen that transitive G spaces are precisely the coset spaces G/H and remembering that when $H = \{e\}$ so that $S = G/H = G$ Haar's theorem (and its extensions) provide us with an essentially unique invariant measure in S whenever G is locally compact, it is natural to inquire into the existence and uniqueness of G invariant measures on G/H when G is locally compact and H is a subgroup of G other than the trivial subgroup $\{e\}$. For simplicity we shall confine ourselves to the case in which G also has the second countability property. In that case weak hypotheses about an assumed Borel structure in S imply that H must be closed. Moreover it is rather easy to prove that if G/H admits an invariant measure this measure is unique up to a multiplicative constant and that it admits one when both H and G are unimodular and more generally when they both fail to be unimodular in "a mutually compatible" manner. While it is easy to find examples of closed subgroups H of second countable locally compact groups G such that G/H has no non-trivial invariant measures this circumstance is less limiting than it might seem. There is a slight generalization of the notion of an invariant measure that allows one to do harmonic analysis on G/H for *any* closed subgroup H of G . Let μ be Haar measure on G and let f be any positive real valued Borel function on G such that $\int f d\mu < \infty$. Let $v(E) = \int_E f d\mu$. Then v is a measure on G having the same sets of measure zero as μ does. Let Φ denote the mapping that takes each element x of G into the coset Hx in G/H . We may then use Φ to define a measure $\tilde{v}$ in G/H by setting $\tilde{v}(E) = \tilde{v}(\Phi^{-1}(E))$. The measure $\tilde{v}$ in G/H is not in general invariant but it is "quasi invariant" in the sense that for every group element x the measures $E \to \tilde{v}(E)$ and $E \to \tilde{v}(Ex)$ have the same sets of measure zero. This implies the existence of a "density"

- 579 -

GEORGE W. MACKEY   217

s $\to$ p(x,s) the "Radon−Nikodym derivative" of E $\to$ $\tilde{\nu}$(Ex) with respect to E $\to$ $\tilde{\nu}$(E) the integration of which converts one measure
into the other. Using this one can compensate for the non-invariance
of $\tilde{\nu}$ and define an analogue of the representation U . In precise
terms one defines $U_X(g)(s)$ for each g in the Hilbert space
$L^2(G/H,\tilde{\nu})$ by the formula

$$U_X(g)(s) = \sqrt{p(x,s)}\, g\,([s]x)$$

and verifies that each U_X is a unitary operator and that $x \to U_X$
defines a countinuous unitary representation of G . This definition
may seem to suffer from the fact that it depends upon choosing an
arbitrary positive integrable function f in order to define ν
and $\tilde{\nu}$. However, if f_1 is a second such function and ν_1 and $\tilde{\nu}_1$
are the associated quasi-invariant measures in G and G/H one
sees at once that $\tilde{\nu}$ and $\tilde{\nu}_1$ have the same null sets and hence
that there exists a positive measurable function $p_{0,1}$ on G/H
such that $\tilde{\nu}_1$ may be obtained from $\tilde{\nu}$ by integrating ρ . The

linear operator $g \to \sqrt{\rho}\, g$ is then more or less obviously a unitary

map of $L^2(G/H,\nu)$ onto $L^2(G/H,\nu_1)$ which sets up an equivalence
between unitary representations of G defined by the above formula
and the two quasi-invariant measures ν and ν_1 . Thus to within
equivalence we have a unitary representation U of G canonically
associated to each coset space G/H where H is a closed subgroup
of G and this representation coincides with that defined above
whenever G/H admits an invariant measure.

The above considerations about quasi-invariant measures in coset spaces G/H suggest the first of the two mild generalizations
alluded to above in our "definition" of harmonic analysis. We need
not confine ourselves to measures μ in the G space S which are
invariant under G . It suffices that they be quasi invariant. At

- 580 -

the same time one need not distinguish between two different quasi-
invariant measures so long as they have the same sets of measure
zero. One may think of being given not a G invariant measure in
the G space S but an invariant *measure class* where by a measure
class C we mean the set of *all* measures having the same sets of
measure zero as any one of its members. As an example of a "natural"
occurrence of an invariant measure class other than that provided by
a coset space G/H consider an arbitrary smooth manifold M satisfying the second axiom of countability. Since M is locally diffeo-
morphic to the interior of the unit ball in E^n for some n , n dimensional Lebesgue measure provides a measure locally whose *class*
is independent of the local diffeomorphism. Blending these one obtains
a well defined canonical measure class for the whole manifold. Evidently every diffeomorphism of M with itself leaves this measure
class invariant. Thus if G is any group of self diffeomorphisms of
M then M will constitute a G space which is supplied with an invariant measure class. Having introduced the notion of measure class
we pause to remark that both left and right Haar measures belong to
the same measure class and this class is both right and left invariant. Conversely once one knows that a group G admits a right (or
left) invariant measure class it is rather easy to deduce that this
class contains a right (left) invariant measure. Thus the essence of
Haar's theorem can be elegantly reformulated as follows: Every locally
compact group G admits one and only one measure class which is both
right and left invariant.

Let us return now to our general framework for harmonic analysis.
We are given a locally compact group G satisfying the second axiom
of countability, a G space S and an invariant measure class C in
S . Choosing a member μ of C we form the Hilbert space $L^2(S,\mu)$
and define a unitary representation U of G by setting $U_x(g)(s) =$
$= \sqrt{\rho(\chi,s)} \, g([s]x)$ where ρ is the appropriate Radon — Nikodym deriva-

tive. The basic problem of harmonic analysis is to decompose the representation U into its irreducible components. We propose now to discuss a characteristic property of those particular representations U of G which arise in this way from a G space S with an invariant measure class C. For each measurable subset E of S let P_E denote the operator in $L^2(S,\mu)$ which maps g into $\phi_E g$ where $\phi_E(s) = 1$ for $s \in E$ and $\phi_E(s) = 0$ for $s \notin E$. Then P_E is self-adjoint and idempotent, a "projection" whose range is the closed subspace of $L^2(S,\mu)$ consisting of those members g which are zero outside of E. Let B denote the range of the mapping $E \to P_E$ that is the set of all projections which are of the form P_E for some measurable E in S. Then B is a so called "complete Boolean algebra of projections" in the sense that it has the following properties

(a) If $P_1 \in B$ then $I - P_1 \in B$ where I is the identity operator.

(b) If $P_1 , P_2 \in B$ then $P_1 P_2 = P_2 P_1 \in B$.

(c) If $P_1, P_2, P_3, \ldots$ are all in B and $P_i P_j = 0$ for $i \neq j$ then $P_1 + P_2 + P_3 + \ldots \in B$. Moreover it is obvious that B commutes with U in the sense that for all x in G $U_x B U_x^{-1} = B$.

It turns out that given only U as an abstract unitary representation of G and the complete Boolean algebra of projections B one can reconstruct the G space S and the invariant measure class C. This suggests that the existence of a U invariant complete Boolean algebra of projection might actually characterize those representations of G which occur in harmonic analysis as we have, so far, defined it. In other words it suggests that one might hope to prove a converse stating that whenever U is a unitary representation of G which admits a complete Boolean algebra B of projections in its Hilbert space such that $U_x B U_x^{-1} = B$ for all x in G then the pair U, B

arises from some G space action as above. Such a converse is *not*
in fact true but it becomes true if we generalize our definition of
harmonic analysis in a subtle but quite natural way.

In order to understand the nature of this generalization as well
as our need for it, it will be useful to concentrate our attention
for a bit on the special case in which B is an "atomic" complete
Boolean algebra of projections. Let U be any unitary representation
of our group G in a Hilbert space $H(U)$ and let $H(U) = H_1 \oplus H_2 \oplus \ldots$
be any direct sum decomposition of $H(U)$ as a direct sum of orthogonal subspaces. Then for each finite or infinite subset $\{i_1 < i_2 < \ldots\} =$
$= E$ of 1,2,... let P_E be the projection whose range is the
closed subspace $H_{i_1} \oplus H_{i_2} \oplus \ldots$. The set of all P_E is then a complete Boolean algebra of projections which is "atomic" in the sense
that every element is a sum of "atoms" i.e. elements which have no
properly smaller subelements except 0 . Conversely every atomic B
can be so obtained from some direct sum decomposition of $H(U)$.

We observe next that the existence of a U invariant complete
Boolean algebra of projections B *which is also atomic* has more or
less obvious important implications about the structure of the representation U of G . Indeed if $H(U) = H_1 \oplus H_2 \oplus \ldots$ is the decomposition of $H(U)$ which defines B then to say that $B = U_x B U_x^{-1}$
is clearly the same as to say that U_x maps each H_j onto some
H_{j^1} where j^1 depends only on x and j . In this way G acts
as a group of permutations of the H_j . Let us look at this permutation group. For each H_j let us form the direct sum of all the H_{j^1}
which are of the form $U_x(H_j)$ for some x and call it $\tilde{H}_j$. It is
clear that $\tilde{H}_j$ and $\tilde{H}_k$ are either orthogonal or identical and that
each of them is carried into itself by all U_x . Thus whenever G
does not permute the H_j *transitively* we have a direct sum decomposition of U which is also (in an obvious sense) a direct sum decomposition of B . Moreover in each $\tilde{H}_j$ the action of G on the H_k

- 583 -

inside it *is* transitive. It follows that we may concentrate our attention on the case in which the action of G on the H_j is transitive — since the general case may be built up from this one by a
transparent construction.

Assuming transitivity then let H_j denote the subgroup of G
consisting of all $x \in G$ such that $U_x(H_j) = H_j$. Then restricting
h to lie in H_j and restricting each U_h to H_j we obtain a unitary representation L^j of the subgroup H_j . Now of course to know
one of these representations L^j is to know them all. The subgroups
H_j are not only isomorphic but conjugate and the isomorphism of
H_j with H_k set up by any x such that $xH_jx^{-1} = H_k$ carry L^j
into a representation of H_k equivalent to L^k . Choosing a particular j , say j = 1 , for convenience we have a unitary representation L^1 of the subgroup H_1 canonically associated with the pair
U , B . It is easy to see moreover that given the subgroup H_1 and
the representation L^1 the pair U , B can be reconstructed.

Suppose now we look at those pairs U , B which arise as above
from a G space with an invariant measure class C . B will be
atomic if and only if S is countable and discrete and in that case
the H_j will be parametrized by the points of S . However, it is
easy to see that all of the H_j will be *one dimensional* and that in
each transitive component the defining representation L will be the
one-dimensional identity representation. Clearly not every pair U , B
can arise from harmonic analysis as defined so far. The fact that
H_j are all one dimensional is a consequence of the fact that we considered only complex valued functions on S and that the complex
numbers are a one-dimensional vector space. If we replace $L^2(s,\mu)$
by $L^2(s,H_0,\mu)$ where H_0 is an arbitrary fixed Hilbert space and
do everything else as before we will find that the H_j all have the
dimension of H_s . However, replacing complex valued functions by

- 584 -

vector valued functions does not go clearly far enough. Although the representations L will no longer be one dimensional they will still be trivial; we will still have $L_h = I$ for all $h \in H$.

The generalization that "works" is quite simple and natural but perhaps not obvious. We must not only use vector valued functions but must permit the vector space in which they lie to vary from point to point. In other words we must replace vector valued functions by what the topologists and differential geometers call the "cross sections of a vector bundle". Consider for example a vector field on the surface S of a sphere in three space; that is an assignment to each point p of a vector $f(p)$ lying in the tangent plane T_p to S at p . Each T_p is a two-dimensional real vector space and so T_p and T_q are isomorphic for all p and q . However, there is no *canonical* isomorphism of one of these vector spaces with the other. There is no meaning in saying that $v \in T_p$ and $w \in T_q$ are the "same" vector. One calls the set of all pairs p , v with $v \in T_p$, $p \in S$ a "vector bundle" with base S and fibres T_p and one calls each vector field on S ; i.e. each function f defined on S such that $f(p) \in T_p$ for all $p \in S$ a "cross section" of this bundle.

Guided by the above we now drop our atomicity assumption and formulate our second and final generalization of our definition of harmonic analysis. We begin with a second countable locally compact group G and a G space S with an invariant measure class C as before. We suppose further that we are given a vector bundle W with base S whose fibers H_s are Hilbert spaces and that G acts not only on S but on the bundle W and that each x defines a so called "bundle automorphism". By the latter we mean that each $x \in G$ maps the subset of all s , v with $v \in H_s$ onto the subset of all $[s]x,w$ with $w \in H_{[s]x}$ in such a way that the associated map $W(x,s)$ of H_s on $H_{[s]x}$ is unitary. If f is any cross section of this Hilbert bundle then $\|f(s)\|^2$ is a well defined real valued function on S which one may integrate with respect to a measure

- 585 -

in C and thus speak of "square integrable cross sections" just as
we spoke of square integrable functions earlier. Similarly we may
translate "cross sections" and define a generalization of the unitary
representation U . If f is a square integrable cross section then
$U_x(f)$ is defined to be the cross section g defined by the equation
$g(s) = \sqrt{\rho(\chi,s)}\ W(\chi,s)\ f([s]\chi)$.

     We have omitted various mild regularity hypotheses and other
technical details but believe we have described the essential ideas
in generalizing our earlier definition of harmonic analysis to one
in which square integrable complex valued functions on a G space
S are replaced by square integrable cross sections of a G invariant "Hilbert bundle" whose base is S . The projection operators
P_E (defined on the base S of the bundle) may be introduced just
as before and the set B of all of them is a complete Boolean algebra of projections which is U invariant. In this more general
setting harmonic analysis is still the decomposition of a certain unitary group representation U and this unitary group representation
is special in that its space admits a complete Boolean algebra of
projections B such that $U_x B U_x^{-1} = B$ for all x in G . Now however there is a converse. One can prove the following

THEOREM. *Let* U *be a unitary representation of the second
countable locally compact group* G . *Suppose that there exists a complete Boolean algebra* B *of projections in the Hilbert space* H(U)
of U *such that* $U_x B U_x^{-1} = B$ *for all* x ∈ G . *Then there exists a*
G *space* S *with an invariant measure class* C *and a* G *invariant
Hilbert bundle* W *with base* S *such that if* U^1 *and* B^1 *are the
unitary representation of* G *and the complete Boolean algebra of
projection associated with* S , C *and* W *as indicated above there
exists a unitary map* V *of* $H(U^1)$ *on* H(U) *such that* $VU^1_x V^{-1} = U_x$
for all x *in* G *and* $VB^1V^{-1} = B$.

The proof of this theorem is quite straightforward in the special case in which B is atomic and is more or less implicit in the motivating analysis of that case given above. The proof in the general case involves a lot of measure theory and is far from trivial.

In the special case in which B is atomic the condition that $U_x B U_x^{-1} = B$ can be given a simple interpretation which is suggestive enough to be worth discussing. Let P_s be the atoms in B where $s \in S$ and S is some indexing set. Then every $P \in B$ is uniquely of the form $\sum_{s \in E} P_s$ where E is some subset of the countable set S. Moreover if M_s is the closed subspace of $H(U)$ which is the range of P_s then the fact that $I = P_s = \sum_{s \in S} P_s$ tells us that $H(U)$ is a direct sum of the orthogonal subspaces M_s. It is trivial to verify that $U_x P U_x^{-1} = B$ if and only if for each M_s, and each $x \in G$, $U_x(M_s) = M_{s^1}$ for some s^1 in S (where of course s^1 depends upon x and s). In other words to say that there exists an *atomic* complete Boolean algebra of projections B such that $U_x B U_x^{-1} = B$ for all x is the same as to say that there exists a direct sum decomposition $H(U) = \sum_{s \in S} M_s$ such that the U_x permute the M_s among themselves.

Now in the theory of permutation groups it sometimes happens that the set A of objects being permuted may be divided into disjoint subsets $A = A_1 \cup A_2 \cup \ldots \cup A_r$ such that each permutation π in the group maps each A_i into some A_j. A division into subsets having this property is classically known as a "system of imprimitivity" for the permutation group. The analogy with the properties of the direct sum decomposition of $H(U)$ is obvious and one accordingly speaks of this direct sum decomposition as a "system of imprimitivity" for the unitary group representation U. But since B and the de-

- 587 -

composition determine one another uniquely one can equally well speak of B as a system of imprimitivity for U and we shall do so. Finally when B is a complete Boolean algebra which is not atomic one has no such immediate relationship to a direct sum decomposition but one does have something analogous. There is a suitably defined "continuous direct sum" or "direct integral" decomposition canonically associated with B . We are thus led to the following:

DEFINITION. Let U be a unitary representation of the second countable locally compact group G . Then by a *system of imprimitivity* for U we shall mean a complete Boolean algebra of projections B in the Hilbert space $H(U)$ of U such that $U_x B U_x^{-1} = B$ for all x in G . The representation U will be said to be *primitive* if it has no non-trivial systems of imprimitivity. (The B consisting only of 0 and I is of course a system of imprimitivity for every representation.) Otherwise we shall say that it is *imprimitive*.

With this terminology available and taking account of the above theorem it is possible to state the connection between harmonic analysis (as defined here) and unitary group representations in a quite concise fashion: Harmonic analysis is concerned with the decomposition into irreducibles of those unitary group representations which are imprimitive and with the implications of this decomposition for the space of Hilbert bundle cross sections associated with the system consisting of the representation in question and a particular system of imprimitivity for it.

Having said what harmonic analysis "is" we turn our attention to the fact that in a sense the definition we have given is too general. More precisely there is a natural division of it into two stages the first of which is more geometrical than analytical and one can reasonably think of only the second stage as being harmonic analytic in a stricter sense. We shall speak of this stricter sense as "harmonic analysis proper". For simplicity of discussion we shall illustrate this point in the special case of the harmonic analysis of the complex valued functions in $L^2(S,\mu)$ where μ is an invariant measure in the G space S. There is no difficulty in carrying everything over to the general case.

It often happens that if one looks at the geometry of the action of a group G on a G space S with an invariant measure μ one finds that one can write $S = S_1 \cup S_2$ where $S_1 \cap S_2 = 0$, $\mu(S_1) \neq 0$, $\mu(S_2) \neq 0$ and every $x \in G$ carries S_1 into S_1 and S_2 into S_2. It is clear that in such an event the G space S "falls apart" into two independent G spaces S_1 and S_2 each with an invariant measure and that S itself can be reconstructed from S_1 and S_2 in an obvious and trivial manner. It is also evident that one has a corresponding decomposition of the group representation U defined in $L^2(S,\mu)$ by $U_x(f)(s) = f([s]x)$ as a direct sum of two subrepresentations U^1 and U^2 where U^j is equivalent to the representations defined from S_j as G was defined from S. The subspaces S_1 and S_2 may be further decomposed in the same manner and one may hope to continue until one reaches components which are indecomposable. To the extent that this is possible one obtains a partial direct sum or "direct integral" decomposition of U into components each of which is again the unitary representation U^1 associated with a G space S^1 as above but now each S^1 is indecomposable or "ergodic" in the sense that S^1 has no measurable G invariant subsets except possibly sets of measure zero and their complements. An example may help make the situation clearer. Let S be the unit disk in the Euclidean plane and let G be the additive group T of the real line. For each x,y in S let $[x,y]t = (x \cos t + y \sin t, -x \sin t + y \cos t)$ for all $t \in T$. Let μ be the "area" measure in S. Given $0 < r_1 < r_2 < 1$ let A_{r_1,r_2} denote the annulus $r_1^2 < x^2 + y^2 < r_2^2$. Evidently each

- 589 -

A_{r_1, r_2} is an invariant set of positive measure whose complement is
of positive measure and the action of G in S is far from ergodic.
On the other hand there are no ergodic invariant subsets of positive
measure so that one cannot hope to write S as a *sum* of invariant
ergodic pieces. However, there is an obvious way of writing S as
an *integral* of invariant subsets of S each equipped with an invariant measure under which the action of G is ergodic. There is one
such subset for each r with $0 < r < 1$; namely the circle
$x^2 + y^2 = r^2 = S_r$ and the invariant measure μ_r is the "arc length"
measure. The μ measure of any subset E of S is easily computed
from the μ_r measures of the sets $E \cap S_r$ by integrating them with
respect to r . $\mu(E) = \int_0^1 \mu_r(E \cap S_r)dr$.

This simple example is typical of what happens in general. There
is a theorem stating that given any G space S equipped with an
invariant measure μ and satisfying mild regularity hypotheses one
can fiber it into invariant subsets analogous to the S_r above. Each
of these is equipped with an invariant measure μ_r from which μ
may be obtained by integration as with the S_r . Moreover the action
of G on these subsets is ergodic. This decomposition into ergodic
parts is essentially unique. Finally one does not have to have a
measure μ which is actually invariant. The whole theory works
when μ is replaced by an invariant measure class C in S .

In terms of this decomposition theorem one can be a bit more
precise about the division of harmonic analysis into two stages. In
the first stage one decomposes the G space S with invariant
measure class C into ergodic parts. This carries with it a decomposition of the unitary group representation U as a direct sum
or direct integral of pieces each of which is the U associated with
an invariant Hilbert bundle in some *ergodic* G space S , C . In the
second stage one uses completely different methods to decompose these
pieces. It is this stage that we call "harmonic analysis proper". In

other words harmonic analysis proper is the special case of harmonic
analysis as defined above in which the relevant G action is ergodic. This definition is easily translated into Boolean terms. If
U is a unitary representation of the second countable locally compact group G and B is a system of imprimitivity for U then the
corresponding G space action in S is ergodic if and only if no
member P of B except for O and I have the property that
$U_x P U_x^{-1} = P$ for all x in G ; that is no non-trivial member of
B commutes with all U_x . One can thus define a system of imprimitivity B for U to be ergodic if it has this property. Moreover
with this definition one may define harmonic analysis proper as the
special case of harmonic analysis as defined earlier in this section in which the system of imprimitivity B is ergodic.

The decomposition of a G space with an invariant measure into
ergodic parts can be very difficult but it is a problem of a rather
different character than that of decomposing a group representation
associated with an ergodic action.

Restricting our attention now to the study of harmonic analysis
proper we note a fundamental dichotomy. Given a group G , a G
space S with an invariant measure class C and given that the G
action is ergodic we consider the "orbits" of the G action; that
is the subsets of S of the form [s]G where s is a fixed element of S . Here of course [s]G denotes the set of all [s]x
with s fixed and x an arbitrary element of G . It is obvious
that the orbits associated with the various s are disjoint when
they are not identical. Moreover under mild regularity hypotheses
one shows that they are all measurable. Suppose now that some orbit
$[s_0]G$ has *positive* measure. It follows at once from the ergodicity
hypotheses not only that all other orbits must have measure zero
but that the complement of $[s_0]G$ must have measure zero. The fundamental dichotomy then is simply this. Either every orbit is of
measure zero or exactly one orbit is of positive measure and then

the union of all other orbits is of measure zero. In the former
case one says that the action of G on S , C is "properly ergodic"
and in the latter that it is "essentially transitive". Since invariant sets of measure zero can be removed or added without changing
the system B , U or anything else that matters the essentially
transitive case can always be reduced to the transitive one; i.e.
the case in which there is only one orbit. Correspondingly one
speaks of B as being a "properly ergodic" system of imprimitivity
for U or a "transitive" system of imprimitivity for U .

In the example given above of a decomposition of an action into
ergodic parts the ergodic parts all turned out to be transitive. A
very simple modification of this example, however, leads to one in
which the ergodic parts are properly ergodic. One has only to replace the additive group of the real line by the discrete subgroup
of all integer multiples of some fixed real number λ where λ/π
is irrational. The fact that an irrational rotation of the code
generates an *ergodic* action of the infinite cyclic group — in spite
of the fact that all orbits are countable and hence of measure
zero — was pointed out in an important paper of G. D. Birkhoff and
Paul Smith published in 1928. This was about three years before von
Neumann and Birkhoff published their celebrated ergodic theorems and
may be regarded as the true birth date of ergodic theory as a new
branch of mathematics. If properly ergodic actions did not exist
the decomposition theorem would be a decomposition into transitive
parts and would reduce most of the questions of ergodic theory to
trivialities or to problems in other parts of mathematics.

Harmonic analysis proper has important applications in both
the transitive and the properly ergodic cases. However the former
is much better understood and much more widely known than the latter.
We shall conclude this paper with a brief discussion of each.

In the transitive case, as explained earlier in this section,
one may always suppose that S is a coset space G/H where H is

- 592 -

a closed subgroup of G and is uniquely determined up to conjugacy
by the action of G on S . Moreover in this case the invariant
measure class C is unique. When the associated system of imprimitivity is atomic (which happens precisely when H is both open and
closed so that G/H is a countable discrete space) the motivating
analysis given near the beginning of this section shows that the
different possibilities for the pair U , B , corresponding to different G invariant Hilbert bundles with base S , correspond one
to one to the different unitary representations L of the subgroup
H . Although the technical problems are considerably more formidable
— mainly because the subspaces H_s are "infinitesimal" and have to
be defined indirectly — a parallel analysis can be given in the
general case. The most general pair B , U arising from the transitive case of harmonic analysis proper can be constructed in a canonical way from an essentially uniquely determined unitary representation L of an essentially uniquely determined closed subgroup H
of the second countably locally compact group G . We proceed to a
description of this construction.

   Let H be a closed subgroup of the second countable locally
compact group G . Let L be an arbitrary unitary representation
of H and let μ be any member of the unique invariant measure in
the coset space G/H . Let F^L denote the vector space of all Borel
functions f from G to the Hilbert space $H(L)$ of the representation L which have the following two properties:

   (a) $f(hx) = L_h(f(x))$ for all $h \in H$ and all $x \in G$.

   (b) $\int_{G/H} (f(x) \cdot f(x)) \, d\mu(x) < \infty$.

(Here (b) derives its meaning from the fact that $(f(hx) \cdot f(h(x))) =$
$= L_h f(x) \cdot L_h(f(x)) = (f(x) \cdot f(x))$ so that $(f(x) \cdot f(x))$ is constant
on the right H cosets and so can be regarded as a function on
G/H .) Defining the "norm" of an element of F^L to be

$\sqrt{\int_{G/H} (fx) \cdot f(x) d\mu(x)}$ and identifying elements of F^L which are equal almost everywhere F^L becomes a Hilbert space. This Hilbert space is the Hilbert space of a unitary representation G which we denote by U^L and define as follows

$$U^L_x(f)(s) = \sqrt{\rho(x,s)}\, f([s]x) \ .$$

Here $\sqrt{\rho(x,s)}$ is the correction factor for the non-invariance of μ defined earlier and is identically one when μ is invariant. For each Borel subset E of G/H the operator $f \to \varphi_E f$ (where $\varphi_E(s) = 1$ or 0 according as $s \in E$ or $s \notin E$) is a projection operator in F^L which we denote by P^L_E . The set of all projections P^L_E for fixed L and variable E is a complete Boolean algebra of projections which we denote by B^L . One checks without difficulty that B^L is a transitive system of imprimitivity for the representation U^L . While this definition may seem to depend upon the choice of μ it does not really do so. If μ_1 and μ_2 are two measures in C then multiplication by the square root of the Radon — —Nikodym derivative of one with respect to the other sets up a unitary map of one Hilbert space on the other which carries the pair U^L, B^L associated with μ_1 to that associated with μ_2 . In terms of these definitions we may now state a precise theorem describing the most general pair U, B which arises in the transitive case of what we have called harmonic analysis proper.

THEOREM. *Let U be a unitary representation of the second countable locally compact group G and let B be any complete Boolean algebra of projections in the Hilbert space $H(U)$ of U which is a transitive system of imprimitivity for U . Then there exists a closed subgroup H of G , a unitary representation L of H and a unitary map W of $H(U)$ in $H(U^L)$ such that*

$WU_xW^{-1} = U_x^L$ *for all* x *in* G *and* $WBW^{-1} = B^L$. *The subgroup* H *is unique up to conjugacy and the representation* L *is unique up to equivalence.*

This theorem is known as the imprimitivity theorem. Actually the theorem with that name is more commonly given a slightly different formulation in which one considers the mapping $E \to P_E$ rather than its range B as the fundamental object. Because this mapping has the property that $P_{\{E_1 \cup E_2 \cup \ldots\}} = P_{E_1} + P_{E_2} + \ldots$ whenever the E_j are mutually disjoint Borel sets it is called a "projection valued measure". Quite generally one defines a projection valued measure to be any mapping $E \to P_E$ from the Borel subsets of a space S to the projection operators in a Hilbert space H which has the property just stated and the following further properties

(1) $\qquad P_\phi = 0$,

(2) $\qquad P_S = I$,

(3) $\qquad P_{E \cap F} = P_E P_F = P_F P_E$.

A corollary of the above discussion is the existence of a canonical mapping $L \to U^L$ from the unitary representations L of a closed subgroup H of a second countable locally compact group G to the unitary representations of G . The representation U^L of G is called the representation of G *induced* by the representation L of H .

For the case of a *finite* group G this mapping (formulated in terms of the characters of L and U^L) was introduced by Frobenius in one of his very first papers on group representations. The role it plays in the applications of the theory of group representations to number theory found by Artin in 1923 is mentioned in Section II. For Frobenius the inducing construction was an important tool for finding and defining irreducible representations which are *not* one dimensional. For many finite groups a large part if not all of the

irreducible unitary representations are of the form U^L where L
is a one-dimensional representation of a suitable subgroup H .
This application of the inducing process is equally important for
determining the irreducible unitary representations of general
locally compact groups.

From this last remark it appears that the inducing process
plays a dual role in harmonic analysis. Not only do the induced representations U^L constitute a large fraction of the representations which the applications ask us to decompose; they also constitute a large fraction of the irreducible pieces into which one seeks
to decompose them.

Before concluding this paper with a brief discussion of the
nature of harmonic analysis in the properly ergodic case we give a
few indications concerning the nature of the applications of the
general theory — first to physics and then to number theory. Let us
specialize our G space S to the case in which S is physical
space and G is the (transitive) group of all isometries of space.
Here we may choose any three dimensional Riemannian space with a
transitive group of isometries as our model for space. Let H be
the Hilbert space of states for a quantum mechanical particle and
to avoid having to discuss so called "projective" representations
let us replace G by its simply connected covering group $\tilde{G}$. For
each Borel subset E of S let P_E denote the self-adjoint operator corresponding to the observable which is one when the particle
is in E and zero when it is not. One can then argue that $E \to P_E$
must be a projection valued measure and that the self-adjoint operators corresponding to all coordinate observables can be constructed
from this projection valued measure. On the other hand it follows
from the assumption of the independence of physical laws of position
and orientation of space that each member α of G is canonically
associated with an automorphism of H which is implemented either
by a unitary or an anti-unitary operator. From this one is able to

argue the existence of a canonical unitary representation $\alpha \to U_\alpha$
of $\tilde{G}$ in the Hilbert space H *and* that U must be related to P
in such a manner that

$$(*) \qquad U_\alpha P_E U_\alpha^{-1} = P_{[E]\alpha}$$

for all α and E . Just as one can construct the self-adjoint
operators corresponding to coordinate or position observables from
P so one can construct the self-adjoint operators corresponding to
momentum observables from U . They are the infinitesimal generators of the restrictions of U to the one parameter subgroups of
G . In other words knowledge of the pair U , P implies knowledge
of all position and momentum operators. On the other hand the imprimitivity theorem provides an overview of all possible pairs
U , P which satisfy $(*)$. Let K be a closed subgroup of $\tilde{G}$ such
that $S = \tilde{G}/K$; that is let K be the group of all "rotations"
about an origin s_0 of space. Then to within unitary equivalence
any pair U , P must be the pair U^L , P^L defined by some unitary
representation L of K . Whether S is flat or curved and if
curved whether of finite or infinite volume the group K is the
compact Lie group $SU(2)$ so that there is a discrete countable infinity of possibilities for L and hence for the pair U , P . For
the more basic particles L is always irreducible and hence one of
the representations $D_0, D_{\frac{1}{2}}, D_1, D_{\frac{3}{2}}, \ldots$ where D_j is $2j+1$ dimensional. The index j is a fundamental invariant called the "spin"
of the particle. To complete the quantum mechanical description of
a particle of spin j one must specify a unitary representation
$t \to V_t$ of the additive group of the real line which then determines how the states of the particle vary in time. General symmetry
principles demand that $U_\alpha V_t = V_t U_\alpha$ for all $\alpha \in \tilde{G}$ and all real t
and this condition limits $t \to V_t$ rather sharply. In particular
using the general theory of induced representations one can give a

complete analysis of all possibilities.

Starting with the above analysis of the quantum mechanics of a particle one can go on to analyze systems of particles and the electromagnetic field in the same spirit. When one does so one finds that one can build much if not all of quantum physics naturally and elegantly into the general framework of the theory of unitary group representations with induced representations playing a prominent role.

The 1927 paper of Hermann Weyl mentioned at the end of Section II was concerned with the beginning of the program just described. He proposed using the theory of group representations to understand why the operators corresponding to the position and momentum observables are what they are and emphasized the significance for this purpose of the conjectured uniqueness of the solutions of the Heisenberg commutation relations. This uniqueness, first rigorously stated and proved by Stone and von Neumann is in fact the special case of the imprimitivity theorem in which G is the additive group of the direct sum of a finite number of replicas of the real line and H is the identity subgroup. The relationship $U_\alpha P_E U_\alpha^{-1} = P_{[E]\alpha}$ can be regarded as a far-reaching generalization of the Heisenberg commutation relations. The reader will find further details in the author's paper in honour of Weyl cited at the end of Section II.

In the applications to number theory the induced representations that occur differ from those arising in quantum physics in that the inducing representation L is usually a finite dimensional representation of an infinite discrete group rather than a compact Lie group. In a typical and important example L is the one-dimensional identity representation of a subgroup Γ of $SL(2,R)$ which is of finite index in the intermediate subgroup $SL(2,Z)$. The analysis of U^{I_Γ} can be facilitated by using two general theorems in the theory of induced representations. One asserts that $U^{L^1 \oplus L^2} = U^{L^1} \oplus U^{L^2}$ and

that more generally "inducing" commutes with the taking of direct
sums and direct integrals. The other deals with pairs of closed subgroups H_1 and H_2 with $H_1 \subset H_2 \subset G$. It says that if L is a
unitary representation of H_1 and M is the representation of H_2
induced by L then the representation of G induced by L is equivalent to the representation of G induced by M. In the case at
hand the representation M of $SL(2,Z)$ induced by I_Γ will be a
finite direct sum of finite dimensional unitary representations
L^j of $SL(2,Z)$ and U^{I_Γ} is accordingly a direct sum of the induced
representations U^{L^j}. This is a useful partial reduction of the
problem because $SL(2,Z)$ has a "more regular structure" than its
subgroups of finite index.

In order to explain how harmonic analysis at this level is relevant to number theory it will be useful to begin with some general
remarks about the nature of number theory. Up through the publication
of Gauss' celebrated treatise "DISQUISITIONES ARITHMETICAE" in 1801
number theory was dominated by the theory of Diophantine equations
of the form $Ax^2 + 2Bxy + Cy^2 = m$ where A, B, C and m are given
integers and x and y are integers to be determined. This theory
is more or less equivalent to the theory of those algebraic number
fields which are second degree extensions of the rationals. One
natural generalization is to a theory of algebraic number fields of
arbitrary finite degree over the rationals. Another is to the theory
of homogeneous Diophantine equations of the second degree in more
than two unknowns; i.e. to the number theory of n-ary quadratic
forms as opposed to binary quadratic forms. These two theories dominated number theory during most of the nineteenth century; the first
being motivated by the work of Gauss and Kummer on the higher reciprocity laws as well as attempts to prove the famous "last theorem"
of Fermat. In our discussion of the work of Artin on algebraic number fields in Section II we indicated the central role played in

that theory by the functions Φ_F defined on the positive integers.

Similarly much of the theory of n-ary quadratic forms reduces to the study of the functions Φ_Q where $\Phi_Q(m)$ for each positive definite quadratic form Q with integer coefficients and each positive integer m is defined to be the number of integer solutions of the equation $Q(x_1,x_2,\ldots,x_n) = m$. (When Q is indefinite one can still define Φ_Q by counting "equivalence classes" of solutions). Just as one finds it useful to introduce the Dirichlet series $\sum\limits_{n=1}^{\infty} \dfrac{\zeta_F(n)}{n^s} = \zeta_F(s)$ in studying the functions Φ_F so one finds it useful to study the functions φ_Q through their associated "generating functions" or "theta series" $\sum\limits_{n=0}^{\infty} \varphi_Q(n)\ell^{\pi i n z} = \hat{\varphi}_Q(z)$.

A large part of the usefulness of introducing the auxiliary functions $\zeta_F(s)$ and $\hat{\varphi}_Q(z)$ may be attributed to the fact that they turn out to be analytic functions of the complex variables s and z respectively and that these analytic functions have rather special properties. The functions $L_F(s)$ for instance, although ostensibly defined only for s in the right half plane $\mathrm{Re}(s) > 1$, have analytic continuations to the entire plane which are meromorphic functions with only one pole (it is at $s = 1$) and have growth properties at ∞ which imply that they are essentially uniquely determined by their zeros. In addition there are "functional equations" relating the behaviour in the left half plane to that in the right which allow one to locate all the zeros except those lying in the "critical strip" $0 < \mathrm{Re}(s) < 1$. The latter are at least known to be symmetric about the central line $\mathrm{Re}(s) = 1/2$ and a celebrated conjecture says that they lie on it. The corresponding facts about the functions $\hat{\Phi}_Q(z)$ defined and analytic in the upper half plane $\mathrm{Im}(z) > 0$ are (in the case of positive definite Q) that $\hat{\Phi}_Q(z)$ has simple transformation properties under $z \to \dfrac{az+b}{cz+d}$ for all

- 600 -

$\begin{pmatrix} a & b \\ c & d \end{pmatrix}$ in some subgroup Γ of finite index of the subgroup $SL(2,Z)$ of $SL(2,R)$. Specifically

$$(*) \qquad \hat{\varphi}_Q\left(\frac{az+b}{cz+d}\right) = (cz+d)^{\frac{k}{2}} \, \hat{\varphi}_Q(z)$$

where k is the number of variables in Q and Γ depends on the coefficients of Q . Moreover, in addition to satisfying $(*)$ $\hat{\varphi}_Q$ has certain growth properties implying that it is what is known as a *modular form of weight* $k/4$ *for* Γ . Note that $(*)$ implies that the zeros of $\hat{\varphi}_Q$ in the upper half plane are invariant under the action of Γ . Moreover the growth properties imply that there can be only a finite number of zeros in any "fundamental domain" for the action of Γ and that these zeros essentially determine $\hat{\varphi}_Q$. Using these facts one can largely reduce the study of the functions $n \to \varphi_Q(n)$ to the study of modular forms and their "Fourier coefficients". Because of what the theory of functions of a complex variable is able to tell us about modular forms this has turned out to be a very fruitful reduction.

Having summarized the nature of certain major parts of number theory in this way we indicate the relevance of harmonic analysis by a series of loosely related remarks.

(1) The fact that the Fourier series of a "reasonably behaved" periodic function converges to the function turns out to have a number of surprising identities as consequences. These identities result from choosing the function f in such a fashion that its Fourier coefficients can be explicitly evaluated and evaluating both sides of $f(x) = \sum_{n=-\infty}^{\infty} c_n e^{inx}$ at suitably chosen x . A wider class of such identities can of course be obtained by considering Fourier series of periodic functions of several variables. The facts outlined above the analytic properties of the functions $\hat{\Phi}_Q(z)$ and $L_F(s)$ are all corollaries of special cases of such identities and

thus have the Fourier—Dirichlet theorem in n dimensions as their ultimate source.

(2) For deriving the number theoretic identities of (1) it is convenient to express the Fourier—Dirichlet theorem in an equivalent form known as the Poisson summation formula. The Poisson summation formula may be derived from the theory of group representations as follows. Let Γ be a discrete subgroup of the separable locally compact commutative group G. Then the induced representation U^{I_Γ} is easily seen to be the direct sum of all the one-dimensional representations defined by those characters $\chi \in \hat{G}$ with $\chi(\gamma) = 1$ for all $\gamma \in \Gamma$. The set of all these characters if a discrete subgroup of $\hat{G}$ called the annihilator $\Gamma^\perp$ of Γ. One has $U^{I_\Gamma} =$

$= \sum\limits_{\chi \in \Gamma^\perp} \chi$. Now for each complex valued function f on G which is summable with respect to Haar measure one may form $U_f^{I_\Gamma} =$

$= \int f(x) U_x^{I_\Gamma} dx$, obtain a well defined operator and deduce that it is a direct sum of the constant operators defined by the numbers $\int f(x)\chi(x)dx$ for $\chi \in \Gamma^\perp$. Taking the trace of both sides (when f is such that it exists) one finds $\sum\limits_{\gamma = \Gamma} f(\gamma) = \sum\limits_{\gamma \Gamma^\perp} \int f(x)\chi(x)dx$.

Now for all χ, $\int f(x)\chi(x)dx = \hat{f}(x)$ is the so called Fourier transform $\hat{f}$ of f evaluated at χ . Thus one arrives at the elegant formula

$$\sum\limits_{\gamma \in \Gamma} f(\gamma) = \sum\limits_{\chi \in \Gamma^\perp} \hat{f}(\chi) \ .$$

When $G = R^n$, the additive group of all n tuples of real numbers, this reduces to the classical Poisson summation formula in n dimensions. The point of deducing the Poisson summation formula in this way is that the procedure has an obvious extension to the case in which G is a not necessarily commutative separable locally com-

pact group. This way of extending the Fourier−Dirichlet theorem
to non-commutative groups yields a host of new identities some of
which are of considerable number theoretical interest. This source
of number theoretical identities was introduced by A. Selberg in
an important paper published in 1957. One speaks of the associated
generalization of the Poisson summation formula as the Selberg trace
formula. Selberg concerned himself mainly with the special case in
which G is SL(2,R) and Γ is a subgroup of finite index of
SL(2,Z) . However, he indicated generalizations to higher dimensional Lie groups.

(3) Artin's use of the representation theory of finite groups
to study the zeta function $\zeta_F(s)$ as explained in Section II has
a counterpart in Hecke's use of this theory in 1928 to study modular
forms. Notice that a modular form of weight k for a subgroup Γ
of finite index of SL(2,Z) is also a modular form of weight k
for every subgroup Γ^1 of Γ which is of finite index in Γ .
Notice further that every Γ of finite index contains a normal subgroup of finite index. Suppose then that f is any modular form of
weight k for a normal subgroup Γ of finite index in SL(2,Z) .
It is easy to see that the functions $z \rightarrow f\left(\dfrac{az+b}{cz+d}\right)(cz+d)^{-2k}$ as
$\begin{pmatrix} a & b \\ c & d \end{pmatrix}$ varies over all elements of SL(2,Z) span a finite dimensional vector space. Call it V_f . For each g in V_f and each
$\begin{pmatrix} a & b \\ c & d \end{pmatrix} \in SL(2,Z)$ let $L_{\begin{pmatrix} a & b \\ c & d \end{pmatrix}}(g)$ denote the function $z \rightarrow$

$\rightarrow \quad g\left(\dfrac{az+b}{cz+d}\right)(cz+d)^{2k}$. It is easy to check that $g \rightarrow L_{\begin{pmatrix} a & b \\ c & d \end{pmatrix}}(g)$ is

a linear transformation and that $\begin{pmatrix} a & b \\ c & d \end{pmatrix} \rightarrow L_{\begin{pmatrix} a & b \\ c & d \end{pmatrix}}$ is a represen-

tation of SL(2,Z) which reduces to the identity in the normal
subgroup Γ . Hence it defines a representation of the finite quo-

tient group $SL(2,Z)/\Gamma$ on V_f and this representation can be decomposed into its irreducible constituents. All members of V_f are modular forms of weight k for Γ and one thus obtains a natural decomposition of f as a sum of modular forms which are special in being intrinsically associated with irreducible representations of $SL(2,Z)/\Gamma$. Hecke found this decomposition a useful tool in studying modular forms. A closely related technique involves introducing vector valued modular forms in which the vectors lie in the space of some irreducible representation L of $SL(2,Z)$ and the basic modular form identity is replaced by $f\left(\dfrac{az+b}{cz+d}\right) = L_{\left(\begin{smallmatrix} a & b \\ c & d \end{smallmatrix}\right)}((f(z))(cz+d)^{2k}$.

To a considerable extent one can then replace the study of modular forms for proper subgroups of $SL(2,Z)$ to the study of vector valued modular forms for the whole group associated with the various finite dimensional irreducible representations of $SL(2,Z)$ — more particularly those defined by representations of its finite quotient groups.

(4) As apparently first observed by Gelfand and Fomin in 1951 there is a significant and close connection between modular forms and the unitary representation theory of $SL(2,R)$. Except for the one-dimensional identity the irreducible unitary representations of this group are all infinite dimensional and fall naturally into three families. The members of one of these, the so called "discrete series", have the distinguishing property of occurring as discrete direct summands in the decomposition of the regular representation of $SL(2,R)$. The remaining representations occur continuously as part of a direct integral decomposition and belong to the family known as the "principal series". Let K denote the maximal compact subgroup of $SL(2,R)$ consisting of all matrices of the form
$\begin{pmatrix} \cos\theta\,, & \sin\theta \\ -\sin\theta, & \cos\theta \end{pmatrix}$ where θ is real. then K is commutative and its

- 604 -

characters are parametrized by the integers; the most general being χ_r where r is an integer and $\chi_r\begin{pmatrix} \cos\theta & \sin\theta \\ -\sin\theta & \cos\theta \end{pmatrix} = e^{i\theta r}$. The members of the principal and discrete series restrict to K in sharply different ways. When L is a member of the principal series, L restricted to K is either $\sum\limits_{j=-\infty}^{\infty} \chi_{2j}$ or $\sum\limits_{j=-\infty}^{\infty} \chi_{2j+1}$ depending upon whether L is trivial on the two element center of $SL(2,R)$ or not. On the other hand when L is a member of the discrete series there exists a unique non-zero integer ℓ such that the restriction of L to K is either $\sum\limits_{j=0}^{\infty} \chi_{\ell+j}$ or $\sum\limits_{j=0} \chi_{\ell-j}$ depending upon whether $\ell > 0$ or $\ell < 0$. Moreover L is determined by ℓ to within unitary equivalence so that ℓ may be used as a label. We denote the member of the discrete series with label ℓ by the symbol W^{ℓ} .

The connection between the unitary representation theory of $SL(2,R)$ and the theory of modular forms may now be explained as follows. Let Γ be a discrete subgroup of $SL(2,Z) \subseteq SL(2,R)$ which is of finite index in $SL(2,Z)$. Let I_{Γ} denote the one-dimensional identity representation of Γ and form the induced representation $U^{I_{\Gamma}}$ of $SL(2,R)$. Then the multiplicity with respect to which $U^{I_{\Gamma}}$ contains the discrete series member W^{ℓ} as a discrete direct summand is always less than the dimension of the space of all modular forms for Γ of weight $|\ell|/2$. It is equal to the dimension of the subspace of all so called "cusp forms". This last fact is a consequence of the important fact that there is a canonical isomorphism of the vector space of all "intertwining operators" for $U^{I_{\Gamma}}$ and W^{ℓ} with the vector space of all modular cusp forms of weight $|\ell|/2$ for Γ . We remind the reader that an intertwining operator for two representations U and V of the same group G is a bounded linear operator T from the space $H(U)$ of U to the space $H(V)$ of V such

that $TU_x = V_x T$ for all x in G and that when U is irreducible
the dimension of the vector space of all intertwining operators for
U and V is equal to the multiplicity with respect to which V
contains U as a discrete direct summand. As to the concept of
"cusp form" it was shown by Hecke, in generalization of a construction of Eisenstein, that associated with each "cusp" of a fundamental region for the action of Γ on the upper half plane one has an
infinite series containing k as a parameter whose sum is a modular
form of weight k and whose "Fourier coefficients" can be explicitly computed. One speaks of the Eisenstein series of weight k for
the cusp in question. Every modular form of weight k for Γ is
uniquely the sum of a finite linear combination of Eisenstein series
and a modular form of weight k which, in a sense we shall not explain, "vanishes at all the cusps". The latter forms are the cusps
forms and it is the study of their Fourier coefficients which presents the more challenging problem. In view of the above one can
say that a cusp form of weight k for Γ "is" an intertwining
operator for U^{I_Γ} and W^{2k} .

(5) The idèle group notion introduced into the theory of algebraic number fields by Chevalley in 1933 and discussed in Section
III above has a non-commutative generalization in which the multiplicative group of a number field is replaced by a more or less
general algebraic group. This generalization was introduced in the
late 1950's by Ono and Tamagawa and is called the idèle group G_A
of the algebraic group G . It contains a canonical countable closed
subgroup Γ isomorphic to G and the analysis of the unitary representation U^{I_Γ} of G_A induced by the identity representation
I_Γ of Γ turns out to be closely connected with many problems in
number theory. In particular when $G = SL(2,Q)$ where Q is the
rational number field one finds, using certain general theorems
about induced representations, that U^{I_Γ} restricted to $SL(2,R)$

(which occurs as a direct factor of G_A) is a direct sum of *all*
induced representations of SL(2,R) of the form U^L where L varies over those finite dimensional unitary irreducible representations of SL(2,Z) which occur in the theory of modular forms. The
consequences of this fact for the theory of modular forms may be
compared to the consequences for algebraic number theory of Chevalley's idea of putting *all* finite algebraic extensions of a field F
having commutative Galois groups into one infinite extension with a
compact Galois group. In the late 1930's and early 1940's Hecke made
important new advances in the theory of modular forms and applied
them to the number theory of quadratic forms in a very significant
way. If one looks at this work from the adele point of view sketched
above one sees the possibilities of vast generalizations through replacing SL(2,Q) by other algebraic groups. R. P. Langlands has been
an important pioneer in developing this program. As one might imag-
ine, the Selberg trace formula is an important tool.

Most of this section has been devoted to the transitive side
of the dichotomy separating harmonic analysis into two cases according to whether the associated group action is transitive or properly
ergodic. In fact the properly ergodic side is much harder to analyze
and consequently much less well developed. In particular one cannot
reduce the classification of the properly ergodic actions of a group
G to anything as "simple" as classifying its (conjugacy classes of)
closed subgroups. Indeed, even when G is the infinite cyclic group
or the additive group of the real line one is still very far from
anything approaching a complete classification of its properly ergodic actions. On the other hand the analogy between the two sides
of the dichotomy is a close one and through the concept of "virtual
subgroup" one can exploit the analogy and what is known of group
theory to gain new insights into ergodic theory and the problem of
classifying ergodic actions.

The chief applications, so far, of harmonic analysis in the

properly ergodic case are to probability theory and statistical mechanics. We now present a brief sketch of the nature of the aplications to probability theory. Let μ be a measure in the standard Borel space S such that $\mu(S) = 1$. One defines a "random variable" to be a real valued measurable function g on S. If $g_1,\dots,g_n$ are random variables then $s \to g_1(s)\dots g_n(s)$ is a measurable function $\bar{g}$ from S to R^n and setting $\bar{\mu}_n(E) =$

$= \mu(\bar{g}^{-1}(E))$ defines a probability measure $\bar{\mu}_n$ in R^n called the "joint distribution" of the random variables $g_1,\dots,g_n$. Questions of independence, correlation etc. between the random variables $g_1,\dots,g_n$ come down to properties of the measure $\overline{\mu_n}$. An infinite sequence $\dots,g_{-2},g_{-1},g_0,g_1,\dots$ of random variables is called a "discrete parameter stochastic process" and a family of random variables $t \to g_t$ parametrized by the real numbers is called a "continuous parameter stochastic process" provided that $g_t(s)$ is measurable as a function of both variables. If $g_\alpha(s)$ is a stochastic process and α is either an integer or a real parameter then the functions $\alpha \to g_\alpha(s)$ for the various values of s are called the "sample functions" of the process. A stochastic process is said to be "stationary" if the joint distribution $g_{\alpha_1+\beta}, g_{\alpha_2+\beta}, \dots, g_{\alpha_N+\beta}$ is independent of β for all choices of $\alpha_1,\dots,\alpha_n$. It is a fundamental theorem that every discrete parameter stationary stochastic process is canonically associated with a measure preserving action of the additive group Z of the integers on S in such a way that $f_n(s) = f_0([s]n)$ for all s and n. Analogously every continuous parameter stationary stochastic process is canonically associated with a single measurable function on S and a measure preserving action of the additive group R of the real line on S. One can show that there is no loss of generality in assuming that the associated measure preserving actions of Z or R are ergodic and that these actions will be *properly ergodic* except in the probabilistically trivial case in which "almost all" sample functions

- 608 -

are periodic. Thus one has to deal with a properly ergodic action
of Z or R and a real valued measurable function on the measure
space S,μ on which Z or R acts. The restriction that f be
contained in $L^2(S,\mu)$ is equivalent to the quite natural probabilistic requirement that the random variables of the process should
have finite dispersions. An important technique in the probabilistic
study of stationary stochastic process consists in relating properties of the sample functions of the process to the properties of the
function f on S whose translates are the random variables. In
particular the harmonic analysis of f in the spirit of this paper
carries with it a parallel analysis of the sample functions. This
way of harmonically analyzing those particular functions on the real
line which occur as sample functions of continuous parameter stationary stochastic processes (in the case in which $f \in L^2(S,\mu)$) is
in essence the "Generalized Harmonic Analysis" of Norbert Wiener's
celebrated paper of 1930.

V.

One cannot work for long in the theory of induced representations without feeling that one is engaged in a fascinating blend of
measure theory and group theory. This blend is a natural outgrowth
of Alfred Haar's famous theorem of 1933. In this paper, written to
honour the memory of Professor Haar on the 100th anniversary of his
birth, I have tried to give a comprehensive picture of the way in
which group theory and measure theory have interacted to produce a
far reaching generalization of the harmonic analysis of Fourier
— and whose extensive applications cover a large part of mathematics.

G. W. Mackey
Mathematics Department
Harvard University
Cambridge, Mass. 02138
USA

WEYL'S PROGRAM AND MODERN PHYSICS

George W. Mackey
Harvard University
Department of Mathematics
Science Center, One Oxford Street
Cambridge, Massachusetts 02138
USA

ABSTRACT. The author is presently engaged in trying to fit as much as possible of modern elementary particle physics into a group theoretical framework described at some length in the reference cited at the end of section IV. This article is in part an exposition of the relationship of this program to one formulated by Weyl in 1927, in part a review and revision of material from the reference mentioned above and in part a description of ideas and incomplete work relevant to carrying the program further.

I. Weyl's program.

Wigner and Weyl introduced group theoretical ideas into quantum physics in independent papers published in 1927 (Zeitschrift für Physik 40 and 46). While Wigner was concerned with simplifying calculations in the quantum theory of atomic spectra, Weyl used group theory in an attempt to clarify the foundations. In his introduction Weyl singles out two basic foundational questions: (1) How does one arrive at the self adjoint operators which correspond to various concrete physical observables? (2) What is the physical significance of these operators; i.e., how does one deduce physical statements? He goes on to say that question (2) has been more or less satisfactorily dealt with by von Neumann and that he proposes to concern himself chiefly with question (1). He asserts that question (1) is the deeper question and "Hier glaube ich mit Hilfe der Gruppentheorie zu eine tieferen Einsicht in der wahren Sachverhalt gelangt zu sein".

Weyl's answer to question (1) rested on his global reformulation of the Heisen-

11

K. Bleuler and M. Werner (eds.), Differential Geometrical Methods in Theoretical Physics, 11–36.
© 1988 by Kluwer Academic Publishers.

berg commutation relations in terms of associated unitary representations of the real line. If P and Q are self adjoint operators then $t \to e^{itQ}$ and $s \to e^{isP}$ are unitary representations of the additive group of the real line. Moreover it is easy to verify, at least on a formal level, that $QP - PQ = iI$ if and only if the associated group representations satisfy the commutation relation

$$e^{itQ}e^{isP} = e^{ist}e^{isP}e^{itQ}$$

for all s and t. More generally if $Q_1, Q_2 \cdots Q_n$ are mutually commuting self adjoint operators then $t_1, t_2, \cdots t_n \to e^{it_1 Q_1} e^{it_2 Q_2} \ldots e^{it_n Q_n} = A_{t_1, t_2, \cdots t_n}$ is a unitary representation of the additive group of an n dimensional real vector space. Let $s_1, s_2 \cdots s_n \to e^{is_1 P_1} e^{is_2 P_2} \ldots e^{is_n P_n} = B_{s_1, s_2, \cdots s_n}$ be a second such representation where $P_1, P_2 \cdots$ is another n tuple of mutually commuting self adjoint operators. Then it follows from the case $n = 1$ considered above that the Q_i and P_j satisfy the classical Heisenberg commutation relations if and only if $A_t B_s = e^{i(t \cdot s)} B_s A_t$ where $(t \cdot s) = t_1 s_1 + \cdots t_n s_n$, $s = s_1, s_2, \cdots s_n$ $t = t_1, t_2, \cdots t_n$. These commutation relations satisfied by the A_t and the B_s are sometimes referred to as the Heisenberg commutation relations in Weyl or global form.

Suppose now that $t \to A_t$ $s \to B_s$ are unitary representations of the above form and let $W_{t,s} = A_t B_s$. As Weyl observed (and as is easy to verify) to say that A and B satisfy the Heisenberg commutation relations in global form is equivalent to saying that

$$W_{t+t', s+s'} = e^{-i(t' \cdot s)} W_{t,s} W_{t',s'}.$$

But this is precisely the statement that the mapping $t, s \to A_t B_s$ is a so-called *projective unitary representative* of the commutative group of all $2n$ tuples of real numbers under addition with the function $\sigma(t, s; t', s') = e^{-i(t' \cdot s)}$ as "multiplier". Recall that, quite generally, one says that a mapping $x \to U_x$ of a group G into the unitary operators in some Hilbert space $\mathcal{H}(U)$ is a projective unitary representation if $U_{xy} = \sigma(x, y) U_x U_y$ for all x and y in G, where $\sigma(x, y)$ is a complex number of modulus one (which of course depends upon x and y). The function σ is called a projective multiplier.

As Weyl emphasized in his paper, the generalization of the concept of group representation to include projective representations is especially important for quantum mechanics. This is because unitary operators occur in quantum mechanics as automorphisms of the state space and two unitary operators U^1 and U^2 define the same automorphism whenever $U^2 = cU^1$ where c is a complex number of modulus one. This means that in a homomorphism of a group G into a group of automorphisms of a state space the unitary operators are determined only up to a constant factor of modulus one and if one chooses this factor at random one can only deduce that $U_{x_1 x_2}$ is a constant multiple of $U_{x_1} U_{x_2}$ *not* that $U_{x_1 x_2} = U_{x_1} U_{x_2}$. Of course one can attempt to choose the arbitrary constant in each U_x in such a manner that one does have $U_{x_1 x_2} = U_{x_1} U_{x_2}$ but this is not always possible. Indeed if $U_{x_1 x_2} \equiv \sigma(x_1, x_2) U_{x_1} U_{x_2}$ for some unitary representation of group G and we define $V_x = r(x) U_x$ where r is an arbitrary function from G to the complex numbers of modulus one then a simple calculation shows that $V_{x_1 x_2} \equiv \sigma'(x_1, x_2) V_{x_1} V_{x_2}$ where $\sigma'(x_1, x_2) \equiv \frac{\sigma(x_1, x_2) r(x_1 x_2)}{r(x_1) r(x_2)}$. Evidently one can choose r so that $\sigma'(x_1, x_2) \equiv 1$ if and only if $\sigma(x_1, x_2)$ can be written in the form $\frac{r(x_1) r(x_2)}{r(x_1 x_2)}$. It is easy to give examples in which one may prove that no such r exists. When it does exist one says that the multiplier is *trivial*. More generally one says that two multipliers are *equivalent* if their ratio is trivial. This is because if σ_1 and σ_2 are equivalent multipliers then every projective representation U with multiplier σ_1 may be made into one with multiplier σ_2 by simply multiplying each U_x by a suitable constant.

Consider now the multiplier $\sigma(t, s, t', s') = e^{-i(t' \cdot s)}$ introduced above. Weyl was much impressed by the fact that, to within equivalence, this is the *only* multiplier on the additive group R^{2n} of all $2n$ tuples of real numbers having a certain property of "non degeneracy". Moreover, although he did not give a rigorous proof, Weyl had reason to believe that every irreducible projective representation of R^{2n} with this particular multiplier is unitarily equivalent to every other. In other words there is essentially only one non degenerate projective multiplier σ for the commutative group R^{2n} and essentially only one irreducible unitary projective representation having σ as multiplier. Moreover this projective representation is just the global counterpart of a set of $2n$ self adjoint operators satisfying the Heisenberg commutation relations.

The fact that sets of operators satisfying the Heisenberg commutation relations may be characterized in such a simple and fundamental group theoretical manner was stated by Weyl as "The kinematical structure of a physical system is expressed by an irreducible Abelian group of unitary ray representations in system space". It seemed to him to be a fact of deep significance and was his answer to question (1).

Weyl also showed that the smallest and simplest commutative group having a non trivial projective multiplier σ is the four element group $Z_2 \times Z_2$. This group also admits an essentially unique irreducible projective representation. It is two dimensional and the three two by two matrices representing the non identity elements may be chosen to be $\begin{pmatrix} 1 & 0 \\ 0 & -1 \end{pmatrix}$, $\begin{pmatrix} 0 & 1 \\ 1 & 0 \end{pmatrix}$ and $\begin{pmatrix} 0 & i \\ -i & 0 \end{pmatrix}$. These of course are the well known Pauli spin matrices.

While Weyl's answer to question (1) is stimulating and suggestive it is far from conclusive or logically compelling. It is significant, however, as being perhaps the first step in the program of deriving the basic equations and relationships in quantum mechanics from group theoretical symmetry principles. It is perhaps therefore appropriate to refer to it as Weyl's program.

II. Stone's Paper on one Parameter Unitary Groups.

Weyl wrote in the spirit of a physicist and made no attempt to give rigorous mathematical proofs of all of the mathematical theorems that played a role in his analysis. These gaps were largely filled in 1930, in a note published in the Proceedings of the U.S. A. National Academy of Sciences, by Marshall Stone. In it he announced the proofs of two theorems about unitary groups in Hilbert space implicitly conjectured by Weyl. One of these states that every (strongly continuous) unitary representation of the additive group of the real line is of the form $t \rightarrow e^{itQ}$ where Q is a uniquely determined self adjoint operator and the other is the theorem that the Heisenberg commutation relations in Weyl form have (to within unitary equivalence) a unique irreducible solution. The first of these theorems is well known as "Stone's theorem" and the second as the theorem of "Stone and von Neumann". The reason for the terminology in the second case is that von Neumann published a proof shortly after Stone's paper appeared and Stone did not publish a proof.

Actually Stone formulated his theorem about unitary representations of the additive group of the real line as an analogue of Hilbert's spectral theorem and not as a theorem connecting such representations with self adjoint operators. Since this formulation will be important for what follows we give it in detail. Let S be any set in which there is given a Borel structure, that is, a distinguished family of subsets containing the whole space and closed under countable unions, countable intersections and the taking of complements. By a projection valued measure P on S, one means a mapping $E \rightarrow P_E$ of the distinguished subsets E (usually called the Borel subsets) of S into projection operators P_E in some separable Hilbert space $\mathcal{H}(P)$ such that the following conditions hold.

(1) $P_{\{0\}} = 0$ and $P_S = I$.

(2) $P_{E \cap F} = P_E P_F$ for all Borel sets E and F.

(3) $P_{E_1 \cup E_2 \cup \cdots} = P_{E_1} + P_{E_2} + \cdots,$

whenever $E_1, E_2 \cdots$ are Borel subsets of S such that $E_i \cap E_j = 0$ whenever $i \neq j$.

Since this concept is not as familiar to physicists as it could be – and perhaps ought to be, I pause to give some examples and make some remarks.

Example 1. Let H be a self adjoint operator with a discrete spectrum. Let $\lambda_1, \lambda_2, \cdots$ be its distinct eigenvalues and let M_{λ_j} be the closed subspace of the underlying Hilbert space spanned by all eigenvectors with eigenvalue λ_j. For each Borel subset E of the real line let P_E denote the projection operator whose range is the closed linear subspace spanned by the union of all M_{λ_j} with λ_j in E. One verifies easily that $E \rightarrow P_E$ is a projection valued measure on the real line. It is easy to see that given P one can reconstruct H and its decomposition into eigenspaces.

Example 2. Let μ be a measure in S which is a finite in the sense that $S = E_1 \cup E_2 \cdots$ where $\mu(E_j) < \infty$ for all j. Let $\mathcal{H}$ denote the Hilbert space $\mathcal{L}^2(S, \mu)$ of all complex valued functions on S which are measurable and **square summable**

with respect to μ. For each Borel subset E of S let us define P_E^μ to be the operator which takes f in $\mathcal{L}^2(S, \mu)$ into $\phi_E f$ where ϕ_E is the function which is 1 or E and 0 on $S - E$. One checks at once that P_E^μ is a projection operator and that $E \to P_E^\mu$ is a projection valued measure.

The projection valued measures of example 2 come close to being universal. To state this precisely one needs the notion of the direct sum of two projection valued measures. Let P and P' be any two projection valued measures on the same Borel space S. Then their direct sum $P \oplus P'$ is the projection valued measure whose Hilbert space is the direct sum $\mathcal{H}(P) \oplus \mathcal{H}(P')$ and $(P \oplus P')_E$ is the operator taking f_1, f_2 into $P_E(f_1), P_E'(f_2)$. Similarly one defines the direct sum of any finite or countably infinite family $P, P', P'' \cdots$ of projection valued measures on S.

Example 2 is almost universal in the sense that *every* projection valued measure on S can be shown to be a finite or countably infinite direct sum of projection valued measures of the form P^μ where μ varies over different measures on S.

It may be helpful to remark that whenever S is countable and every one point set is a Borel set, then $\mathcal{H}(P)$ is simply the direct sum of the ranges of the $P_{\{s\}}$ and this direct sum decomposition completely determines P. A projection valued measure on S *is* a direct sum of Hilbert spaces parametrized by the points of S. More generally it is intuitively useful to think of a general projection valued measure as a positively continuous direct sum or direct integral of "infinitesimal" Hilbert spaces parametrized by the points of S.

In terms of projection valued measures Stone's theorem may be formulated as follows. For every strongly continuous unitary representation U of the additive group of the real line there exists a unique projection valued measure P defined on all Borel subsets E of the real line such that $\mathcal{H}(P) = \mathcal{H}(U)$ and for all elements f of this Hilbert space $(U_x(f) \cdot f) = \int_{-\infty}^\infty e^{ixt} d(P_t(f) \cdot f)$. Moreover every projection valued measure P on the real line occurs for some U and the correspondence $U \to P$ is one-to-one and onto between unitary equivalence classes. In understanding the meaning of the integral in the above formula one should note that for each f the mapping $E \to (P_E(f) \cdot f)$ is an ordinary measure on the line and the integral is just the integral of the function $t \to e^{ixt}$ with respect to this measure.

III. The Imprimitivity Theorem.

Using Stone's theorem it is easy to rewrite the Stone-von Neumann theorem in a form which admits far reaching generalizations. Observe first that there is no difficulty in generalizing Stone's theorem to apply to unitary representations of the additive group of all k tuples of real numbers. One simply replaces projection valued measures on the line by projection valued measures on k space and the fundamental equation by

$$(U_{x_1, x_2, \cdots x_k}(f) \cdot f) = \int e^{i(x_1 t_1 + x_2 t_2 + \cdots x_n t_n)} d(P_{t_1, t_2, \cdots t_n}(f) \cdot f).$$

Now consider the Weyl form of the Heisenberg commutation relations.

$$(*) \qquad A_{t_1,t_2,\cdots t_k} B_{s_1,s_2,\cdots s_k} = e^{i(t_1 s_1 + \cdots + s_k t_k)} B_{s_1,s_2\cdots s_k} A_{t_1,t_2\cdots t_k}$$

where A and B can be regarded (in view of Stone's theorem) as otherwise unrestricted strongly continuous unitary representations of the additive group of all k tuples of real numbers.

The basic trick in putting the Stone-von Neumann theorem in a more generalizable form consists in applying the generalized Stone's theorem to one, but *not both*, of the two representations A and B. Let P denote the projection valued measure corresponding to B by this theorem. A straightforward calculation then shows that A and B satisfy $*$ if and only if A and P satisfy the commutation relations

$$(**) \qquad A_{t_1,t_2\cdots t_n} P_E = P_{E-(t_1,t_2\cdots t_n)} A_{t_1,t_2\cdots t_n}$$

where $E - (t_1 \cdots t_n)$ is the set of all k-tuples of the form $x_1 - t_1, x_2 - t_2 \cdots x_n - t_n$ with $x_1, x_2, \cdots x_n \in E$. It follows that seeking all possible solutions of $(*)$ is completely equivalent to seeking all possible solutions of $(**)$.

The reason that $(**)$ is more generalizable than $(*)$ is that the factor $e^{i(t_1 s_1 + \cdots + t_k s_k)}$ has disappeared and only the group structure of the common domain of A and P plays a role. Let G be any group with a Borel structure and let A, P be a pair consisting of a unitary representation A of G and a projection valued measure P defined on the Borel subsets of G. Then one can write down the following commutation relation which A and P might satisfy

$$(***) \qquad A_x P_E = P_{[E]x^{-1}} A_x$$

where $[E]x^{-1}$ is the set of all yx^{-1} with y in E. Evidently $(***)$ reduces to $(**)$ when G is the additive group of all k tuples of real numbers.

The present author noticed these things some forty years ago, after reading Stone's paper and wondering how one might go about proving the uniqueness theorem. He was then pleased to discover that he could prove a corresponding uniqueness theorem when the additive group of k tuples of real number was replaced by an arbitrary separable locally compact group G. The results were published in 1949 in the Duke Mathematical Journal.

Finding a particular solution of $(***)$ for any separable locally compact group G is quite easy. By a celebrated theorem of Haar there is an essentially unique invariant σ finite measure μ defined on all Borel subsets E of G and one obtains a unitary representation U of G in the Hilbert space $\mathcal{L}^2(G,\mu)$ by setting

$U_x(f)(y) = f(yx)$. This is the so-called regular representation of G. Also using the Haar measure μ one obtains the projection valued measure P^μ defined in II. A trivial calculation now shows that

$$U_x P^\mu_E = P^\mu_{[E]x^{-1}} U_x \text{ for all } x \text{ and } E$$

so that U and P^μ satisfy $(\ast\ast\ast)$. One can also show that no non trivial closed subspace of $\mathcal{L}^2(\sigma, \mu)$ is invariant under all U_x and all P^μ_E so that the pair is irreducible. Our generalization of the Stone-von Neumann theorem now states that for every separable locally compact group G the solution of $(\ast\ast\ast)$ first constructed is (up to unitary equivalence) the only one that exists.

We are not finished however. The regular representation of a separable locally compact group G has an obvious generalization which also has an associated projection valued measure which satisfies $(\ast\ast\ast)$. It is natural to investigate uniqueness in this case as well. Let H be any closed subgroup of G and let G/H denote the set of all right H cosets; that is, the set of all closed subsets of G of the form Hx where $x \in G$. Two such cosets are either disjoint or identical and so define a partition of G. If Hx is a right H coset and y is an element of G then $(Hx)y$ is by definition $H(xy)$. Each y thus defines a permutation of G/H. G/H inherits a Borel structure from that in G. Let us suppose that G/H admits a σ-finite measure μ which is preserved by all the transformations $Hx \to Hxy$. Then taking $\mathcal{L}^2(G/H, \mu)$ instead of $\mathcal{L}^2(G, \mu)$ we produce a unitary representation U as before by setting $U_x(f)(y) = f(yx)$ where now y is a right coset instead of a group element. P^μ is defined just as before but now E is a Borel subset of G/H rather than G. The identity $(\ast\ast\ast)$ holds just as before where now $[E]x^{-1}$ is the set of all transforms by x^{-1} of a set E of right H cosets. To have $(\ast\ast\ast)$ make sense it is not necessary that the domain of P be subsets of G but only subsets of a space S on which G acts.

Our question now is this. Let U' be any unitary representation of G and let P' be any projection valued measure defined on G/H such that U' and P' are collectively irreducible and satisfy $(\ast\ast\ast)$. Is the pair U', P' unitarily equivalent to the U, P^μ defined above?

The answer is no! Given any unitary representation L of the closed subgroup H of G one can define a generalization of the construction of U, P^μ and all of these pairs satisfy $(\ast\ast\ast)$. Two such pairs are unitarily equivalent if and only if the corresponding representations L are unitarily equivalent and such a pair is irreducible if and only if L is irreducible. One no longer has a uniqueness theorem but a classification theorem. The solutions of $(\ast\ast\ast)$ for a given coset space G/H correspond one-to-one modulo unitary equivalence to the unitary representations of the subgroup H. One gets a uniqueness theorem only when H has only one equivalence class of irreducible unitary representations, e.g., when H contains only the identity element e and $G/H = G$.

To motivate the generalized construction it is useful to note that $\mathcal{L}^2(G/H, \mu)$ can also be described as a space of functions on G which satisfy the identity

$$(\dagger) \qquad f(\xi x) = f(x) \text{ for all } \xi \in H \qquad x \in G.$$

It consists in fact of all measurable such functions such that $|f(x)|^2$ when regarded as a function on G/H has a finite integral with respect to μ. In our generalization we replace complex valued functions f by vector valued functions whose values are taken in the Hilbert space $\mathcal{H}(L)$ of the representation L of H and the identity $(\dagger)$ is replaced by

$$(\dagger\dagger) \qquad f(\xi x) = L_\xi(f(x)).$$

For such functions the self scalar product $(f(x) \cdot f(x))$ satisfies $(\dagger)$ and can be integrated with respect to μ as before.

Thus one is led to define a Hilbert space $\mathcal{H}'$ whose elements are measurable functions f from G to $\mathcal{H}(L)$ which satisfy the identity $(\dagger\dagger)$ and are square summable in the sense that

$$\int_{G/H} (f(x) \cdot f(x)) d\mu(x) < \infty.$$

Of course one identifies two functions when they are equal almost everywhere with respect to μ. If f is any element in $\mathcal{H}'$ then for all x its translate $y \to f(yx)$ is also clearly in $\mathcal{H}'$ and one obtains a (strongly continuous) unitary representation of G in $\mathcal{H}'$ by setting $U_x^L(f)(y) = f(yx)$. One associates a projection valued measure P^L with U^L by simply noting that the definition given above applies equally well to vector valued functions. It is easy to check now that the pair U^L, P^L satisfies the commutation relations $(* * *)$.

We may now give a formal statement of the theorem alluded to in the title of this section.

THEOREM. *Let U be a unitary representation of the separable locally compact group G and let H be a closed subgroup of G. Let P be any projection valued measure in G/H whose values are projections in $\mathcal{H}(U)$ such that U and P satisfy the commutation relations $(* * *)$. Then there exists a unitary representation L of H such that the pair U^L, P^L is unitarily equivalent to the pair U, P. Moreover if L and M are two unitary representations of H then the pair U^L, P^L is unitarily equivalent to the pair U^M, P^M if and only if L and M are unitarily equivalent. Finally the pair U^L, P^L is irreducible if and only if L is irreducible and any direct sum decomposition of a reducible L implies an associated direct sum decomposition of U^L, P^L. $U^{M^1 \oplus M^2}, P^{M^1 \oplus M^2} \simeq U^{M^1}, P^{M^1} \oplus U^{M^2} \oplus P^{M^2}$.*

This theorem was first stated and its proof sketched by the present author in 1949 in a note in the U.S.A. National Academy of Sciences Proceedings. Strictly speaking, it only makes sense as stated when G/H has an invariant measure μ. However, using so-called quasi invariant measures, this limitation is easily removed. It is to be noticed that one obtains as a by-product a natural way of passing from an arbitrary irreducible unitary representation L of an arbitrary closed subgroup H of a separable locally compact group G to a unitary representation U^L of the whole group G. This representation U^L of G is called the representation L induced by L. It turns out that very many infinite dimensional unitary representations can be obtained as induced representations U^L where L is finite dimensional and even one dimensional.

In view of the role that projective representations play in quantum mechanics it is natural to ask whether the imprimitivity theorem may be extended so as to apply to projective unitary representations. The answer is yes and the formulation of the generalized theorem is obvious once we have extended the notion of inducing, $L \to U^L$, to apply when L is a projective unitary representation of H. Let σ be any multiplier for the whole group G and let L be any unitary projective representation of H whose multiplier is the restriction of σ to H. In the definition of U^L for ordinary representations, given above, replace the identity $f(\xi x) = L_\xi f(x)$ by $f(\xi x) = \sigma(\xi, x)L_\xi f(x)$ and the defining equation $U_x^L(f)(y) = f(yx)$ by $U_x^L(f)(y) = f(yx)/\sigma(y, x)$. With these changes U^L becomes a projective unitary representation with multiplier σ – in brief a unitary σ representation of G. Thus the inducing construction generalizes to allow us to construct a σ unitary representation of G for each σ unitary representation of H.

IV. The Imprimitivity Theorem and Weyl's Program.

In generalizing from the uniqueness theorem for solutions of the Heisenberg commutation relations to the imprimitivity theorem it may appear that we have been indulging in a purely mathematical digression which has taken us far away from any conceivable application to physics. We shall now show that this is definitely not the case and that in fact the imprimitivity theorem is just what is needed to give a much better answer to Weyl's question (1) than could possibly have been given at the time Weyl's paper was written.

We shall proceed in steps as follows.

(1) We shall recall the conventional quantum mechanical model for the unconstrained motion of a spinless particle of mass m.

(2) We shall reformulate this model in a simple and natural manner which is more intrinsic and coordinate free and which at the same time involves a pair U, P satisfying $(***)$ in an essential way.

(3) We shall observe that the existence of a pair U, P satisfying $(***)$ follows from simple and natural axioms involving little more than group theoretical invariance principles.

(4) Using the imprimitivity theorem we shall find the most general possible pair satisfying the axioms and observe that the Heisenberg commutation relations

are satisfied in all cases. Solutions of $(* * *)$ other than that provided by the conventional model will include the conventional model for particles with spin and other kinds of "internal degrees of freedom".

In the conventional model for a free spinless particle one chooses a rectangular coordinate system in space so that the points of space are in one-to-one correspondence with triples of real numbers x, y, z. The Hilbert space of quantum states is then $\mathcal{L}^2(x, y, z)$ where the relevant measure is the usual Lebesgue or volume measures $dx\, dy\, dz$. The self adjoint operators X, Y, Z corresponding to the observables defined by the three coordinates are then $X(\psi) = x\psi \quad Y(\psi) = y\psi \quad Z(\psi) = z\psi$ and those corresponding to the x, y, z components of the momentum P_X, P_Y, P_Z are

$$P_X(f) = \frac{1}{i}\frac{\partial f}{\partial x} \quad P_Y(f) = \frac{1}{i}\frac{\partial f}{\partial y} \quad P_Z(f) = \frac{1}{i}\frac{\partial f}{\partial z}.$$

Here we have chosen our units so that Planck's constant h is 2π.

In reformulating this model our main observation is that the unbounded operators X, Y and Z may be usefully replaced by a single projection valued measure on space. For each Borel subset E of $S(=$ set of all $x, y, z)$ let $Q_E\psi = \psi\phi_E$ where $\phi_E(x, y, z) = 1$ when $x, y, z \in E$ and $\phi_E(x, y, z) = 0$ when x, y, z is not in E. Using the spectral theorem for self adjoint operators one can easily construct the Q_E when X, Y and Z are given and vice versa. Actually, given the Q_E one can immediately describe the self adjoint operator corresponding to any real valued Borel function of x, y and z. As functions of E the Q_E evidently constitute a projection valued measure as defined in II. Moreover each Q_E has a simple physical interpretation. It is the self adjoint operator corresponding to the observable which is one when the particle is in the set E and zero otherwise.

One can describe the operators P_X, P_Y, P_Z in a coordinate free way using a rather different maneuver. Let $\mathcal{E}$ denote the group generated by the translations and rotations in space. Then for each α in $\mathcal{E}$. The mapping $x, y, z \to (x, y, z)\alpha$ is volume preserving and hence Lebesgue measure preserving. Hence the linear operator U_α such that $U_\alpha(f)(x, y, z) = f((x, y, z)\alpha)$ is a unitary operator and one can show that $\alpha \to U_\alpha$ is a unitary representation of $\mathcal{E}$. Indeed the fact that this is so is a special case of a much more general theorem stated in III. Consider now the restriction of U to the subgroup of all members of $\mathcal{E}$ of the form $x, y, z \to x+\lambda, y, z$ that is the subgroup of all translations in the x direction. By Stone's theorem this group may be written in the form $\lambda \to e^{iA\lambda}$ where A is a self adjoint operator and a simple calculation shows that $A = P_X$. We construct P_Y and P_Z analogously. Just as giving the projection valued measure Q gives all possible coordinate observables so giving the group representation U gives not only P_X, P_Y, and P_Z but the linear momentum in any direction. Actually giving U gives still more. In addition to the one parameter subgroups of translation $\mathcal{E}$ has other one parameter subgroups; in particular one parameter subgroups of rotations about the origin. Dealing with these in an analogous fashion leads to calculation of the self adjoint operators corresponding to the angular momentum observables. The connection with the imprimitivity theorem is now immediate. It follows by specializing a calculation already mentioned in III that U and Q satisfy the identity $(***)$.

With the foregoing considerations at our disposal let us now reconsider Weyl's question (1). Weyl wanted to know why the position and momentum operators take the special form that they do as described explicitly at the beginning of this section. We have now refined the question somewhat by replacing the position and moment operators by the pair U, Q consisting of a unitary representation of $\mathcal{E}$ and a projection valued measure Q on space. We repair an earlier omission here by pointing out the easily verified fact that space may be identified with $\mathcal{E}/R$ where R is the closed subgroup of $\mathcal{E}$ consisting of all rotations about $0,0,0$, i.e., all α in $\mathcal{E}$ which carry $0,0,0$ with itself.

Let us suppose that we have a quantum mechanical particle moving in a space S and let $\mathcal{H}$ be its Hilbert space of states. We no longer assume at the outset that $\mathcal{H} = \mathcal{L}^2(S)$ or that $\mathcal{H}$ has any other special representation. In addition we do not assume any particular form for the operators corresponding to the position and momentum observables. Instead we merely assume that for each Borel set E of S we are given a self adjoint operator Q_E in $\mathcal{H}$, that for each α in $\mathcal{E}$ the isometry group of space that we are given a unitary operator U_α in $\mathcal{H}$ and that these satisfy certain natural conditions which we now explain. Q_E is the self adjoint operator corresponding to the observable which is one when the particle is in E and zero when it is not. Since this observable takes on only the values 0 and 1, Q_E must be a projection operator. Moreover with a little reflection it is easy to convince oneself that, in its dependence on E, it is natural to assume that it is a projection valued measure. We do. In other words we assume that the position observables of or system are described by some (completely unspecified) projection valued measure on S. Our second assumption is that the physics of our system is invariant under all rigid motions in space and hence that every α in $\mathcal{E}$ defines an automorphism of $\mathcal{H}$ which we may assume to be described by a unitary operator U_α. The mapping $\alpha \to U_\alpha$ would be a representation of $\mathcal{E}$ if it were not for the fact that for every complex c with $|c| = 1$, U_α and cU_α define the same automorphism. We may assume only that $\alpha \to U_\alpha$ is a projective representation of $\mathcal{E}$ with some multiplier σ. Our only other assumption is that Q and U must be related in a way that is forced on us by the very definition of invariance. If T is the operator corresponding to some observable $\mathcal{O}$ and T' is the operator corresponding to the transform by α then T' must equal the transform of T by U_α. In particular we must have

$$U_\alpha Q_E U_\alpha^{-1} = Q_{[E]\alpha^{-1}}$$

for all α in $\mathcal{E}$ and all Borel subsets E of $S = \mathcal{E}/R$.

But this last identity is just $(***)$.

In other words, we have found an *a priori* argument leading to a description of the quantum mechanics of a free particle by a pair U, Q consisting of a projective unitary representation U of $\mathcal{E}$ and a projection valued measure Q on $S = \mathcal{E}/R$; this pair necessarily satisfying the crucial identity $(* * *)$ of the imprimitivity theorem. The imprimitivity theorem itself then allows us to survey *all* possibilities for U, Q. It tells us in particular that to within unitary equivalence there is only a discrete countable infinity of possibilties – one for each equivalence class of unitary projective representations of the compact subgroup R of $\mathcal{E}$.

The simplest cases are those in which the inducing projective representation L of R is irreducible. It turns out that $\mathcal{E}$ has just two equivalence classes of projective multipliers and that the one not containing the identity can be chosen so as to take on only the values 1 and -1. These multipliers remain inequivalent when restricted to R. Thus there are two fundamentally different kinds of particles depending upon whether L is an ordinary representation or a projective representation with a non trivial multiplier. In the first case there is precisely one equivalence class of irreducible representations whose dimension is d for $d = 1, 3 \cdots$ and in the second case there is just one equivalence clss of irreducible projective representations of dimension d for $d = 2, 4, 6 \cdots$. It is customary to denote the projection representation whose dimension is d by the symbol $D_{\left(\frac{d-1}{2}\right)}$ whether its multiplier is trivial or not. The case in which L is irreducible and equal to $D_{\frac{d-1}{2}}$ coincides exactly with the model for a free particle of "spin $\frac{d-1}{2}$" which was introduced in the early days of quantum mechanics.

To summarize: Weyl based a suggestive, if not logically compelling, answer to his question (1) on the conjectured uniqueness of the solutions to the Heisenberg commutation rules and his elegant reformulation of this uniqueness theorem in terms of projective unitary representations of finite dimensional vector groups. We have just seen that a mathematically inspired generalization of the uniqueness theorem may be used to go much further in the direction Weyl wished to go.

It is possible to go further still. Just how far remains to be seen. Some steps in this direction were taken in the author's Oxford lectures during the academic year 1966-67. (See Chapters 17-24 in the published version of these lectures, "Unitary Group Representations in Physics, Probability and Number Theory", Benjamin Cummings 1978). The author is currently at work on extending the scope of the material in these chapters with the hope of making significant contact with a large part of present day elementary particle physics.

In the remaining sections we shall outline some of our ideas as well as relevant material from the book cited in the last paragraph.

V. The dynamics of a free particle and velocity operators.

Up to now we have only considered what might be called quantum statics. Nothing has been said about how the Hilbert space vectors describing the states of our particle change with time. It follows easily from an axiomatic approach to the Hilbert space formulation of von Neumann that there must be a unitary representation $t \rightarrow V_t$ of the additive group of the real line (thought of now as the group of translations in time) such that if ϕ defines the state of the system at some instant of time then $V_t(\phi)$ defines the state of the same system t times units later. In order to describe a "particle" completely we must not only be given Q and U but the unitary representation V. Weyl's question (1) must be amplified to include the question: (1') Given Q and U what can we say about V? Are there principles with some degree of *a priori* justification which limit the possibilities in some useful way? In the case of a "free" particle it is easy to convince one's self that the mathematical counterpart of freeness is the assumption that U_α and V_t must commute for all α and t – at least up to a multiplicative factor. Thus the mapping $\alpha, t \rightarrow U_\alpha V_t = W_{\alpha,t}$ must be a projective unitary representation of the group $\mathcal{E} \times T$ where T denotes the time translation group. Actually an analysis of the possible projective multipliers on $\mathcal{E} \times T$ shows that we may assume without loss of generality that U_α and V_t actually commute.

By Stone's theorem V_t may be written uniquely in the form $V_t = e^{itH}$ where H is a self adjoint operator and one checks easily that $V_t U_\alpha = U_\alpha V_t$ for all t and α if and only if $U_\alpha H = H U_\alpha$ for all α. Finding the most general possible V is equivalent to finding the most general self adjoint operator H which commutes with all U_α. This is a problem in the theory of unitary representations which standard theory enables one to solve very completely by exploiting the fact that U is induced by a representation L of G. The answer is especially simple when L is irreducible and one-dimensional so that we are dealing with a free particle of zero spin. In that case one cam forget group representations and prove that H is necessarily of the form $\rho\left(-\left(\frac{\partial^2}{\partial x^2} + \frac{\partial^2}{\partial y^2} + \frac{\partial^2}{\partial z^2}\right)\right)$ where ρ is an arbitrary real valued Borel function defined on the positive real axis. (We assume of course that we have realized our Hilbert space as $\mathcal{L}^2(x, y, z)$ so that the operators corresponding to the coordinate and momentum observables have their conventional form.) To define ρ we observe that the Fourier transform of the operator $-\left(\frac{\partial^2}{\partial x^2} + \frac{\partial^2}{\partial y^2} + \frac{\partial^2}{\partial z^2}\right)$ is multiplication by $u^2 + v^2 + w^2$ so that $\rho\left(-\left(\frac{\partial^2}{\partial x^2} + \frac{\partial^2}{\partial y^2} + \frac{\partial^2}{\partial z^2}\right)\right)$ is the inverse Fourier transform of the operator of multiplication by $\rho(u^2 + v^2 + w^2)$.

Without making further assumptions one can conclude no more. *Any* choice of ρ will lead to a V satisfying all assumptions. One can select out the conventional choices for ρ by imposing space time invariance. The group $\mathcal{E} \times T$ occurs in a canonical way as a closed subgroup of the group $\mathcal{G}$ of all automorphisms of space time in the sense of Galileo and also as a closed subgroup of the Poincarè group $\mathcal{P}$ of all automorphisms of space time in the sense of Einstein. One can show that to have our particle space time invariant in the sense of Galileo (respectively Einstein) is equivalent to assuming that W has an extension to a (possibly projective) unitary representation $\widetilde{W}$ of $\mathcal{G}$ (respectively $\mathcal{P}$). Here the projective multiplier for $\widetilde{W}$ must restrict on $\mathcal{E} \times T$ to the given projective multiplier of $\mathcal{E} \times T$. It turns out that in the case of $\mathcal{G}$, $\widetilde{W}$ will have a non trivial projective multipier even when W is an ordinary representation of $\mathcal{E} \times T$.

Assuming Galilean relativity one can conclude that H must be multiplication by a constant λ which without loss of generality may be taken to be positive. Thus H is an operator of the form $-\lambda \left(\frac{\partial^2}{\partial x^2} + \frac{\partial^2}{\partial y^2} + \frac{\partial^2}{\partial z^2} \right)$ where λ turns out to depend upon the so-called mass m of the particle. $\lambda = \frac{1}{2m}$. On the other hand, one can deduce from the assumption of Einstein's relativity that ρ must be a function of the form $r \to \rho(r) = \sqrt{M + c^2 r}$ where c is a universal constant equal to the velocity of light and M is a constant which depends on the particle and can be shown to be equal to $m^2 c^4$ where m is the so-called "rest mass" of the particle.

If we investigate the possibilities for V (or H) when L remains irreducible but is no longer one-dimensional we encounter a somewhat wider range of possibilities. Let ℓ denote the dimension of the space of L. The most general possible H is then specified by means of, not one, but ℓ independent functions ρ. To be more precise we note that the Hilbert space of our particle may be conveniently represented as the direct sum $\mathcal{H}_0 \oplus \mathcal{H}_0 + \cdots \oplus \mathcal{H}_0$ of ℓ replicas of $\mathcal{H}_0 = \mathcal{L}^2(x, y, z)$ in such a way that the position and linear momentum operators are simply direct sums of replicas of the corresponding operators Q_E^0 and U_α^0 (α a translation) in $\mathcal{H}_0$. It will be convenient to introduce some notation to describe operators formed in this way from operators in $\mathcal{L}^2(x, y, z)$. If T is any linear operator in $\mathcal{H}_0$ we shall denote the operator which takes $\phi_1, \phi_2, \cdots \phi_\ell$ in $\mathcal{H}_0 \oplus \mathcal{H}_0 + \cdots \mathcal{H}_0$ into $T(\phi_1), T(\phi_2) \cdots T(\phi_\ell)$ by $\tilde{T}$. Also if $A = \{a_{ij}\}$ is any $\ell \times \ell$ matrix of complex numbers we shall use $\hat{A}$ to denote the operator in $\mathcal{H}_0 \oplus \mathcal{H}_0 \oplus \cdots \mathcal{H}_0$ which takes $\phi_1, \phi_2, \cdots \phi_\ell$ into $\psi_1, \psi_2 \cdots \psi_\ell$ where $\psi_i = \sum_{j=1}^{\ell} a_{ij} \phi_j$. It is obvious that $\hat{A}$ and $\tilde{T}$ commute for all A and T and easy to show that the set of all $\tilde{T}$ is precisely the commutator of the set of all $\hat{A}$ and vice versa.

In these terms what one can prove is the following. There exist ℓ self adjoint operations $H_1, H_2, \cdots H_\ell$ in $\mathcal{H}_0 \oplus \mathcal{H}_0 \oplus \cdots \mathcal{H}_0$, each of which commutes with all U_α and hence is a possible H, such that the most *general* possible H is uniquely of the form

$$\hat{T}_{\rho_1} H_1 + \hat{T}_{\rho_2} H_2 + \cdots + \hat{T}_{\rho_\ell} H_\ell$$

where $T_{\rho_j} = \rho_j \left(-\left(\frac{\partial^2}{\partial x^2} + \frac{\partial^2}{\partial y^2} + \frac{\partial^2}{\partial z^2} \right) \right)$ and each ρ_j is a Borel function.

To complete this description of the possible operators H one must specify the operators $H_1, H_2, \cdots H_\ell$ for each $\ell = 1, 2, 3, \cdots$. We shall content ourselves with a statement of the facts in the special case in which $\ell = 2$. Here H_1 is the identity operator and H_2 is the operator

$$\frac{1}{i} \widetilde{\frac{\partial}{\partial x}} \hat{S}_1 + \frac{1}{i} \widetilde{\frac{\partial}{\partial y}} \hat{S}_2 + \frac{1}{i} \widetilde{\frac{\partial}{\partial z}} \hat{S}_3$$

where S_1, S_2 and S_3 are the well known Pauli spin matrices $S_1 = \begin{pmatrix} 0 & 1 \\ 1 & 0 \end{pmatrix}$, $S_2 = \begin{pmatrix} 0 & -i \\ i & 0 \end{pmatrix}$ and $S_3 = \begin{pmatrix} 1 & 0 \\ 0 & -1 \end{pmatrix}$. Since the operators $\tilde{T}_\rho$ are obvious generalizations of the possible H's for the spinless case, H_1 is the essentially new constituent in the spin 1/2 case. A straightforward computation shows that $H_2^2 = \hat{T}_0$ where $T_0 = -(\frac{\partial^2}{\partial x^2} + \frac{\partial^2}{\partial y^2} + \frac{\partial^2}{\partial z^2})$. Thus H_2 is a square root of the Laplace operator of a quite different character from that obtained by Fourier transforms or the operational calculus.

Just as in the spin zero case one can sharply limit the possibilities for ρ_1 and ρ_2 and hence for H by imposing space time invariance. One finds that Galilean invariance holds if and only if $\rho_2 = 0$ and ρ_1 is multiplication by a constant and that Einsteinian invariance holds if and only if $\rho_1 = 0$ and $\rho_2(r)$ is of the form $\pm\sqrt{\frac{m^2 c^4}{r} + c^2}$ where c is once again the velocity of light and m is the rest mass of the particle.

There is a logical gap in the analysis of section IV in that we have tacitly identified the self adjoint operators associated with the restrictions of U to the one-parameter subgroups of $\mathcal{E}$ with those defining the momentum observables. The most obvious justification for the procedure is that the identification holds in the classical model of a spinless quantum particle. We recall however that our aim was to derive the classical model from axioms, so this justification is not acceptable. In order to repair this gap, as well as for later use in discussing particle interactions, we close this section with a discussion of velocity operators in quantum mechanics and their relationship to momentum operators.

In classical mechanics the linear momentum of a particle is defined as the mass of the particle multiplied by its velocity. Moreover the reason for the usefulness of the momentum concept is the invariance principle known as the law of conservation of momentum. One could approach the problem of defining linear momentum in quantum mechanics either by seeking an analogous invariance principle or by first seeking the operators which define the velocity observables and multiplying them by a suitably defined mass. It is gratifying to find that the two approaches lead to the same result.

Let H denote the dynamical operator discussed above and let X, Y, Z denote the self adjoint operators defining the x, y and z coordinates. Then as a function of time the expected value of the x coordinate on the trajectory $e^{iHt}(\phi)$ will be $(Xe^{iHt}(\phi) \cdot e^{iHt}(\phi))$ with analogous expressions for the expected values of the y and z coordinates. If these expected values have time derivatives at $t = 0$ their values will clearly be the expected values in the state ϕ of the self adjoint operators $\frac{1}{i}(HX - XH)$, $\frac{1}{i}(HY - YH)$ and $\frac{1}{i}(HZ - ZH)$. Whether these exist or not evidently depends upon H and when they do they must clearly be the operators corresponding to the x, y and z components of the particle velocity. Using what we have learned about the nature of H we may compute the velocity operators explicitly. In the case in which the particle has spin zero so that $H = \rho(-(\frac{\partial^2}{\partial x^2} + \frac{\partial^2}{\partial y^2} + \frac{\partial^2}{\partial z^2}))$ one computes that

$$\frac{1}{i}(HX - XH) = +\frac{2}{i}\rho'(-(\frac{\partial^2}{\partial x^2} + \frac{\partial^2}{\partial y^2} + \frac{\partial^2}{\partial z^2}))\frac{\partial}{\partial x}$$

where ρ' is the derivative of the function ρ. Of course the argument fails when ρ fails to have an almost everywhere defined derivative. Similarly

$$\frac{1}{i}(HY - YH) = \frac{2}{i}\rho'(-(\frac{\partial^2}{\partial x^2} + \frac{\partial^2}{\partial y^2} + \frac{\partial^2}{\partial z^2}))\frac{\partial}{\partial y}$$

$$\frac{1}{i}(HZ - ZH) = \frac{2}{i}\rho'(-(\frac{\partial^2}{\partial x^2} + \frac{\partial^2}{\partial y^2} + \frac{\partial^2}{\partial z^2}))\frac{\partial}{\partial z}.$$

If we denote the self adjoint operator $2\rho'(-(\frac{\partial^2}{\partial x^2} + \frac{\partial^2}{\partial y^2} + \frac{\partial^2}{\partial z^2}))$ by M we see that the operators defining the velocity observables may be written in the form $M(\frac{1}{i}\frac{\partial}{\partial x})$, $M(\frac{1}{i}\frac{\partial}{\partial y})$, $M(\frac{1}{i}\frac{\partial}{\partial z})$ where M is a self adjoint operator which commutes with all U_α and hence with H, $\frac{1}{i}\frac{\partial}{\partial x}$, $\frac{1}{i}\frac{\partial}{\partial y}$, $\frac{1}{i}\frac{\partial}{\partial z}$.

Note that in the special case in which our particle is Galilean invariant the operator M is a constant operator $\frac{1}{m}I$ where m is the classical mass of the particle. Hence in this case the abstract (group theoretical) definition of linear momentum agrees with the classical definition according to which the momentum is the mass times the velocity.

More generally the operator M will not be a constant operator but it will still be true that the group theoretically defined linear momentum operators $\frac{1}{i}\frac{\partial}{\partial x}$, $\frac{1}{i}\frac{\partial}{\partial y}$, $\frac{1}{i}\frac{\partial}{\partial z}$ may be written in the form

$$\frac{1}{M}(V_x), \quad \frac{1}{M}(V_y), \quad \frac{1}{M}V_z$$

where V_x, V_y, V_z are the velocity operators. What is, then, the meaning of the non constant self adjoint operator $\frac{1}{M}$? To answer this question let us consider the spectral resolution of $\frac{1}{M}$. Using it, the Hilbert space may be written as a discrete direct sum of time invariant subspaces in each of which $\frac{1}{M}$ has its spectrum confined to a very short interval. In each of these invariant subspaces $\frac{1}{M}$ will differ only slightly from a constant operator and the motion of our particle will differ only slightly from that of Galilean particle whose constant mass is in the very short interval just alluded to. *In other words our particle will move just like a Galilean particle whose mass is not constant but varies with its velocity in a manner indicated by the preceding discussion.*

It is interesting to compare this mass velocity relationship with the well known one predicted by applying Einstein's special theory of relativity to classical mechanics. First of all we see that some such variation is predicted not just by

Einsteinian relativity but by *any* denial of Galilean relativity. Secondly it is easy to check that the particular form of the variation predicted by the present method, in the case when ρ has the special form demanded by Einsteinian invariance, coincides with the one found by Einstein.

To keep this paper from getting too long we shall *not* at this time investigate what happens in the case of higher spin.

VI. Reducibility of L and connections with particle multiplets.

In section V and the latter part of section IV we have confined attention to the special case in which the pair U P is irreducible or equivalently (once the imprimitivity theorem has been invoked to write $U = U^L$ $Q = Q^L$) the case in which L is irreducible. It is time to explore what happens when this restriction is removed.

The simplest case is that in which L is the direct sum of two irreducible representations L^1 and L^2 so that U^L is $U^{L^1} \oplus U^{L^2}$ and in which the direct sum decomposition (it will not be unique if L^1 and L^2 are equivalent) can be so chosen that the V_t leave $\mathcal{H}(U^{L^1})$ and $\mathcal{H}(U^{L^2})$ separately invariant. In that case the system is simply the direct sum of two systems U^{L^1}, V^1 and U^{L^2}, V^2 in which L^1 and L^2 are both irreducible. Since L^1 and L^2 are irreducible the component systems describe particles with definite spins as explained above. How shall we interpret the direct sum of two systems each of which describes a particle? A little reflection makes it clear that the corresponding physical system can be thought of as a "particle" which has two "natures" or two "identities". Depending upon whether the state vector ϕ is in $\mathcal{H}(U^{L^1}), \mathcal{H}(U^{L^2})$ or is a sum $\phi_1 + \phi_2$ of two non zero vectors with ϕ_1 in $\mathcal{H}(U^{L^1})$ and ϕ_2 in $\mathcal{H}(U^{L^2})$ the particle will exhibit its first nature, exhibit its second nature or exhibit each nature with a certain probability. The extension to the case in which L is the direct sum of three or more irreducible representations should be obvious.

There is nothing very profound in all this. We have merely seen that it is possible to maintain a consistent particle interpretation for any system satisfying our axioms – even when L is reducible – provided that V happens to be compatible with the reduction of the system U^L, P^L in the manner described above. Actually it is sometimes both technically and conceptually convenient to group two or more particles together and consider them as a single entity in this way. In the case of the neutron and the proton this circumstance was first exploited by Heisenberg in his 1932 paper on nuclear structure. His point of view of course was quite different from ours. In the neutron-proton case the new joint entity has even been given a name. It is called the nucleon.

One can be led to such "groupings" of particles in a natural and suggestive way if one supposes *a priori* that the physical world is invariant under a somewhat larger group than the group $\mathcal{E}$ of all isometries; specifically a group of the form $\mathcal{E} \times K$ where K is some compact group which acts trivially on space. The analysis of sections four and five goes through without essential change, and leads to conclusions of the same sort, except that now projective representations of the rotation group R are replaced by projective representations of the product group $R \times K$. For simplicity let us assume that it is such that the only projective multipliers of $\mathcal{E} \times K$ are direct products of projective multipliers of $\mathcal{E}$ with the identity multiplier on K. Then the most general possible irreducible projective unitary representation of $R \times K$ will be of the former $L \times M$ where L is one of the irreducible unitary projective representations of R already discussed and M is some irreducible unitary representation of K. The unitary projective representation $U^{L \times M}$ describing the symmetry of our "particle" is easily shown to be equivalent to $U^L \times M$. From the point of view of an observer unaware of the symmetry group K this representation appears as its restriction to $\mathcal{E} \times e$ which is just U^L repeated as many times as the dimension of $\mathcal{H}(M)$. For those "particles" with $\mathcal{H}(M)$ one-dimensional there is nothing new, but for those for which $\mathcal{H}(M)$ has a dimension two or greater the unaware observer sees several different particles which are more properly regarded as the "same" particle in different identities or different "internal states" – related to symmetry under the group K in much the same manner as the different spin states are related to symmetry under the group R.

Although the technical description becomes more elaborate the same considerations apply when the symmetry is only "approximate". The neutron-proton pairing is a case in point. Here the group K is $SU(2)$ which is a double covering of R so that the internal degrees of freedom are quite similar to those associated with spin. Indeed when this symmetry was recognized the term "isotopic spin" was introduced to emphasize its similarity to spin. One says that the nucleon is a "particle" with isotopic spin $1/2$. After the discovery of the pi mesons and their relationship to nucleon nucleon scattering it became clear that the three differently charged pi mesons fitted into the same scheme and could be regarded as components of an isotopic triplet. The pion is a "particle" with isotopic spin 1. Later still when the properties of the new so-called strange particles began to be known it was pointed out by Gell-Maun and Ne'eman that these could be best understood by assuming that the symmetry group $\mathcal{E} \times SU(2)$ is a subgroup of a (less exact) symmetry group $\mathcal{E} \times SU(3)$. Now if M is any irreducible unitary representation of $K = SU(3)$, and one looks at the "particle" associated with $L \times M$ for some irreducible projective unitary representation L of R, one can consider it not only from the point of view already described in which the observer knows only about $\mathcal{E}$ but also from the point of an observer who knows about $\mathcal{E} \times SU(2)$ but not about $SU(3)$. This observer will restrict $U^{L \times M} = U^L \times M$ to $\mathcal{E} \times SU(2)$ and get $U^L \times M{\restriction}$ where $M{\restriction}$ is the restriction of M to $SU(2)$ and may be reduced to a direct sum of irreducible representations of $SU(2)$. If $M{\restriction} = M^1 \oplus M^2 + \cdots M^n$ where $M^1, M^2, \cdots M^n$ are irreducible representations of $SU(2)$, then $U^L \times M\uparrow = U^L \otimes M^1 \oplus U^L \otimes M^2 \oplus \cdots U^L \times M^n$. Thus our observer will see n isotopic multiplets having perhaps different isotopic spins. Taking $SU(3)$ into account these are all part of a single "supermultiplet" in the same sense that the individual particles are parts of multiplets.

Gell-Mann and Ne'eman (independently) observed that $SU(3)$ has an eight-dimensional irreducible representation whose restriction to $SU(2)$ is the direct sum $D_0 + 2D_{1/2} + D_1$ where D_j is the unique irreducible representation of dimension $2j + 1$. Taking this eight-dimensional representation for M one would have a supermultiplet made up of one isotopic singlet, two isotopic doublets and one isotopic triplet. This just fitted the then known particles of spin $1/2$ and mass greater than or equal to that of the proton. In addition to the nucleon one had the Λ hyperon, the Ξ hyperon and the Σ hyperon which were respectively an isotopic singlet, an isotopic doublet and an isotopic triplet. Because of the dimension of the appropriate M one spoke of the theory as the "eightfold way". As more particles (or resonances) were found these too could be fitted into supermultiplets but with M sometimes having a dimension other than eight. The failure of a three-dimensional representation to occur led to the famous "quark" hypothesis of Gell-Mann and Zweig in 1964.

VII. Particle interactions and the gauge principle.

Up to this point we have said nothing about how quantum mechanics deals with interactions between particles. This of course is a serious omission since isolated particles do not exist in nature and, if they did, would not do anything very interesting. On the other hand, Weyl said nothing about extending his program to explain why interactions take the general form that they do. Presumably this is because once one knows the quantum mechanical formalism for free particles it is easy in many cases to guess the quantum mechanical counterpart of known classical interactions.

Nevertheless it is interesting and illuminating to put classical mechanics aside and attempt to understand interactions between quantum mechanical particles from first principles in the spirit of Weyl's original program. After all classical mechanics is presumably only a limiting form of the truer and richer quantum mechanics and important features can easily be blurred or eliminated in the classical limit. We shall describe here some incomplete but encouraging efforts in this direction.

The first remark to be made is that if U^{L^1}, V^1 and U^{L^2}, V^2 define two distinct particles with Hilbert spaces $\mathcal{H}^1$ and $\mathcal{H}^2$ and one constructs the tensor product $\mathcal{H}^1 \otimes \mathcal{H}^2$ and the associated tensor products of the operators $U_\alpha^{L^1}, V_t^1, U_\alpha^{L^2}, V_t^2$ one obtains a system which has an immediate and obvious interpretation as the quantum mechanics of a system of two particles each of which is moving as though the other were absent. The question before us then becomes: What changes must be made when the particles interact so that the motion of each is affected by the behavior of the other. It is natural to assume that the position and momentum observables as described by U^{L^1} and U^{L^2} remain unchanged but that $t \to V_t^1 \otimes V_t^2$ must be replaced by some other one-parameter unitary group $t \to V_t$; the exact nature of which depends upon the particular particles.

One plausible assumption, with interesting consequences, is that the interaction leaves unchanged, not only the position and momentum observables, but also the velocity observables as defined in section 5. Let H^1, H^2, and H respectively denote the self-adjoint generators of V^2, V^2 and V. Then $H^1 \times I + I \times H^2$ is the self-adjoint generator of $V_t^1 \times V_t^2$ and H is of course completely determined by H^1, H^2 and $J = H - (H^1 \times I + I \times H^2)$. But our hypotheses and the definition of velocity observable imply immediately that J, the so-called interaction operator, must commute with all $P_E^{L^1}$ and all $P_E^{L^2}$. In the simplest special case, that in which L^1 and L^2 are one-dimensional identity representations, so that the particles are spinless, one concludes without difficulty that the Hilbert space $\mathcal{H}^1 \otimes \mathcal{H}^2$ may be realized as $\mathcal{L}^2(x_1, y_1, z_1, x_2, y_2, z_2)$ so that J is a multiplication operator of the form $\psi \to V\psi$ where V is some real-valued function defined on six space. Using invariance under simultaneous translation and rotation of both particles by the same amount one shows that V depends only on the distance $\sqrt{(x_1 - x_2)^2 + (y_1 - y_2)^2 + (z_1 - z_2)^2}$ so that $V(x_1, y_1, z_1, x_2, y_2, z_2)$ $= v(\sqrt{(x_1 - x_2)^2 + (y_1 - y_2)^2 + (z_1 - z_2)^2})$ for some real function v defined on the positive real axis.

The Schrödinger equation which results in the special case in which the two constituent free particles are subject to Galilean invariance is just that customarily assigned to two interacting particles whose classical equations of motion are determined by the potential energy function

$$V(x_1, y_1, z_1, x_2, y_2, z_2) = v(\sqrt{(x_1 - x_2)^2 + (y_1 - y_2)^2 + (z_1 - z_2)^2}\,)^2.$$

Now, however, we have a justification for this equation which is independent of classical mechanics. Indeed one can argue in reverse to show that certain features of classical mechanics that seem arbitrary are in fact consequences of the fact that classical mechanics is a limiting form of quantum mechanics – in a sense are "quantum effects".

When the particles have spin, a parallel analysis may be carried out to just as complete an extent but the results are more complicated and have no familiar classical analogues. The key difference is that a self-adjoint operator which commutes with all coordinate observables can no longer be proved to be multiplication by a real-valued function of the coordinates. Instead it is determined by an operator valued function of the coordinates in the following precise sense. Recall that one can realize the members of $\mathcal{H}^1$ and $\mathcal{H}^2$ respectively as vector valued functions on space having values in $\mathcal{H}(L^1)$ and $\mathcal{H}(L^2)$ respectively so that members of $\mathcal{H}^1 \otimes \mathcal{H}^2$ may be realized as functions of $x_1, y_1, z_1, x_2, y_2, z_2$ having values in $\mathcal{H}(L^1) \otimes \mathcal{H}(L^2)$. The interaction operation J will be defined by a function F of $x_1, y_1, z_1, x_2, y_2, z_2$

whose values are linear operators in the vector space $\mathcal{H}(L^1) \otimes \mathcal{H}(L^2)$. One can think of F as defining a "multiplication operator" as in the spinless case provided that one interprets the "product". A θ of an operator A with a vector θ is the transform $A(\theta)$ of θ by A. The analogue of the reduction from V to v in the spinless case can be carried out but involves a more elaborate discussion than seems appropriate in an article of this character. The final results can be interpreted as stating that in addition to an exact analogue of the "potential" found in the spinless case one must expect "potentials" corresponding to "forces" which depend upon the relative orientation of the spins of the particles. Such additional "forces" have been found experimentally and are especially important in the interactions of nucleons with one another.

While the immediately preceding considerations provide a certain amount of insight and illumination they can hardly be regarded as a final answer to the question. Our basic axiom about velocity observables is plausible but hardly logically compelling. Indeed it is clearly at most approximately true because it rules out interactions corresponding to so-called "velocity dependent forces" in classical mechanics. These occur in the interactions between charged particles and can only be neglected when the relevant velocities are small. It is thus gratifying to discover that a slight and plausible weakening of this axiom is also susceptible of a complete analysis and in the spinless case lets in just the missing velocity dependent "forces". Moreover when one examines the higher spin cases one finds the sort of new particle interactions implied by the introduction of "gauge fields" in the spirit of the work initiated in 1954 by Yang and Mills and continued so successfully by many others in the 1960's and after.

Our new assumption states not that the velocity operators are the *same* as they would be if the interaction were not there but only that the *commutation relations* between the velocity operators on the one hand and the position operators on the other are the same. In the case in which both particles are spinless this implies that each of the velocity operators for the two particles must differ from what it reduces to where there is no interaction by a multiplicative operator. In other words, there must exist six real-valued functions $A_1, B_1, C_1, A_2, B_2, C_2$ of the coordinates $x_1, y_1, z_1, x_2, y_2, z_2$ of the two particles such that the operator describing the x component of the velocity observable for the first particle is the sum of the operator $\frac{2}{i}\rho'\left(-\left(\frac{\partial^2}{\partial_1^2}+\frac{\partial^2}{\partial y_1^2}+\frac{\partial^2}{\partial x_1^2}\right)\frac{\partial}{\partial x_1}\right.$ and the operator which takes ψ into $A_1\psi$ and similarly for the other five velocity components. Once the six functions $A_1, B_1, C_1, A_2, B_2, C_2$ are known, the interaction operator is uniquely determined up to addition of an operator which commutes with all coordinate operators and hence is a multiplication operator in the above sense. In other words, under our new axiom the dynamics of two interacting spinless particles is determined by specifying not just one function V (the potential energy) but seven functions $A_1, B_1, C_1, A_2, B_2, C_2, V$. Of course invariance constraints put severe and useful constraints on the functions $A_1, B_1, C_1, A_2, B_2, C_2, V$ but time and space do not permit us to explore the matter here.

Passing on to the more difficult but important case in which the particles have nonzero spin one finds closely analogous results but in which the functions $A_1, B_1, \cdots$ are operator valued just as was the single function V in our earlier analysis. One is led into lengthy considerations involving linear algebra and group representations but finding all possible interactions is a doable problem. The author hopes to examine it in greater detail on a future occasion.

It is well known that in classical mechanics and in Galilean invariant quantum mechanics one can reduce the study of a pair of interacting particles to the study of a single particle moving in a "force field". It thus seems reasonable to attempt to gain insight into the above considerations by way of a parallel axiomitization which relates the velocity and position operators of a single free particle to those of a single particle which is not assumed to be free.

Consider a free spinless particle in the special case of Galilean invariance so that the operator H is of the form $-\frac{1}{2m}\left(\frac{\partial^2}{\partial x^2} + \frac{\partial^2}{\partial y^2} + \frac{\partial^2}{\partial z^2}\right)$. The velocity operators are then $v_x = \frac{m}{i}\frac{\partial}{\partial x}$, $v_y = \frac{m}{i}\frac{\partial}{\partial y}$ and $v_z = \frac{m}{i}\frac{\partial}{\partial z}$ and one computes at once that the velocity operators and the position operators X, Y, Z all commute with one another except that

$$v_x X - X v_x = \frac{m}{i}I, \quad v_y Y - Y v_y = \frac{m}{i}I, \quad v_z Z - Z v_z = \frac{m}{i}I.$$

Conversely suppose that we ask for the most general self-adjoint operator H' having these properties; that is, such that if v_x^1, v_y^1 and v_z^1 are defined by the equations

$$v_x^1 = \frac{1}{i}(H'X - XH'), v_y^1 = \frac{1}{i}(H'Y - YH'), v_z^1 = \frac{1}{i}(H'Z - ZH')$$

then $v_x^1 X - X v_x^1 = v_y^1 Y - Y v_y^1 = v_z^1 Z - Z v_z^1 = \frac{m}{i}I$ and all other pairs commute. One deduces at once that $v_x^1 - v_x$, $v_y^1 - v_y$ and $v_z^1 - v_z$ all commute with X, Y and Z. Hence each of these three operators is a multiplication operator. Hence $v_x^1 = \frac{m}{i}\frac{\partial}{\partial x} + \tilde{A}_1$, $v_y^1 = \frac{m}{i}\frac{\partial}{\partial y} + \tilde{A}_2$, $v_z^1 = \frac{m}{i}\frac{\partial}{\partial z} + \tilde{A}_3$ where A_1, A_2, A_3 denote real valued functions of x, y, z and $\tilde{A}_j$ denotes the operator $\psi \to A_j\psi$.

Now consider the self-adjoint operator

$$H_0 = \frac{1}{2m}\left[\left(\frac{1}{i}\frac{\partial}{\partial x} + \tilde{A}_1\right)^2 + \left(\frac{1}{i}\frac{\partial}{\partial y} + \tilde{A}_2\right)^2 + \left(\frac{1}{i}\frac{\partial}{\partial z} + \tilde{A}_3\right)^2\right].$$

One computes at once that H_0 and H' lead to exactly the same velocity operators and hence that $H' - H_0$ commutes with all position observables and so is a multiplication operator. In short the most general possible H' is $H_0 + \tilde{V}$ where $\tilde{V}$ is multiplication by the real-valued function V. But the operator $H_0 + \tilde{V}$ is precisely the one that occurs in the quantum mechanics of a charged spinless particle moving in an electric field of potential V and a magnetic field of vector potential $A = A_1, A_2, A_3$.

Notice that if we had made the stronger assumption that the velocity operators are the *same* for free and unfree particles then the magnetic field would have been forced out. Notice also that the physics is completely determined by the quadruple consisting of H', X, Y and Z. This means that if H'', X, Y, Z and H', X, Y, Z are unitarily equivalent quadruples H'' and H' yield the same physical predictions. But with a spinless particle the most general unitary operator which carries X into X, Y into Y and Z into Z is a unitary operator which commutes with X, Y, Z; that is, a multiplication operator $\tilde{U}$ in which U is a function of the form $x, y, z \to e^{i\sigma(x,y,z)}$ where σ is real valued. One computes at once that for such a U

$$U^{-1}(\frac{1}{i}\frac{\partial}{\partial x} + \tilde{A}_1)U = U^{-1}(\frac{1}{i}\frac{\partial}{\partial y})U + \tilde{A}_1 = \frac{1}{i}\frac{\partial}{\partial x} + \frac{\widetilde{\partial\sigma}}{\partial x} + \tilde{A}_1$$

with the obvious analogous formulas for $\frac{1}{i}\frac{\partial}{\partial y} + \tilde{A}_2$ and $\frac{1}{i}\frac{\partial}{\partial z} + \tilde{A}_3$. Thus $U^{-1}H'U$ will have the same form as H' and will differ from it only in that the vector potential A_1, A_2, A_3 will be replaced by $A_1 + \frac{\partial\sigma}{\partial x}$, $A_2 + \frac{\partial\sigma}{\partial y}$, $A_3 + \frac{\partial\sigma}{\partial z}$. This is interesting because in the classical theory of electricity and magnetism the components of the magnetic field defined by a given vector potential A_1, A_2, A_3 have the components $\frac{\partial A_3}{\partial x_2} - \frac{\partial A_2}{\partial x_3}$, $\frac{\partial A_1}{\partial x_3} - \frac{\partial A_3}{\partial x_1}$, $\frac{\partial A_2}{\partial x_1} - \frac{\partial A_1}{\partial x_2}$ so that A_1, A_2, A_3 and $A_1 + \frac{\partial\sigma}{\partial x}$, $A_2 + \frac{\partial\sigma}{\partial y}$, $A_3 + \frac{\partial\sigma}{\partial z}$ define the same magnetic field. What is interesting about it is the emergence of the result from abstract axioms. It is the embryo of the principle of gauge invariance and the multiplicative unitary operators U are called *gauge* transformations. The principle of gauge invariance as introduced by Weyl in his celebrated book of 1928, "Gruppentheorie und Quantummechanik" (page 100 of the English translation) differs from the above chiefly in allowing U as well as the functions A_1, A_2, A_3 and V to be time dependent.

Now consider the same problem with the restriction removed that the inducing representation L of the free particle be one-dimensional. In the simplest case L is just the two-dimensional identity representation and the free H is just as before, except that it now acts on functions which have two vectors as values instead of complex numbers. The velocity operators will change in the same way and obey the same commutation relations with X, Y and Z as before. Now, however, when one seeks the velocity operators associated with the most general H' having velocity operators satisfying these same commutation relations one finds that they are defined not by a triple A_1, A_2, A_3 of real-valued functions but by a triple A_1, A_2, A_3 of 2×2 self-adjoint matrix valued functions and that V is similarly replaced by a self-adjoint matrix-valued function. This suggests a generalization of an electro-magnetic field in which once again one has analogues of vector and scalar potentials, velocity dependent forces and a principle of gauge invariance.

However the gauge transformations will be defined by operator valued functions rather than scalar valued functions and will not commute with one another.

The author has not yet done the detailed calculations required to check this, but would be very surprised if pursuing this line of thought did not lead to something very close to, and possibly more general than, the modern theory of non Abelian gauge fields which originated in the celebrated work of Yang and Mills. If so, it is interesting that one could have been led to it by systematically seeking to extend Weyl's program in the manner indicated in this section. On page 216 of the author's book (mentioned at the end of section 4) one will find the words "Although more general cases are interesting and important, we shall consider here only the case in which A'_j, B'_j and C'_j are all zero so that there are no 'velocity dependent forces'". This sentence was written down in the fall of 1965 but unfortunately the author was not moved to look further in the direction indicated until quite recently.

VIII. Super symmetry.

Let us return to the theory of a single free particle and consider the case in which the particle has spin 1/2. We saw in section VI that the most general possible dynamical operator is of the form

$$\rho_1\left(-\left(\frac{\partial^2}{\partial x^2} + \frac{\partial^2}{\partial y^2} + \frac{\partial^2}{\partial z^2}\right)\right) I + \rho_2\left(-\left(\frac{\partial^2}{\partial x^2} + \frac{\partial^2}{\partial y^2} + \frac{\partial^2}{\partial z^2}\right)\right) H_2$$

where H_2 is the operator

$$\frac{1}{i}\frac{\widetilde{\partial}}{\partial x}\hat{S}_1 + \frac{1}{i}\frac{\widetilde{\partial}}{\partial y}\hat{S}_2 + \frac{1}{i}\frac{\widetilde{\partial}}{\partial z}\hat{S}_3 \quad \text{and} \quad S_1, S_2, S_3$$

are the Pauli spin matrices. Let us choose $\rho_1(r) = r$ and $\rho_2(r) \equiv 0$ so that our dynamical operator is $-(\frac{\partial^2}{\partial x^2} + \frac{\partial^2}{\partial y^2} + \frac{\partial^2}{\partial z^2}) = (P^2 + Q^2 + R^2) = T_0$ where for convenience we have denoted $\frac{1}{i}\frac{\widetilde{\partial}}{\partial x}$, $\frac{1}{i}\frac{\widetilde{\partial}}{\partial y}$, and $\frac{1}{i}\frac{\widetilde{\partial}}{\partial z}$ by P, Q and R respectively. Recall that $H_2^2 = F_0$ so that H_2 commutes with T_0. Also it is obvious that $\hat{S}_1, \hat{S}_2, \hat{S}_3$ commute with P, Q, R and hence with T_0.

We now focus our attention on the four dimensional vector space spanned by $\hat{S}_1, \hat{S}_2, \hat{S}_3, H_2$. Its self-adjoint members are all "symmetries" of the system in the sense that they commute with T_0. However this vector space has only zero in common with the usual Lie algebra of symmetries generated by $P, Q, R, T, 0$ and the angular momentum operators. Moreover it is not a Lie algebra in the sense of being closed under the usual bracket operation. Instead it is closed under a slightly different operation; namely the operation $A, B \to AB^2 - B^2A$. Given that $(B+C)^2 = B^2 + BC + CB + C^2$ one sees at once that closure under $A, B \to AB^2 - B^2A$ is equivalent to closure under the ternary operation $A, B, C \to A(BC+$

$CB) - (BC + CB)A$. This ternary operation may be regarded in turn as a sort of composite of the commutator $AB - BA$ and the anti-commutator $AB + BA$.

These extra symmetries are called super symmetries and the vector space generated by them and the usual Lie algebra of symmetries is called a Lie superalgebra. The two vector spaces, whose direct sum it is, are called the odd and even parts respectively. They are related in an interesting way which it is instructive to explore in the abstract.

Let V be any vector space of linear operators which is such that $AB^2 - B^2A$ is in V whenever A and B are in V. Let $\tilde{V}$ denote the vector space of all linear combinations of squares of elements of V. Then the following mildly surprising lemma has a very short and simple proof.

LEMMA 1. $\tilde{V}$ is a Lie algebra in the sense that $AB - BA$ is in $\tilde{V}$ whenever A and B are in $\tilde{V}$.

PROOF: It will clearly suffice to show that $A^2B^2 - B^2A^2$ is in $\tilde{V}$ whenever A and B are in V. But $A^2B^2 - B^2A^2 = A^2B^2 - AB^2A + AB^2A - B^2A^2 = A(AB^2 - B^2A) + (AB^2 - B^2A)A$. By hypothesis $AB^2 - B^2A = C$ is in V so $A^2B^2 - B^2A = AC + CA = (A + C)^2 - A^2 - C^2$ where C is in V. But since A and C are in V, $(A + C)^2 - A^2 - C^2$ is in V^2 and the proof is complete.

Since $\tilde{V}$ is in an obvious sense the *square* of V it is natural to call V a *square root* of $\tilde{V}$ and to define a Lie algebra square root to be any V closed under $A, B \to AB^2 - B^2A$.

In the example with which this section started V is the four-dimensional vector space generated by $\hat{S}_1, \hat{S}_2, \hat{S}_3$ and $H_2 = (\hat{S}_1P + \hat{S}_2Q + \hat{S}_3R)$. One checks easily that $\tilde{V}$ is the five-dimensional vector space generated by P Q R T_0 and I and is an invariant subalgebra of the Lie algebra generated by these and the angular momentum operators. Actually one can verify that V itself is invariant under commutation with the angular momentum operators and then the fact that $\tilde{V}$ is an invariant subalgebra is a consequence of a simple extension of Lemma 1.

LEMMA 2. *Let V be a Lie algebra square root and let $\mathcal{L}$ be a Lie algebra of operators in the same vector space such that $AB - BA$ is in V whenever A is in V and B is in $\mathcal{L}$. Then $\tilde{V} + \mathcal{L}$ the linear sum of $\tilde{V}$ and $\mathcal{L}$ is a Lie algebra and contains $\tilde{V}$ as an invariant subalgebra.*

The proof, which we omit, is very similar to that of Lemma 1.

The example of a Lie algebra square root presented above is quite special in that its square is a commutative Lie algebra. The first example to appear in the physical literature is of quite a different character in that it is two-dimensional and has a three-dimensional square which is not only non commutative but simple. In fact, it is possible to give a complete analysis of all possible two-dimensional Lie algebra square roots V and show that if $\tilde{V}$ is three-dimensional it is either commutative or simple. Obviously $\tilde{V}$ cannot be more than three-dimensional when V is two-dimensional.

The two-dimensional example mentioned above appeared in a paper of Neveu and Schwartz on string theory which appeared in 1971. It was constructed there in a rather complicated way. A much simpler construction was shown to the author by H. Bacry in the spring of 1986. Let P and Q be two self-adjoint operators which satisfy the Heisenberg commutation relations. Then the two-dimensional vector space of linear combinations of P and Q is readily checked to be a Lie algebra square root whose square is three-dimensional and simple.

Because of the fundamental role played by the Heisenberg commutation relations in the early stages of the implementation of Weil's program it is interesting to notice that Weyl's integrated or global form of the commutation relations has an analogue – or better an extension – which gives a global form to the Lie super algebra generated by the linear span of P and Q and its square. Consider the semi direct product $N \odot G$ of a two-dimensional real vector space N and the group G of all automorphisms of N which leave fixed a non degenerate projective multiplier σ on N. Then σ can be extended to a projective multiplier $\tilde{\sigma}$ on $N \odot G$ by setting $\tilde{\sigma}(n_1, x_1, n_2, x_2) = \sigma(n_1, n_2)$ for all n_1 and n_2 in N and all x_1 and x_2 in G. From the general theory of induced projective representations one can then show that there exists a unique irreducible $\tilde{\sigma}$ representation U of $N \odot G$ whose restriction to N is irreducible. The self adjoint generators of U restricted to N then generate a Lie algebra square root whose square is generated by the self adjoint generators of G.

IX.   Concluding Remarks..

The author had originally hoped to discuss the elements of quantum field theory from a point of view consistent with the foregoing. However there turned out to be neither enough time nor enough space for this to be possible. The author is very conscious of many other sins of omission and over sketchiness and asks the reader's indulgence on the grounds that the later sections must be regarded as a premature report of vague ideas and work in progress.

ACKNOWLEDGEMENTS: The material in section VIII was found while the author was a guest at the Max Planck Institut für Mathematik in Bonn in 1985-86 while being supported by a prize from the Alexander von Humboldt foundation. More details will be found in "Some remarks on Lie superalgebras", Czechoslovak Journal of Physics, Vol. B 37, 1987, pp. 373-386.

INDUCED REPRESENTATIONS AND THE APPLICATIONS
OF HARMONIC ANALYSIS

GEORGE W. MACKEY

HARVARD UNIVERSITY, DEPARTMENT OF MATHEMATICS

1. Introductory remarks.

During the past twenty years or so I have gradually shifted my interest from working on the development of harmonic analysis as a technically interesting branch of pure mathematics to studying harmonic analysis as a tool. As I have done so, I have become more and more impressed with the power and universality of this tool, and in particular with the extent to which such fundamental and disparate topics as the quantum mechanical analysis of matter and the theory of Diophantine equations can be organized and understood as applications of fundamental results in harmonic analysis. Indeed I now divide my time in roughly equal proportions between studies in elementary particle physics and studies in number theory – the emphasis in each case being the illumination provided by systematic use of harmonic analysis.

As should be clear from the title, when I speak of harmonic analysis I mean something much more far reaching than classical Fourier analysis; namely the unification of Fourier analysis with the representation theory of finite groups inaugurated in a famous paper by F. Peter and H. Weyl in 1927.

A few high points in the historical development of our subject include: (a) the introduction of the notion of group character in a problem in number theory by Gauss in 1801; (b) Fourier's memoir of 1807 on the use of his series in the theory of heat conduction; (c) the introduction of analytic methods into number theory by Jacobi in 1828; (d) Dirichlet's 1835 use of Fourier analysis in establishing the sign of the Gauss sums in number theory; (e) Dirichlet's implicit use of group characters in his 1839 proof of the infinity of primes in non degenerate arithmetic progressions; (f) Klein's work of 1870 and after in popularizing the group concept; (g) Dedekind's 1879 generalization of the Gauss character notion; (h) Frobenius' 1896 extension of characters to non commutative finite groups; (i) Artin's discovery in 1923 of the importance to number theory of Frobenius' non commutative characters; (j) the extension of Frobenius' theory to compact Lie groups by Schur and Weyl in 1924-1926; (k) the Peter-Weyl theorem of 1927; (ℓ) the applications of Frobenius theory to the new quantum mechanics by E. Wigner and H. Weyl in 1927; (m) the 1933 discovery of Haar measure; (n) the duality theory of locally compact commutative groups discovered by Pontrjagin and von Kampen in 1933 and 1934; (o) Wigner's 1939 determination of the irreducible unitary representations of the Poincaré group; (p) the 1946, 1947 papers of Bargmann and of Gelfand and Neumark inaugurating forty years of intensive activity in the infinite dimensional unitary representation theory of non compact locally locally compact groups; (q) the 1951 paper of Gelfand and Fomin connecting modular forms with the unitary representation theory of $SL(2,R)$; (r) Selberg's paper of 1956 introducing his "trace formula"; (o) the 1967 work of Langlands in generalizing some of Hecke's profound number theoretical work by first translating it into the language of group representations.

2. The central role of induced representations in harmonic analysis.

The basic observation of Peter and Weyl which recognized the group theoretical character of Fourier analysis and suggested its extension to other groups is that the functions $x \to e^{2\pi i n x}$ consist precisely of all continuous functions χ from the real line to the complex numbers of modulus one which are periodic of periodic one and satisfy the identity $\chi(xy) = \chi(x)\chi(y)$. In other words they are the continuous characters on the quotient group R/Z where R is the additive group of the real line and Z is the closed subgroup of all integers. Applied to square summable functions the Fourier expansion theorem becomes the theorem that the regular representation of R/Z is the direct sum of all the one-dimensional representations each occurring just once.

The generalization envisaged by Peter and Weyl replaced R/Z by an arbitrary compact Lie group G. However before very long it became clear that G could be any locally compact group and more generally still that one did not have to restrict attention to functions defined on a group but could apply the same ideas to functions defined on a space S on which a group G acts. Once this last generalization is accepted one sees that the theory of surface harmonics developed by Laplace and Legendre in the 1780's is just harmonic analysis for the case in which S is the unit sphere in three space and G is the rotation group.

A further natural generalization replaces complex valued functions on S by vector valued functions but this generalization is rather trivial unless one lets the vectors be elements of a vector space which varies from point to point of S and assumes that G acts not only on S but on the "vector bundle" whose "cross sections" are the vector valued functions under consideration.

In its most general form a vector bundle is, by definition, a set S and a family $s \to X_s$ of vector spaces X_s parametrized by the points of S. A cross section of the bundle is a vector valued function f defined on S such that for all s in S the vector $f(s)$ lies in X_s. One usually assumes that the vector spaces X_s have the same dimension for all s and a point of the bundle is a pair s, v where $v \in X_s$. S is called the base of the bundle and for each fixed s_0, the set of all s_0, v with v in X_{s_0} is called the fiber over s_0. A one-to-one transformation α of the set of points of a vector bundle onto itself is said to be a bundle automorphism if α maps each fiber onto some other fiber in such a way that it sets up a vector space isomorphism between the corresponding vector spaces. Usually S is more than just a set, being perhaps a measure space, a topological space or a differentiable manifold. One then has such a structure in the bundle also and bundle automorphisms are assumed to preserve this as well. The original motivating example was, no doubt, the tangent bundle to a differentiable manifold wherein X_s is the tangent space to the point s of the manifold S and the "vector fields" on the manifold are precisely the cross sections of this bundle.

In extending classical Fourier analysis in the indicated group theoretical direction one needs a substitute for Haar measure on G when replacing G itself by a space S on which G acts. Simply postulating the existence of a G invariant measure on S will not do as there are too many interesting and important examples in which it can be proved that no such measure exists. Fortunately there is a weaker condition which is strong enough for the needs of the program and holds much more widely. This is the condition that S admits a measure which is *quasi invariant* under the action of G in the sense that G carries sets of measure zero into sets of measure zero. If μ is such a measure then the unitary operator U_x^μ defined in $\mathcal{L}^2(S,\mu)$ by the formula $U_x^\mu(f)(s) = f([s]x)$ (where $[s]x$ denotes the transform of s by x) need not be unitary unless μ is actually invariant. However since μ is quasi invariant the measures $E \to \mu(E)$ and $E \to \mu([E]\,x)$ have the same sets of measure zero. This implies the existence of a "density" $s \to \rho(s,x)$ the so-called "Radon-Nikodym derivative" of $E \to \mu([E]x)$ with respect to $E \to \mu(E)$. If one now replaces the definition $U_x^\mu(f)(s) = f([s]x)$ by $U_x^\mu(f)(s) = \sqrt{\rho(s,x)}\, f([s]x)$ one finds that U_x is unitary just as in the special case in which μ is invariant.

It is important to notice that if μ is quasi invariant and μ' is a second measure with the same null sets then μ' is also quasi invariant and the two unitary representations U^μ and $U^{\mu'}$ are unitarily equivalent. In general one says that two measures with the same null sets lie in the same class and defines a measure class to be the set of all measures lying in the same class as some particular measure. With this terminology a quasi invariant measure (under a group G) may be equivalently defined as a measure whose class is invariant under G. Moreover to define the unitary representation U^μ whose decomposition is the generalization of harmonic analysis we have been describing, it is now clear that we do not need to specify any particular quasi invariant μ but only an invariant measure class.

The fact that invariant measure classes may be used instead of invariant measures is especially interesting in two cases.

Case I. S is a standard Borel space on which the separable locally compact group G acts transitively; i.e., such that the equation $[s_1]x = s_2$ may be solved for x for *every* pair s_1, s_2 of points of S.

Case II. S is a differentiable manifold and G is any group of diffeomorphisms of S.

In case I S admits one and only one invariant measure class and in case II S admits a canonical measure class which is invariant. This canonical measure class is the unique class whose members are locally in the class of Lebesgue measure.

Putting all this together we suppose that we are given a set S on which a separable locally compact group G acts and a vector bundle B with base S. We suppose that the fibers of B are Hilbert spaces and that G acts on the whole bundle as a group of bundle automorphisms in a manner consistent with the action of G on S. Finally we suppose that we are given a G invariant measure class C in S. These things being given, and being assumed to satisfy the technical regularity requirements described more precisely above, we may construct a unitary representation U of G acting in the space of μ square summable cross sections of B for some μ in C. To within unitary equivalence this representation is independent of the choice of μ in C. It is the decomposition of representations U of *this particular form* that we have in mind when we speak of harmonic analysis.

However there is a sense in which this definition of harmonic analysis is a bit too general. Consider the special case in which S is the unit ball in three space (that is, the set of all real triples x, y, z with $x^2 + y^2 + z^2 \leq 1$), G is the group of all rotations about $0, 0, 0$ and B is the trivial one-dimensional bundle so that cross sections are just complex-valued functions on S. Choose μ to be the restriction to S of Lebesgue measure. In this case there is a part of the decomposition of U which has a purely geometrical character and is completely different in spirit from the decomposition produced by Fourier analysis. This is because it is possible in many ways to decompose S as a sum of disjoint measurable subsets each of positive measure and each invariant under G. Given any such decomposition $S = S_1 \cup S_2$ the Hilbert space $\mathcal{L}^2(S, \mu)$ decomposes as the direct sum of $\mathcal{L}^2(S_1, \mu_1)$ and $\mathcal{L}^2(S_2, \mu_2)$ where μ_1 and μ_2 are the restrictions of μ to S_1 and S_2 respectively. These sub Hilbert spaces are U invariant and so define a decomposition of U. This process may be iterated and in the limit one finds that U is a "direct integral" of representations each of which is defined by replacing S by the sphere S^r consisting of all x, y, z with $x^2 + y^2 + z^2 = r^2$ where $0 < r < 1$. Of course $\mu(S^r) = 0$ for all r but μ is an integral over r of measures μ^r concentrated in the S^r and it is these that go into the definition of the "infinitesimal" constituents U^r the direct integral decomposition of U.

The U^r can of course be further decomposed. Indeed they are just the representations that one decomposes in the group theoretical version of the classical theory of spherical harmonics mentioned above. However this decomposition cannot be carried out by examining the geometry of the action of G on the S^r. Indeed G acts transitively on each S^r and so there are no invariant subsets.

It turns out that it is always possible to fiber the underlying space S into sets which cannot be further decomposed and then write the measure μ as an integral over measures in the fibers – just as in the above fibering of the unit ball into spheres S^r and thus to obtain a partial decomposition of U with one component for each fibre. To go further one must use completely different methods. We shall refer to this further decomposition as *harmonic analysis proper*. Evidently harmonic analysis proper is just the general harmonic analysis defined above in the special case in which it is *impossible* to write S as the disjoint union of two *measurable G invariant* sets neither of which is of measure zero. This condition is a condition on the system S, C, G and does not depend on the vector bundle B with base S. The concept which it defines is fundamental in the theory of groups of measure preserving transformations – and extends immediately to groups of measure class preserving transformations. It was introduced in a special case by G.D. Birkhoff and Paul Smith in 1928 where it was labeled "metric transitivity". After its connection with the ergodic problem in statistical mechanics was recognized two or three years later it began to be more common to label it with the shorter word "ergodicity". One says that the action of G on S is *ergodic* if the condition holds. Thus harmonic analysis proper is concerned with the decomposition of the group representation defined by G invariant Hilbert space bundles defined over *ergodic G* spaces.

There is an important dichotomy which divides ergodic G spaces rather sharply into two classes. Consider the "orbits" in S of the G action that is the sets $[s_0]x$ where s_0 is fixed in S and x varies through G. If any one orbit is of positive measure then the union of *all* other orbits is necessarily of measure zero and may be discarded without essential loss of generality. One is reduced to the so-called transitive case in which there is just one orbit. However it also may happen that every orbit is of measure zero. We shall refer to this as the properly ergodic case. It is the existence of properly ergodic actions that makes ergodic theory a deep and interesting subject. There is no way of reducing their study to the transitive case.

This dichotomy of course brings about an associated dichotomy in harmonic analysis proper. The underlying ergodic action may be transitive or it may be properly ergodic. Both cases are equally deserving of intensive study but up to now rather little has been done in the more difficult properly ergodic case. We pass on to the transitive case with the hope that mathematicians will soon begin a concerted attack on the almost virgin field presented by the properly ergodic case. Probability theory is especially rich in potential applications.

In the transitive case it is possible to make a useful classification of the possible invariant Hilbert bundles over the G space S. Fix an arbitrary point s_0 of S as an origin or reference point and let H_{s_0} denote the subgroup of all x in G such that $[s_0]x = s_0$. Under mild regularity conditions on the systems $S, \not{C}$ one can prove that H_{s_0} is a closed subgroup of the separable locally compact group G. It follows easily from the assumed transitivity of G on S that changing s_0 to some other s_1 in S leads to a subgroup H_{s_1} that is *conjugate* to H_0. A more elaborate argument shows that (modulo natural isomorphisms) the possible Hilbert space bundles over S satisfying the G invariance conditions listed earlier correspond one to one to the equivalence classes of unitary representations of H_{s_0}. In particular the unitary representation U of G, whose decomposition is what harmonic analysis of the bundle cross sections consists of, is uniquely determined by the unitary representation of H_{s_0} which defines the bundle. In other words our analysis of possible G invariant Hilbert bundles with base S implies and is determined by a canonical mapping of unitary representations L of the closed subgroups H_{s_0} of G into unitary representations U of G. This mapping is of fundamental importance in the theory of unitary group representations and may be set up directly without explicit mention of bundles. Indeed this is how it was first defined. Here is the definition. Let H be any closed subgroup of the separable locally compact group G. Let L be any unitary representation of H and let $S = G/H$ be the set of all H right cosets Hx. Then setting $[Hx]y = H(xy)$ converts S into a transitive G space. Moreover H is precisely the subgroup of all x in G such that $[H]x = H$. As remarked above, S admits a unique invariant measure class C. Let μ be any measure in C. Let $\mathcal{F}^L$ denote the set of all Borel functions f from G to $\mathcal{H}(L)$ the space of L such that

$$(*) \qquad\qquad f(hx) = L_h(f(x))$$

for all $h \in H$ and all $x \in G$. For each f in $\mathcal{F}^L$ the function $x \to (f(x) \cdot f(x))$ is a constant on the right cosets since $(f(hx) \cdot f(hx)) = (L_h(f(x)) \cdot L_h(f(x))) = (f(x) \cdot f(x))$ for all h in H and all x in G. Hence $f(x) \cdot f(x)$ may be considered a function on $S = G/H$ and we may form $\int_S (f(x) \cdot f(x) d\mu(x)$. If this integral is finite we shall say that f is in $\mathcal{F}^{L,\mu}$. If we identify functions which are equal almost everywhere the functions in $\mathcal{F}^{L,\mu}$ form a Hilbert space with the norm of $f \in \mathcal{F}^{L,\mu} = \sqrt{\int_S (f(x) \cdot f(x)) d\mu(x)}$. For each $f \in \mathcal{F}^{L,\mu}$ and each $x \in G$ let $U_x^L(f) = \sqrt{\rho(y,x)} f(yx)$ where ρ is the Radon Nikodym derivative associated with the quasi invariant measure μ as above. Then $x \to U_x^L$ is a unitary representation of G. By definition *it is the unitary representation of G induced by the* unitary representation L of H. Modulo equivalence it is identical with the unitary representation whose reduction is what we mean by harmonic analysis for the cross sections of the vector bundle over S canonically associated with L. In order to make connections with the less abstract considerations with which we began the reader is advised to convince him(or her)self that in the special case in which μ is invariant and L is the one dimensional identity representation U^L is just the representation of G in $\mathcal{L}^2(S,\mu)$ defined by the action of G on S.

We have thus seen that the induced representations U^L are precisely the unitary representations which must reduce in order to do harmonic analysis proper in the transitive case of the dichotomy explained above.

This fact is a major part of "the central role" announced in the title to this section. However there is another part. In order to carry out harmonic analysis we must analyze our given induced representation U^L into its irreducible constituents and this requires finding all or a large part of the irreducible representations of G. When G is commutative as it was with Fourier this is just a matter of finding the characters of G and is fairly straightforward for those groups that arise in applications. When G is not commutative, finding the irreducible representations is usually a non trivial problem itself. It is an interesting and important fact that the inducing construction is one of the key tools used in solving this problem. Indeed in many cases one can construct all or most of the infinite dimensional irreducible representations of G as representations induced by finite or even one dimensional representations of subgroups. Moreover there are general theorems which enable one, in many cases, to find all of these induced irreducible representations systematically.

We sum up by stating that induced representations play a curious dual role in harmonic analysis proper. They are the representations that one wants to decompose and to a considerable extent they provide the irreducible constituents into which one wishes to decompose them.

3. The role of induced representations in the foundations of quantum physics. Perhaps the simplest physical system is that defined by a particle of mass m moving freely in three dimensional Euclidean space. In classical mechanics its motion is described by the differential equations $m\frac{d^2 x}{dt^2} = 0$, $m\frac{d^2 y}{dt^2} = 0$, $m\frac{d^2 z}{dt^2} = 0$ where m is the so called *mass* of the particle and x, y, z are its coordinates on some rectangular coordinate system. The general solution of these equations may be written down at once $x = v_1 t + x_0$, $y = v_2 t + y_0$, $z = v_3 t + z_0$ where $x_0, y_0, z_0, v_1, v_2, v_3$ are six constants. We see that the equations of motion imply that the particle moves in a straight line with constant velocity.

In quantum mechanics this simple picture becomes considerably more complicated. This is because it turns out to be impossible in principle to assign a definite position x_0, y_0, z_0 and a definite velocity v_1, v_2, v_3 to a particle at the same time. One can only assign probability distributions to these six numbers and the more concentrated the probability distribution is for x_0, y_0, z_0 the more spread out is that for v_1, v_2, v_3. In order to have a theory at all one must take these probability distributions rather than actual numerical values of the variables to be the fundamental objects and replace the classical equations of motion by differential equations which determine how these probability distributions change with time.

A scheme for doing this was worked out between 1924 and 1927 with inputs from a half dozen or so physicists. In the elegant mathematically rigorous form given it by von Neumann one has a complex Hilbert space $\mathcal{H}$ whose elements of unit norm define pure states of the system. A pure state changes with time in such a way as to satisfy a differential equation of the form $\frac{1}{i}\frac{d\phi}{dt} = H(\phi)$ where H is a self-adjoint operator. For each "observable" of the system there is a corresponding self-adjoint operator A and there is a definite algorithm from which one can compute the probability distribution of the observable corresponding to A when the system is in the pure state corresponding to ϕ. This algorithm is easiest to state in the special case in which A has a discrete spectrum so that $\mathcal{H}$ has an orthonormal basis $\phi_1, \phi_2, \ldots$ each member of which is an eigenvector $A(\phi_j) = \lambda_j \phi_j$. In that case the probability that the observable corresponding to A will yield a value in the set E when measured in the state defined by ϕ is $\sum_{\lambda_j \in E} |(\phi \cdot \phi_j)|^2$.

When A has a spectrum which is in part continuous the algorithm is best stated in terms of the concept of a projection valued measure. Since we shall need this concept in the ensuing discussion we pause to explain it. Returning for the moment to the above formulation in the discrete case let us consider for each Borel subset E of the real line the closed linear span of all ϕ_j for which λ_j is in E and define P_E to be the projection operator whose range is this closed linear span. One verifies at once that the mapping $E \to P_E$ has the following properties:

(a) $P_0 = 0$ and $P_R = I$ where 0 and R denote the empty set and the whole real line respectively.

(b) $P_E P_F = P_F P_E = P_{E \cap F}$ for all E and F.

(c) If $E_1, E_2, E_3 \ldots$ are mutually disjoint then $P_{E_1 \cup E_2 \cup \ldots} = P_{E_1} + P_{E_2} + \cdots$.   *weak convergence!*

It is easy to see that our algorithm can be restated in terms of the mapping $E \to P_E$ without mentioning the ϕ_j and λ_j because $\sum_{\lambda_j \in E} |\phi \cdot \phi_j|^2 = (P_E(\phi) \cdot \phi)$ for all ϕ.

Quite generally if S is any set in which a notion of Borel set has been defined and $\mathcal{H}$ is a separable Hilbert space then a mapping $E \to P_E$ from Borel subsets of S to projection operators in $\mathcal{H}$ is said to be a *projection valued measure* on S if it satisfies (a), (b) and (c) above (with R replaced by S of course). Notice that whenever $E \to P_E$ is a projection valued measure on S and ϕ is a unit vector in the underlying Hilbert space then $E \to (P_E(\phi) \cdot \phi)$ is a probability measure in S.

The importance of this notion for our immediate purposes is that canonically associated with *every* self adjoint operator A (bounded or unbounded) there is a projection valued measure P^A on the real line which uniquely determines A by way of the equation $(A(\phi) \cdot \phi) = \int_{-\infty}^{\infty} x \, d(P_x^A(\phi) \cdot \phi)$ where the integral on the right is to be interpreted as the integral of x with respect to the measure $E \to (P_E(\phi) \cdot \phi)$. Actually *every* projection valued measure on the line occurs and the indicated correspondence is bijective. These facts constitute the celebrated spectral theorem first proved by Hilbert for bounded self adjoint operators and later extended to the general case by Stone and von Neumann. Using the correspondence $A \to P^A$ set up by the spectral theorem it is obvious how to extend the above algorithm to all self adjoint operators.

Having described the general scheme of quantum mechanics let us return to the classical picture of a free particle and examine its quantum mechanical version. The simplest case is that in which the so-called "spin" of the particle is zero. In that case one takes the Hilbert space to be $\mathcal{L}^2(E^3, \mu)$ where μ is Lebesgue measure in Euclidean 3-space E^3 and then the self adjoint operators X, Y and Z corresponding to the x, y and z coordinates are $\psi \to x\psi$, $\psi \to y\psi$ and $\psi \to z\psi$ respectively. Those corresponding to the three velocity components are $\frac{\hbar}{im} \frac{\partial}{\partial x}$, $\frac{\hbar}{im} \frac{\partial}{\partial y}$ and $\frac{\hbar}{im} \frac{\partial}{\partial z}$ where $\hbar = \frac{h}{2\pi}$ and h is the fundamental constant introduced by Planck.

The Schrödinger equation; that is the differential equation describing how the stated vector ψ changes with time is

$$\frac{\partial \psi}{\partial t} = -\frac{1}{i} \frac{\hbar}{2m} \left(\frac{\partial^2 \psi}{\partial x^2} + \frac{\partial^2 \psi}{\partial y^2} + \frac{\partial^2 \psi}{\partial z^2} \right).$$

It may be rewritten in the form $\frac{\partial \psi}{\partial t} = \frac{\hbar}{i} H(\psi)$ where H is the positive self adjoint operator

$$\frac{m}{2}\left[\left(\frac{\hbar}{im}\frac{\partial}{\partial x}\right)^2 + \left(\frac{\hbar}{im}\frac{\partial}{\partial y}\right)^2 + \left(\frac{\hbar}{im}\frac{\partial}{\partial z}\right)^2\right]$$

and as such is related to the operators corresponding to the velocity observables just as the classical energy of a free particle is related to the velocity components. It is the operator corresponding to the energy observable in quantum mechanics.

Simple and elegant as this model is, it appears at first sight to be quite arbitrary and *ad hoc*. In particular it is difficult to understand how anyone could have guessed it and by no means obvious how to modify it to fit a model for space different from E^3. We now propose to recast it in more group theoretic terms and show how this reformulation

(a) Removes much of the mystery.

(b) Generalizes in a straightforward way to any model for space with a separable locally compact group of isometries.

(c) Relates in an extremely intimate way to the considerations of section 2.

There are two major steps in the reformulation. In the first we replace the operators X, Y, and Z by a single projection valued measure on E^3. In the second we replace the velocity observables and the Schrödinger equation by a unitary representation W of the group $\mathcal{E} \times T$ where T is the time translation group and $\mathcal{E}$ is the group generated by the translations and rotations of E^3.

We define the relevant projection valued measure Q as follows. For each Borel subset E of E^3, Q_E is the operator $\psi \to \phi_E\psi$ where ϕ_E is the function which is one on the set E and zero outside it. It is trivial to check that $E \to Q_E$ is a projection valued measure. Given Q let f be any real valued Borel function on E_3 and consider the mapping $E \to Q_{f^{-1}(E)}$ where E varies over the Borel subsets of the real line. This is easily seen to be a projection value measure on the real line. By the spectral theorem it corresponds to a self adjoint operator A_f. It can be shown that A_f is precisely the operator $\psi \to f\psi$. Thus *all* coordinates (rectangular or not) which are real valued Borel functions on space have operators which are uniquely determined by giving Q. Q itself is defined in a coordinate free manner. Moreover it has a direct physical interpretation. For each E, Q_E is the self adjoint operator corresponding to the observable which is one when the particle is in E and zero when it is not in E.

We define the unitary representation W of $\mathcal{E} \times T$ in terms of unitary representations U and V of $\mathcal{E}$ and T respectively. To define U we observe that $\mathcal{E}$ acts naturally on E^3 and preserves Lebesgue measure μ. U is just the unitary representation of $\mathcal{E}$ defined by the formula $U_\alpha(\psi)(x, y, z) = \psi((x, y, z)\alpha)$. To define V one takes advantage of Stone's theorem which sets up a natural one-to-one correspondence between self adjoint operators and unitary representations of the additive group of the line. The unitary representation corresponding to the self adjoint operator A is the representation $t \to e^{iAt}$. Conversely (modulo technical questions about the domains of operators) one can obtain A from $t \to e^{iAt}$ by noticing that $\frac{d}{dt}e^{iAt}$ at $t = 0$ is $i\,A$. This being said we define $V_t = e^{-i\hbar t H}$. Evidently giving V is the same as giving H. Moreover for each ψ in $\mathcal{L}^2(E^3, \mu)$, $t \to V_t(\psi)$ is a solution of Schrödinger's equation. Thus giving V amounts to giving Schrödinger's equation in integrated form. Stone's theorem is also involved in recovering the velocity operators from U. Consider the restriction of U to the subgroup of $\mathcal{E}$ consisting of all elements of the form $x, y, z \to x + a, y, z$. This unitary representation of the additive group of the real line can then be written in the form $a \to e^{iAt}$ where A is easily seen to be the self adjoint operator $\frac{1}{i}\frac{\partial}{\partial x}$. This is just the constant $\frac{\hbar}{m}$ times the operator corresponding to the x component of the velocity. The y and z velocity operators can be obtained analogously. Passing from H and the velocity operators to U and V has the added advantage of getting rid of unbounded operators and their delicately defined domains. One proves easily that U_α and V_t commute for all α and t so that U and V may be blended into a single representation W of $\mathcal{E} \times T$, $W_{\alpha,t} = U_\alpha V_t$ for all $\alpha, t \in \mathcal{E} \times T$.

Having seen that the position and velocity operators may be given by specifying a projection valued measure Q on space and a unitary representation U of the group $\mathcal{E}$ of isometries of space let us now observe that our constructions of Q and U may be generalized to any system S, G consisting of a separable locally compact group G and a transitive Borel G space S with a G invariant measure μ. Thus one has a plausible extension of the quantum mechanics of a free particle to a wide variety of models for space – at least as far as position and momenum operators are concerned. Here we have to remark that in generalizing it is best to shift the emphasis from velocity to momentum since velocity operators depend upon the choice of H in a complicated way. In general mass will depend upon velocity as in the special theory of relativity and only the momentum operators can be obtained directly from U. Velocity is conceptually more elementary than momentum but the latter is theoretically more fundamental.

We are now ready to explain the fundamental connection between the quantum mechanics of a free particle and the material of #2. Let us observe first that U and Q satisfy a very simple commutation relation. Direct calculation shows that for all α and E

$$(**) \qquad\qquad U_\alpha Q_E U_\alpha^{-1} = Q_{[E]\alpha^{-1}}.$$

This identity is interesting and indeed of central importance for three reasons.

(a) One can deduce it from simple axioms about the quantum mechanics of a particle which have, on physical grounds, a high degree of *a priori* plausibility.

(b) There is a theorem known as "the imprimitivity theorem" which allows one to find all solutions of $(**)$ and they turn out to generalize the solution considered above in precisely the way that harmonic analysis of vector bundles generalizes the harmonic analysis of complex valued functions. Moreover these more general solutions also occur in the physics of particles. Among other things they include the systems in which U and Q define particles with non zero spin. Indeed one could have been led to the spin concept by this analysis

(c) In a certain sense the identity ∗∗ characterizes induced representations. The precise sense in which this is true is made clear in the statement of the imprimitivity theorem to which we now turn our attention.

By an obvious adaptation of the definition of the projection Q_E in the case of a free spinless quantum mechanical particle one assigns a projection valued measure $E \to P_E^L$ to every induced representation U^L. This projection valued mesure is defined on the Borel sets of G/H where U^L is the unitary representation of G induced by the unitary representation L of the closed subgroup H of G. One verifies that U^L and P^L satisfy (∗∗). The imprimitivity theorem is a converse.

THEOREM. *Let H be any closed subgroup of the separable locally compact group G and let U be any unitary representation of G. Suppose that there exists a projection valued measure P defined on G/H such that P and U satisfy*

$$U_\alpha P_E U_\alpha^{-1} = P_{[E]\alpha^{-1}}$$

for all α in G and all Borel sets $E \subseteq G/H$. (One says that P is a system of imprimitivity for U). Then there exists a unitary representation L of H (unique to withn unitary equivalence) such that the pair U, P is unitarily equivalent to the pair $U^L \, P^L$ defined above.

It follows from (a) and (b) above that one can deduce the explicit form of the quantum mechanics of a free particle from plausible axioms. This axiomatic analysis extends without difficulty to higher dimensional not necessarily flat models for space. So extended one finds that the mathematical model for a free quantum mechanical particle almost coincides with the mathematical apparatus of what we have called harmonic analysis proper.

Now, to a very considerable extent, modern particle physics is concerned with interactions of (otherwise) free particles. Thus one can hope to describe it mathematically using the above picture of a free particle as a starting point. This program can be carried quite far and its implementation leans heavily on the theory of induced representations. Since even quantum field theory can be included it seems fair to say that harmonic analysis is not just applicable to the quantum mechanics of elementary particles but can be made to *be* the mathematical apparatus of that theory.

4. Some remarks on the nature and history of number theory.

Before embarking on an exposition of the far-reaching character of the applications of harmonic analysis to number theory it will be useful to provide the non expert with some kind of overview of the nature of number theory itself.

It turns out that an astonishingly large part of that subject may be reduced to the analysis of a class of complex valued functions on the integers which I shall call *number theoretical problem functions*. For example if a is a positive integer and we define $\phi_a(n)$ to be the number of integer pairs x, y such that $x^2 + ay^2 = n$ then ϕ_a is a number theoretical problem function. More generally we may replace

$x^2 + ay^2$ by a general binary positive definite quadratic form Q where $Q(x,y) = Ax^2 + Bxy + Cy^2$ and define $\phi_Q(n)$ as the number of integer pairs x, y such that $Ax^2 + Bxy + Cy^2 = n$. We shall give other examples of number theoretical problem functions below.

Given a number theoretical problem function what does one do with it? The most obvious thing to do is to evaluate it for various values of n by solving the associated number theoretical problem. However in many cases it has proved possible to do much better than this, specifically to find an "explicit" function defined on the positive integers which is identically equal to the given problem function. A simple example of an explicit function is the function $n \to n^2$ or more generally $n \to P(n)$ where P is any polynomial with complex coefficients. If a function ϕ on the positive integers is periodic with period N and one knows $\phi(1), \phi(2), \cdots, \phi(N)$ one can immediately compute $\phi(n)$ for any n by a simple straightforward procedure. Thus periodic functions are explicit. Of course the sum, difference, product, etc., of two explicit functions may also be regarded as explicit.

There is a less obvious way of combining two explicit functions to get a third which plays an extremely important role in number theory and which I now propose to define. Let ϕ_1 and ϕ_2 be any two complex valued functions defined on the positive integers. Then their "convolution" $\phi_1 * \phi_2$ is defined by the formula

$$\phi_1 * \phi_2(n) = \sum_{d|n} \phi_1(d)\phi_2\left(\frac{n}{d}\right).$$

An example of an explicit representation of the problem function ϕ_Q was already known to Fermat in the special cases in which $Q = Q_1, Q_2$ or Q_3 and $Q_a(x,y) = x^2 + ay^2$. Fermat did not formulate his results in this way but, in effect, announced that when $a = 1, 2$ or 3 then $\phi_{Q_a}(n) = c_a(1 * \chi_a)(n)$ where c_a is a constant and χ_a is periodic with period $4a$. Once one knows c_a one need only compute $\phi_{Q_a}(n)$ by trial and error for $n = 1, 2, 3 \cdots 4a$ to obtain an explicit formula for all values of n. One checks easily that $c_1 = 4$ and $c_2 = c_3 = 2$.

This early example is important because so much of subsequent number theory can be understood as finding (or making progress toward finding) closely analogous results for more general number theoretical problem functions: The function χ_a in Fermat's theorem has another property besides its periodicity which it is important to isolate in understanding subsequent work. It is *completely multiplicative* in the sense that $\chi_a(1) = 1$ and $\lambda_a(nm) = \chi_a(n)\chi_a(m)$ for all positive integers n and m. The property of complete multiplicativity of χ_a implies that $1 * \chi_a$ and hence ϕ_{Q_a}/c_a is multiplicative in a somewhat weaker sense. One defines a function ϕ on the positive integers to be *multiplicative* if $\phi(1) = 1$ and $\phi(nm) = \phi(n)\phi(m)$ whenever n and m are positive integers with no common divisors and proves without difficulty that $\phi * \psi$ is multiplicative whenever both ϕ and ψ are multiplicative. It is important to notice however that $\phi * \psi$ is almost never completely multiplicative even when this is true of both ϕ and ψ.

With the above defined notions at hand it is possible (and useful) to divide Fermat's results about

ϕ_{Q_a} $(a = 1, 2, 3)$ into three parts as follows:

I. ϕ_{Q_a} is a constant c_a times a *multiplicative* function ψ_a.

II. $\psi_a(n) \equiv (1 * \chi_a)(n)$ where χ is completely multiplicative.

III. χ_a is periodic of period $4a$.

Parts I and II tells us that to determine $\phi_{Q_a}(n)$ for all n it is sufficient to determine it for all prime values of n. Part III allows us to determine $\phi_{Q_a}(p)$ for all primes when it has been found for a finite number of primes.

Fermat did not publish proofs of his assertions and no proofs by anyone reached the mathematical public until Euler took up the study of number theory a century later. He found I and II relatively easy to prove but asserts that he struggled for seven years before he found a proof of III – even for $a = 1$. The issue in the case $a = 1$ reduces to proving that every prime of the form $4n + 1$ is a sum of two squares.

If one wants to generalize the above cited results of Fermat the first question that suggests itself, is as to what happens when $a \neq 1, 2$ or 3. The case $a = 4$ is easily reduced to the case $a = 1$ since $x^2 + 4y^2 = x^2 + (2y)^2$ but Fermat was baffled by $a = 5$ and so was Euler. Among other things there is no constant c such that $\frac{\phi_{Q_5}}{c}$ is multiplicative.

It was Lagrange, in a paper published in 1773, who made the first big step toward clarification of the difficulty. He began by considering not just special examples of binary quadratic forms as Fermat and Euler had done, but the general binary quadratic form $Ax^2 + Bxy + Cy^2$ where A, B and C are arbitrary integers. His basic insight was that the theory of a particular form $Q(x, y) = Ax^2 + Bxy + Cy^2$ can only be understood by simultaneously considering all forms $Q'(x, y) = A'x^2 + B'xy + C'y^2$ for which $(B')^2 - 4A'C' = B^2 - 4AC$. One calls $D = B^2 - 4AC$ the discriminant of the form and notes that the form is positive definite if and only if $D < 0$. Lagrange observed that two forms Q and Q' will have $\phi_Q(n) \equiv \phi_{Q'}(n)$ whenever the forms are "equivalent" in the sense that there exists a two by two matrix of integers $\begin{pmatrix} a & b \\ c & d \end{pmatrix}$ with $ad - bc = \pm 1$ such that $Q(ax + by, cx + dy) \equiv Q'(x, y)$. Equivalent forms have the same discriminant D but the converse is not true. However he was able to show that for each D there are only *finitely* many inequivalent forms $Ax^2 + Bxy + Cy^2$ with $B^2 - 4AC = D$. This number is one for only a finite number of values of D and only when D has one of these values can one prove that ϕ_Q is a constant times a multiplicative function.

On the other hand if one shifts attention from the individual ϕ_Q to the whole set $\phi_{Q_1}, \phi_{Q_2} \cdots \phi_{Q_h}$ where the Q_j constitute a maximal set of inequivalent forms with the same discriminant as Q one can use Lagrange's results to prove an elegant generalization of Fermat's theorems I and II. For each negative integer D let $Q_1, Q_2 \cdots Q_h$ denote a complete set of inequivalent positive definite forms of discriminant D. Let these be so numbered that for $j = 1, 2, \cdots \nu \leq h$, there exists Q'_j such that Q'_j is equivalent to Q_j under an integer matrix $\begin{pmatrix} a & b \\ c & d \end{pmatrix}$ with $ad - bc = -1$ but not under any such matrix with $ad - bc = 1$

while for $j > \nu$ no such Q_j' exists. Let

$$\phi_D \equiv 2\phi_{Q_1} + 2\phi_{Q_2} + \cdots 2\phi_{Q_r} + \phi_{Q_{r+1}} + \cdots \phi_{Q_h}.$$

One can then prove

I'. ϕ_D is a constant C_D times a multiplicative function ψ_D.

II'. $\psi_D(n) \equiv 1 * \chi_D(n)$ where χ_D is completely multiplicative.

It is natural to conjecture the truth of the following generalization of Fermat's theorem III.

III'. $\chi_D(n)$ is periodic of period $-4D$. This conjecture is true but lies much deeper. It is completely equivalent to the celebrated quadratic reciprocity law which was first proved by Gauss over two decades later. The first published proof appeared in Chapter IV of Gauss' "Disquititiones Arithmeticae" in 1801. Lagrange proved various special cases of III' (although not in this formulation) but did not formally conjecture the full theorem.

While theorems I', II', III' constitute rather beautiful and far-reaching generalizations of theorems I, II, and III of Fermat they tell us nothing about ϕ_Q itself except for very special values of the discriminant D. Further information was a long time in coming and had to await the epoch-making work of Dedekind on "ideals" in algebraic number fields. Dedekind's theory was first published in 1871 and was the final step in a long development which began with Gauss in 1801 and included important contributions of Dirichlet and Kummer.

We shall skip over the intermediate history for the moment and outline the main features of Dedekind's beautiful theory. Once it has been described we can introduce a new class of number theoretical problem functions which includes the ϕ_Q for positive definite binary forms Q as a very special case. In particular we shall obtain substitutes for theorems I and II of Fermat for these more general problem functions which will then include such theorems for the ϕ_Q.

Let $A_0, A_1, A_2 \cdots A_n$ be integers and consider the field $\mathcal{F}$ generated by the rational numbers and *all* the complex roots of the equation $A_n Z^n + A_{n-1} Z^{n-1} + \cdots A_0 = 0$. In this article we shall refer to fields generated in this way as algebraic number fields although this term usually refers to a somewhat wider class of objects. Given an algebraic number field $\mathcal{F}$ one defines $R_{\mathcal{F}}$ the ring of integers of $\mathcal{F}$ to be the set of all elements Z of $\mathcal{F}$ which are "algebraic integers" in the sense that they satisfy an equation of the form $Z^k + B_{k-1} Z^{k-1} + \cdots B_0 = 0$ where $B_0, B_1, \cdots$ are all integers. One shows that $R_{\mathcal{F}}$ is closed under multiplication, addition and subtraction and thus is indeed a ring. It shares many of the properties of the ring of ordinary integers.

In particular there is a notion of congruence and for every congruence relation except the trivial one the number of congruence classes is finite. Any congruence notion is completely determined by the set I of all elements which are congruent to zero. This is because x is congruent to y if and only if $x - y$ is in I. One says that x is congruent to y mod I. Of course an essential property of the congruence notion is that the congruence classes of $x \pm y$ and xy depend only on those of x and y so that the ring structure of $R_{\mathcal{F}}$ may be carried over to the set of congruence classes and define a residue class ring $R_{\mathcal{F}}/I$. One checks easily that this is so far an arbitrary subset I of $R_{\mathcal{F}}$ if and only if I is a subring *and* xy is in I whenever x is an I and y is an $R_{\mathcal{F}}$. Dedekind called such subsets *ideals*. If w is an arbitrary nonzero element of $R_{\mathcal{F}}$ then the set of *all* wx with x in $R_{\mathcal{F}}$ is an ideal called the principal ideal generated by w. We denote it by I_w. One of the most important ways in which an $R_{\mathcal{F}}$ may differ from the ring of ordinary integers is that there may exist ideals which are *not* principal ideals. In other words there may be congruence relations which cannot be described as congruences modulo an *element* of $R_{\mathcal{F}}$.

Principal ideals have the important property that the principal ideal defined by a product $I_{w_1 w_2}$ can be described by I_{w_1} and I_{w_2} without mentioning w_1 and w_2. Indeed $I_{w_1 w_2}$ is simply the set of all sums $x_1 y_1 + x_2 y_2 + \cdots x_v y_v$ where each x_j is in I_w, and each y_j is in I_{w_2}. This suggests that one define the product of *any* two ideals I_1 and I_2 as the set $I_1 I_2$ of all sums $x_1 y_1 + x_2 y_2 + \cdots x_v y_v$ where such x_j is in I_1 and each y_j is in I_2. It is immediate that the product of any two ideals is again an ideal. Of course $I_1 I_2$ is contained in both I_1 and I_2 and it is not difficult to show that both inclusions are proper. In the ring of ordinary integers where every ideal is principal the ideals generated by the primes are precisely those which are *maximal* in the sense that they are properly contained in no other proper ideals. More generally one refers to the maximal ideals as the *prime* ideals.

With these definitions one can now state Dedekind's fundamental theorem: In every $R_{\mathcal{F}}$ every proper ideal I may be written as a product of prime ideals and this decomposition is unique up to a reordering (the multiplication of ideals is commutative and associative). In the special case in which every ideal is principal the unique factorization theorem for ring elements is an immediate corollary. In more general cases Dedekind's theorem is a substitute for an element factorization theorem which no longer exists.

Dedekind also analyzed the extent to which ideals in an $R_{\mathcal{F}}$ can fail to be principal and showed that there is always a *finite number* of non principal ideals such that all other ideals can be built up from these and principal ideals. Indeed let us say that the ideals I_1 and I_2 are *equivalent* if there exist principal ideals I_{w_1} and I_{w_2} such that $I_1 I_{w_1} = I_2 I_{w_2}$ and write $I_1 \sim I_2$. It is easy to see that $\sim$ is an equivalence relation in the technical sense and Dedekind showed that there are only a *finite* number of equivalence classes. It is easy to see that if $I_1 \sim I_1'$ and $I_2 \sim I_2'$ then $I_1 I_2 \sim I_1' I_2'$ so that one can define the *product* of two equivalence classes. Under this product the set $C_{\mathcal{F}}$ of all ideal classes in $R_{\mathcal{F}}$ becomes a finite commutative group called the *ideal class group* of $\mathcal{F}$.

We are now in a position to define our new number theoretical problem function. Recall first that for each ideal $I \neq 0$ in $R_{\mathcal{F}}$, $R_{\mathcal{F}}/I$ is a ring with a finite number of elements. This number is called the *norm*, $N(I)$, of the ideal I. For each positive integer n let $\phi_{\mathcal{F}}(n)$ be the number of ideals I in $R_{\mathcal{F}}$ such that $N(I) = n$ and for each ideal class C in $C_{\mathcal{F}}$ let $\phi_{\mathcal{F},C}(n)$ be the number of ideals I *in the class C*

GEORGE W. MACKEY   289

which have $N(I) = n$. Then $\phi_{\mathcal{F}}(n) = \sum_{C \in C_{\mathcal{F}}} \phi_{\mathcal{F},C}(n)$ and $\phi_{\mathcal{F}}$ and the various $\phi_{\mathcal{F},C}$ are number theoretical problem functions. It follows almost immediately from the fundamental theorem on the factorization of ideals that $\phi_{\mathcal{F}}$ is multiplicative. This property is *not* shared by the component functions $\phi_{\mathcal{F},C}$ but there is a substitute for it which is of great importance. When suitably specialized it leads to the missing generalization of Fermat's theorem I for the individual ϕ_Q.

In 1879 Dedekind introduced the following definition. A function χ from an ideal class group $C_{\mathcal{F}}$ to the complex numbers of modulus 1 is said to be a *character* of $C_{\mathcal{F}}$ if $\chi(C_1 C_2) = \chi(C_1)\chi(C_2)$ for all C_1 and C_2 in $C_{\mathcal{F}}$. Given a character χ of $C_{\mathcal{F}}$ one defines $\phi_{\mathcal{F},\chi}(n)$ as $\sum_{C \in C_{\mathcal{F}}} \chi(C)\phi_{\mathcal{F},C}(n)$ and proves without difficulty that $\phi_{\mathcal{F},\chi}$ is multiplicative for all χ. In the special case in which $\chi(C) \equiv 1$ one checks at once that $\phi_{\mathcal{F},\chi}$ reduces to $\phi_{\mathcal{F}}$. Thus the fact that $\phi_{\mathcal{F},\chi}$ is multiplicative for all χ is a natural generalization of the fact that $\phi_{\mathcal{F}}$ is multiplicative. Now not only is each $\phi_{\mathcal{F},\chi}$ a finite linear combination of the functions $\phi_{\mathcal{F},C}$ but one can prove conversely that each $\phi_{\mathcal{F},C}$ is a finite linear combination of the $\phi_{\mathcal{F},\chi}$ and hence a finite linear combination of multiplicative functions. [The proof is a simple exercise in harmonic analysis on finite commutative groups.] This is the substitute for the missing multiplicity of the functions $\phi_{\mathcal{F},C}$ mentioned above. Such functions are always, and in a canonical fashion, finite linear combinations of multiplicative functions.

Now it is not difficult to prove that if Q is any positive definite binary quadratic form then $\phi_Q(n) \equiv \phi_{\mathcal{F},C}(n)$ for some ideal class C where $\mathcal{F}$ is the algebraic number field generated by the square root of the discriminant D of Q. Hence the truth of a generalization of Fermat's theorem I follows at once. It states that each ϕ_Q is a finite linear combination of multiplicative functions. It is natural to ask whether this result continues to hold when Q has more than two variables and such results were proved in special cases in the latter part of the nineteenth century – especially in the work of Glaisher on numbers of representations as sums of squares. However it is perhaps some indication of the depth of the problem that fairly general theorems along these lines had to await rather profound work of E. Hecke published between 1935 and 1940 which rested in turn on important advances made earlier by the same man. Further details will be found in section 5.

Our discussion so far suggests that a first major goal in the analysis of number theoretical problem functions might be to prove that they have unique or at least canonical decompositions as finite linear combinations of multiplicative functions. These multiplicative constituents then become new problem functions and we must now turn our attention to the question of what one can hope to do in the further analysis of multiplicative problem functions. Turning once more to Fermat for guidance we recall theorem II which states that the multiplicative functions which he encountered may be factored as $1 * \chi$ where χ is *completely* multiplicative. Of course 1 is also completely multiplicative and Fermat's simple result suggests that one should attempt to factor multiplicative problem functions in the form

$\phi = \phi_1 * \phi_2 \cdots * \phi_n$ where each ϕ_j is completely multiplicative. While this is true or "almost" true in a very general setting it needs modification to eliminate the "almost" and even when true is too non unique to be useful. When one limits oneself to suitably "canonical" factorizations it is seldom clear that one can achieve a factorization into factors which are completely multiplicative. One may have to be content with partial factorizations which are canonical. Achieving these is already highly non trivial and those that have been found constitute some of the major results of number theory.

These introductory remarks having been made, we turn our attention to the definition of some basic concepts in the factorization of multiplicative functions ϕ. Let us begin by defining δ to be the function on the positive integers such that $\delta(1) = 1$ and $\delta(n) = 0$ for $n \neq 1$ and observing that $\phi * \delta = \phi$ for all complex valued functions ϕ defined on the positive integers. This suggests introducing the notion of "inverse" with respect to $*$. One verifies easily that there exists a θ such that $\phi * \theta = \delta$ if and only if $\phi(1) \neq 0$ and then θ is uniquely determined by ϕ. When $\phi(1) \neq 0$ we define ϕ^{-1} to be this unique θ. Evidently $(\phi^{-1})^{-1} = \phi$ and ϕ is an explicit function whenever ϕ is. It is easy to prove that ϕ^{-1} is multiplicative if and only if ϕ is multiplicative and that the set M of all multiplicative functions is a group under ϕ.

Next let us recall the obvious fact that a member ϕ of M is completely determined by its values at the prime powers p^k and state the following easily proved lemmas.

LEMMA 1.. *The member ϕ of M is completely multiplicative if and only if for all primes p, $\phi^{-1}(p^k) = 0$ whenever $k > 1$.*

LEMMA 2.. *Let ϕ_1 and ϕ_2 be in M and let p be a prime. Let $\phi_1(p^k) = 0$ for $k > k_1$, $\phi_1(p^{k_1}) \neq 0$. Let $\phi_2(p^k) = 0$ for $k > k_2$, $\phi_1(p^{k_2}) \neq 0$, where k_1 and k_2 are positive integers. Then $\phi_1 * \phi_2(p^k) = 0$ for $k > k_1 + k_2$ and $\phi_1 * \phi_2(p^{k_1 + k_2}) \neq 0$.*

From lemmas 1 and 2 we deduce at once a necessary condition on a member ϕ of M in order that it be equal to a convolution of k_0 completely multiplicative functions. It is that for all p, $\phi^{-1}(p^k) = 0$ for $k > k_0$. With a little more work it can be shown that this condition is also sufficient. Given ϕ satisfying the condition it suffices to find completely multiplicative functions $\phi_1, \phi_2, \cdots \phi_{k_0}$ such that $\phi_1^{-1} * \phi_2^{-1} \cdots * \phi_{k_0}^{-1}(p^k) = \phi^{-1}(p^k)$ for all $k \leq k_0$ (where of course ϕ_j^{-1} is completely known when $\phi_j^{-1}(p)$ is known for all primes p). Working out the convolutions one finds that for each p, $\phi_1^{-1}(p), \cdots \phi_{k_0}^{-1}(p)$ must the k_0 roots of a k_0-th degree polynomial with given coefficients and that these roots may be assigned to $\phi_1^{-1}(p), \cdots, \phi_{k_0}^{-1}(p)$ in any order. Each assignment leads to a factorization of the desired kind. Thus a factorization always exists but no particular one may be specified without making countably many arbitrary choices – one for each prime. The analysis does however give us one useful invariant. The number of factors k_0 is the same for all factorizations – provided that we do not allow δ as a factor. This suggests introducing a notion of "order" for multiplicative functions which measures how far they are from being completely multiplicative. However the most useful notion along these lines is slightly different from that most immediately suggested by the foregoing. In part this is because of the fact that the multiplicative problem functions which occur in practice often have exceptional behaviour at a finite number of primes. In this connection it is important to observe that changing the behaviour at a finite number of prime powers can never destroy the explicitness of a multiplicative function. In particular, if ϕ_1 and ϕ_2 are multiplicative functions which are zero except at a finite number of prime powers then $\phi_1 * \phi_2^{-1}$ is explicit.

We shall say that a member ϕ of $\mathcal{M}$ is of order $k = 0, 1, 2 \cdots$ if for all but a finite number of primes p, $\phi(p^k) \neq 0$ and $\phi(p^k) = 0$ for $\ell > k$. We shall say that ϕ has order $k = -1, -2, \cdots$ if ϕ^{-1} has order $-k$. It follows from lemma 2 that if ϕ_1 and ϕ_2 are multiplicative of orders k_1 and k_2 and k_1 and k_2 are both non negative or both negative then $\phi_1 * \phi_2$ is multiplicative of order $k_1 + k_2$. It is clear that a completely multiplicative function has order -1 but that the converse is true only to the extent that we ignore integers divisible by a certain finite set of primes.

In terms of these notions one can state a generalization of Fermat's theorem II which holds for all binary positive definite quadratic forms ϕ_Q: The multiplicative constituents of ϕ_Q are all of order -2. The theory of Hecke mentioned above implies that this is also true for a broad class of quadratic forms in n variables. Having gotten this far in the theory of the ϕ_Q the next step would be to investigate the p dependence of these multiplicative functions of order -2 as well as their explicit factorizability into multiplicative functions of order -1. It turns out that in the binary case and in those n-ary cases analyzed by Hecke one of the multiplicative constituents (but only one) is of the form $1 * \chi$ where χ is of order -1 and is an explicit function. In the binary case this constituent is the ϕ_D of Lagrange and the fact that χ is periodic and hence explicit is, as already noted, equivalent to the quadratic reciprocity law. In many, if not all, of Hecke's n-ary cases one also has explicit formulas for χ. For the remaining multiplicative constituents one knows much less – even in the binary case. It would be interesting to know more.

The factorization problem for the multiplicative constituents becomes more interesting as well as more difficult when we investigate the $\phi_{\mathcal{F},C}$ instead of the ϕ_Q. Here these multiplicative functions are of order $-n$ instead of -2 where n is the degree of $\mathcal{F}$ over the rational field Q. Here a factorization into functions of order -1 would involve n factors and one can contemplate partial factorizations. Indeed one knows of explicit partial factorizations in important cases in which no explicit complete factorization is known and indeed may not exist. As in the case of the ϕ_Q one of the multiplicative constituents of each $\phi_{\mathcal{F},C}$ plays a special role. It is $\phi_{\mathcal{F}}$, i.e., $\phi_{\mathcal{F},\chi}$ where $\chi(C) \equiv 1$. It has an explicit partial factorization of the form $1 * \psi$ where ψ is multiplicative of order $n-1$ and much more is known about its further explicit factorizations than about the explicit factorizations of the other multiplicative constituents.

We conclude this section with a brief account of how the work of Gauss and Kummer led to

Dedekind's ideal theory and the significance for Diophantine problems of analyzing the multiplicative problem functions $\phi_{\mathcal{F}}$.

In the form in which it is usually stated the quadratic reciprocity law asserts that when p and q are odd primes then p is a square mod q if and only if q is a square mod p unless p and q are both congruent to 3 mod 4. In that case p is a square mod q if and only if q is *not* a square mod p. (We omit the "supplements" dealing with -1 and 2). The discovery and proof of this law was one of Gauss's main achievements. (He was unaware of the earlier statements of Euler and Legendre and gave the first valid proof.) He gave two distinct proofs in his celebrated *Disquistiones Arithmeticae* of 1801 and six more proofs during the next two decades. Of course one can attempt a generalization in which squares are replaced by higher powers and Gauss soon got to work on this problems. He encountered unexpected difficulties, however, and published nothing on the subject until 1824. Even then his results were incomplete and dealt only with cubic and biquadratic reciprocity. It was the further development of these ideas that led to the Dedekind theory and later in the hands of Hilbert to modern class field theory. The main figure between Gauss and Dedekind was Kummer, although Jacobi and Eisenstein made important contributions as well, and in particular completed Gauss's work on cubic and biquadratic reciprocity. Kummer, in work extending over a fifteen year period, (1845-1859), succeeded in formulating and proving reciprocity laws for all powers.

We have already observed (without full explanation) that the quadratic reciprocity law is equivalent to an explicit formula for $\phi_{\mathcal{F}}$ when $\mathcal{F}$ is a certain kind of second degree extension of the rationals. Similarly Kummer's n-th power reciprocity laws are equivalent to explicit formulas for $\phi_{\mathcal{F}}$ whenever $\mathcal{F}$ is the field generated by *all* the n-th roots of some rational a having no rational n-th root. However the explicitness is of a rather more subtle sort than in the quadratic case and cannot be explained here. In the course of his work Kummer encountered the difficulty of non unique factorization of the integers of $\mathcal{F}$ and overcame it by a device which does not work for general fields. This was, no doubt, Dedekind's inspiration.

Actually, although this does not seem to have been much of a motivating force there are cogent Diophantine reasons for studying $\phi_{\mathcal{F}}$ even when $\mathcal{F}$ is not a Kummer field. Consider the equation

$$(*) \qquad A_0 x^n + A_1 x^{n-1} y + \cdots A_n y^n = m$$

where $A_0, A_1, \cdots A_n$ and m are given integers and one seeks all integer pairs x, y which satisfy it. When $A_0 \neq 0$ this equation may be rewritten in the form

$$(A_0 x - \alpha_1 y)(A_0 x - \alpha_2 y) \cdots (A_0 x - \alpha_n y) = m A_0^{n-1}$$

where $\alpha_1, \alpha_2, \cdots \alpha_n$ are integers in the field $\mathcal{F}_0$ generated by the n roots of the equation $A_0 Z^n + A_1 Z^{n=1} + \cdots A_0 = 0$. Thus every solution of our equation is canonically associated with a factorization of $m A_0^{n-1}$

into the integers of the field. Exploiting this fact one may develop a theory of the solutions of $*$ which makes heavy use of Dedekind's theory and the functions $\phi_{\mathcal{F}_0,C}$. In the special case $x^n + y^n = m$, which arises in attempts at proving Fermat's last theorem the field $\mathcal{F}_0$ is a Kummer field. Indeed Kummer used his theory to obtain a proof of Fermat's theorem whenever n is a so-called "regular prime".

In the 1890's Hilbert made an intensive study of the work of Kummer and Dedekind which culminated in his celebrated "Zahlbericht". This is a beautiful synthesis and reworking of the ideas of both men with many simplifications and improvements. Shortly thereafter he found a "relativized" version of the quadratic reciprocity law in which quadratic extensions of the rational field are replaced by quadratic extensions of an arbitrary algebraic number field. This led him to formulate a vast program in which the whole Kummer theory was to be relativized in an analogous fashion. During the first two decades of the present century Hilbert's program was carried out by Fueter, Furtwangler and above all Takagi. In all of this work the relative Galois group was commutative. E. Artin made a very important advance in the late 1920's when he reformulated the main results in an illuminating new way and at the same time pointed the way toward dealing with non commutative relative Galois groups. All of this work can be looked upon as discovering properties of the problem functions $\phi_{\mathcal{F}}$ for algebraic number fields $\mathcal{F}$.

5. The role of harmonic analysis in number theory.

In section 4 we have discussed the history and nature of number theory but have said next to nothing about how its non elementary results might be proved. We now propose to give some indications that harmonic analysis in all of its manifestations is a powerful tool in proving number theoretical theorems. Its full scope is still far from understood and, while it would be rash to predict anything close to universality, the author is sometimes tempted to do so. Even theorems which can be proved by elementary but ingenious and complex arguments can often be given much simpler and more understandable proofs using harmonic analysis.

One important aspect of the applicability of harmonic analysis is not widely appreciated. It has to do with the fact that the Fourier convergence theorem contains concealed within it an infinite number of surprising identities. If f is any sufficiently regular complex valued function on the real line which is defined by "explicit" formulas, has period 1 and Fourier coefficients $\{C_n\}$ which can be explicitly computed then for each real x

$$f(x) = \sum_{n=-\infty}^{\infty} C_n e^{2\pi i n x}$$

is a statement of equality between two numbers which is far from obvious *a priori*. As every student of elementary Fourier series theory soon learns, the identities

$$\frac{\pi^2}{6} = 1 + \frac{1}{2^2} + \frac{1}{3^2} + \cdots$$

$$\frac{\pi}{4} = 1 - \frac{1}{3} + \frac{1}{5} - \frac{1}{7} + \cdots$$

may be proved in this way. The interesting thing for us is that the identities so obtainable include key results in number theory, in particular, the quadratic reciprocity law.

In the interests of historical accuracy we shall henceforth refer to the Fourier convergence theorem as the Fourier-Dirichlet theorem. Correcting earlier work of Cauchy, it was Dirichlet in 1829 who first gave a rigorous proof of this theorem. It has an equivalent form which is more directly useful in obtaining identities which are useful in number theory. Let f be any complex valued function on the real line which goes to zero sufficiently rapidly at ∞. Then $\tilde{f}(x) = \sum_{n=-\infty}^{\infty} f(x + n)$ exists for all x and defines a function $\tilde{f}$ which is periodic of period 1. If $\tilde{f}$ is such that the Fourier-Dirichlet theorem applies then that theorem gives a formula for $\tilde{f}$ which may be written in the form

$$\tilde{f}(x) = \sum_{n=-\infty}^{\infty} e^{2\pi i n x} f^T(n)$$

where $f^T(y) = \int_{-\infty}^{\infty} f(x)e^{-2\pi i y x}dx$ which is of course just the Fourier transform of f. Writing $\tilde{f}$ in terms of f we arrive at the identity

$$(1) \qquad \sum_{n=-\infty}^{\infty} f(x + n) = \sum_{n=-\infty}^{\infty} f^T(n)e^{2\pi i n x}.$$

When $x = 0$ this reduces to

$$(2) \qquad \sum_{n=-\infty}^{\infty} f(n) = \sum_{n=-\infty}^{\infty} f^T(n).$$

The identity (2) is known as the *Poisson summation formula* (although Cauchy seems to have discovered it first). Actually, applying (2) to an arbitrary translate of f gives back (1) so that (1) and (2) are equivalent. Moreover it is elementary to deduce the Fourier-Dirichlet theorem from (1). Thus the Poisson summation formula (2) and the Fourier-Dirichlet theorem are equivalent and lead to the same identities. The Poisson summation formula is more useful in number theory because the functions f which has been found to yield number theoretically useful identities are simpler and more natural than their associated periodifications $\tilde{f}$.

Let p and q be positive integers and let f_{pq} be the function which is $e^{2\pi i y^2 q/p}$ for $0 \le y \le p$ and is zero outside of this interval. The Poisson summation formula applied to $f_{p,q}$ yields an identity which can be manipulated to be an identity involving "Gauss sums". This identity has the quadratic reciprocity law as an immediate consequence.

There are other examples, some of which will be mentioned below, but the full power of the method has yet to be systematically investigated.

The Poisson summation formula has a more or less obvious generalization in which the real line is replaced by an arbitrary separable locally compact commutative group G and the subgroup of all integers by any closed discrete subgroup Γ such that G/Γ is compact. In that case the subgroup $\Gamma^\perp$ of the character group $\hat{G}$ consisting of all χ in $\tilde{G}$ such that $\chi(\gamma) = 1$ for all $\gamma \in \Gamma$ is also discrete and for suitably restricted complex valued functions f on G the generalized Poisson summation formula asserts that

$$\sum_{x \in \Gamma} f(x) = \sum_{\chi \in \Gamma^\perp} f^T(\chi)$$

where $f^T(\chi) = \int_G f(x)\chi(x)dx$ is the Fourier transform of f. The special case in which G is the additive group of all n tuples of real numbers was of course known classically and the identities it implies play an important role in number theory.

There is an alternative way of deriving the Poisson summation formula for the pair G, Γ which involves the concept of an induced representation. Thus derivation has the great advantage that it makes sense for *non commutative* separable locally compact groups and so suggests a non commutative generalization of the Poisson summation formula. Let I_Γ denote the identity representation of Γ and form the induced representation U^{I_Γ} of G. For each L^1 function f on G one can form the operator $U_f^{I_\Gamma} = \int f(x)U_x^{I_\Gamma}dx$. On the other hand one can decompose U^{I_Γ} as a direct sum of irreducibles and, since G is commutative, these are all defined by characters χ of $\hat{G}$. Thus those that occur are easily seen to be just the members of $\Gamma^\perp$. Thus we have

$$U_f^{I_\Gamma} = \sum_{\chi \in \Gamma^\perp} \chi_f I$$

where $\chi_f = \int \chi(x)f(x)dx = f^T(\chi)$. Taking the trace of both sides (when it exists) yields $\mathrm{Trace}(U_f^{I_\Gamma}) = \sum_{\chi \in \Gamma^\perp} f^T(\chi)$. Finally a straightforward analysis shows that the left side is equal to $\sum_{\gamma \in \Gamma} f(\gamma)$.

Most of the above argument can be carried through without assuming the commutativity of G provided that U^{I_Γ} decomposes discretely. However one cannot describe the irreducible constituents of U^{I_Γ} explicitly and the right side of the final formula must be written as $\sum_{j=1}^{\infty} \mathrm{Trace}(L_f^j)$ where the L^j are the (often unknown) irreducible constituents of U^{I_Γ}. The left hand side can be reduced to a form reminiscent of $\sum_{r \in \Gamma} f(\gamma)$ but more complicated in that one must include conjugates in G of the elements of Γ.

That the Poisson summation formula can be usefully generalized in this way was first pointed out in a seminal paper published in 1957 by Atle Selberg. Accordingly one speaks of the Selberg trace formula. Like the Poisson summation formula it has had several impressive successes but its full potential has yet to be realized. Selberg mainly concerned himself with the special case in which $G = SL(2, R)$ and Γ is a subgroup of $SL(2, Z)$ of finite index. In this case $U^{I\Gamma}$ does *not* decompose discretely but Selberg showed how to deal with this by adding an integral to the sum on the right. Also Selberg did not think in terms of induced representations but found another path to his formula.

We turn our attention now to a different kind of application which is more in the spirit of the classical application of Fourier analysis to the partial differential equations of mathematical physics. Just as the classical Fourier transform allows one to solve partial differential equations by converting them into algebraic equations there are slight variants of this classical transform that allow one to convert number theoretical problem functions into analytic functions of a complex variable and then exploit the methods of complex analysis.

Let ϕ be a number theoretical problem function. One cannot apply Fourier analysis directly because the positive integers do not constitute a group in a natural way. However they form a discrete subset of a countable group in at least two ways and in each case one can regard ϕ as defined on the whole group by setting it equal to zero whenever it is not already defined. The two groups with which we shall be concerned are the additive group of all the integers Z and the multiplicative group of all the positive rationals Q^+. In either case we can form the appropriate Fourier transform and obtain a new function on the group $\hat{Z}$ (resp. $\hat{Q}^+$) of all characters. We shall distinguish the two cases by speaking respectively of the *additive* Fourier transform and the *multiplicative* Fourier transform.

The most general character on Z is $n \to e^{2\pi i n x}$ where x is a real number so the additive Fourier transform of ϕ is the periodic function on the real line $x \to \sum_{n=1}^{\infty} \phi(n) e^{2\pi i n x}$. Unfortunately most number theoretical problem functions are such that this series fails to converge. The difficulty can be bypassed by extending the Fourier transform to be defined on non unitary characters. These are the functions $n \to e^{2\pi i n z}$ where z is any *complex* number $x + iy$. The Fourier transform so extended $z \to \sum_{n=1}^{\infty} \phi(n) e^{2\pi i n z}$ we shall call the *additive* Fourier transform $\tilde{\phi}$ of ϕ. For most number theoretical problem functions it is defined for all z in the upper half plane and is easily seen to be an *analytic function* there.

The additive Fourier transform $\tilde{\phi}$ is especially useful when $\phi = \phi_Q$ for some positive definite quadratic form Q and its utility there stems largely from the fact that for all such Q the function $\tilde{\phi}_Q$ can be shown to satisfy a functional equation of the following form: There exists a positive integer λ and a complex constant c_0 such that if $f_Q(z) = \tilde{\phi}_Q(\frac{z}{\lambda}) + c_0$ then

$$(3) \qquad f_Q\left(-\frac{1}{z}\right) \equiv z^{2k} f_Q(z)$$

where $4k$ is the number of variables in Q. It is of course obvious that

$$(4) \qquad f_Q(z + \lambda) \equiv f_Q(z).$$

Now $-\dfrac{1}{z} = \dfrac{az+b}{cz+d}$ with $\begin{pmatrix} a & b \\ c & d \end{pmatrix} = \begin{pmatrix} 0 & -1 \\ 1 & 0 \end{pmatrix}$ and $z + \lambda = \dfrac{az+b}{cz+d}$ with $\begin{pmatrix} a & b \\ c & d \end{pmatrix} = \begin{pmatrix} 1 & \lambda \\ 0 & 1 \end{pmatrix}$. Thus (3) and (4) together imply the existence of a formula expressing $f_Q\left(\dfrac{az+b}{cz+d}\right)$ in terms of $f_Q(z)$ for all $\begin{pmatrix} a & b \\ c & d \end{pmatrix}$ in the subgroup Γ_λ of $SL(2, Z)$ generated by $\begin{pmatrix} 0 & -1 \\ 1 & 0 \end{pmatrix}$ and $\begin{pmatrix} 1 & \lambda \\ 0 & 1 \end{pmatrix}$. This formula is easily worked out and reads

$$(5) \qquad\qquad f_Q\left(\frac{az+b}{cz+d}\right) \equiv (cz+d)^{2k} f_Q(z)$$

The fact that f_Q satisfies (3) and hence (5) is not only very useful but *a priori* quite astonishing. It is thus a significant contribution to the point of view being developed here that the truth of (3) is another consequence of the Poisson summation formula. For certain elementary quadratic forms such as binary forms of the form $x^2 + ay^2$ and sums of squares in any number of variables one gets (3) by applying the classical Poisson summation formula in one variable to e^{-ax^2} (note the similarity to the function used in deriving the quadratic reciprocity law). For more general quadratic forms one must use the Poisson summation formula for discrete subgroups of R^n.

The connection between a special class of analytic functions and the arithmetic of quadratic forms epitomized by (5) was discovered in a special case by Jacobi a full century before the work of Hecke began to clarify the general theory. Moreover Jacobi's discovery was an accidental consequence of his work on elliptic functions. His approach to elliptic functions was to represent them as quotients of certain entire functions which have doubly periodic zeros and are known as theta functions. In his great memoir of 1829, "Fundamenta Nova Theoriae Functionum Ellipticarum", Jacobi developed the theory of theta functions in great detail and comparing two different expansions of the same function was delighted to find that he had derived an explicit formula for the number of solutions in integers of the equation $x^2 + y^2 + z^2 + u^2 = n$. This is just $\phi_Q(n)$ for the case in which $Q(x, y, z, u) = x^2 + y^2 + z^2 + u^2$. Jacobi's formula asserted that for this Q, $\phi_Q(n)$ is 8 multiplied by the sum of all divisors of n which are *not* divisible by 4. This result was interesting for two reasons. In the first place it established a totally unexpected connection between analysis and number theory. In the second place it implied immediately that *every* positive integer is a sum of four squares. This theorem had been suggested by the writings of Diophantus and Fermat had claimed to have a proof. However no proof was known to the mathematical public until Lagrange produced one in 1772. Euler had worked on the problem without success for forty years but did prove an important lemma of which Lagrange made use. (He also simplified Lagrange's proof).

It is trivial to show from Jacobi's formula that ϕ_Q is multipicative of order -2 and that for all odd n, $\phi_Q(n) \equiv 8(1 * n)$. Thus Jacobi's theorem fits into the general scheme we have outlined (2 is an exceptional prime).

As far as the unexpectedness of the connection between analysis and number theory is concerned it seems worthwhile to quote from a letter in which Jacobi explained his formula to Legendre. In appreciating this quotation one should bear in mind that Legendre was 76 years old and had devoted his life to two apparently independent interests: number theory and elliptic integrals. One should also have in mind that Jacobi and his contemporary Abel, having been inspired by Legendre's work on elliptic integrals, had just made it out of date, by inventing elliptic functions. Here is the quotation: "Ne vous fait-il pas plaisir, Monsieur, de voir se rapproche l'une à l'autre deux théories si hétérogènes en apparence et qui se datent en quelque sorte de vos travaux?"

Before saying more about the usefulness of the additive Fourier transform $\tilde{\phi}$ in dealing with quadratic forms it will be convenient to discuss the multiplicative Fourier transform $\hat{\phi}$ and a remarkable relationship between $\tilde{\phi}$ and $\hat{\phi}$. In regarding ϕ as a function defined on Q^+, the multiplicative group of all positive rationals which is zero outside of Z^+, we observe first of all that the composition law $\phi_1 * \phi_2$ defined in section 4 coincides exactly with group theoretical convolution: $f * g = \int f(xy^{-1})g(y)dy = \sum_{y \in G} f(xy^{-1})g(y)$ when the group is discrete. This implies at once that the multiplicative Fourier transform of $\phi_1 * \phi_2$ is the product of their multiplicative Fourier transforms. To see what this multiplicative Fourier transform looks like in concrete terms we must first examine the Pontrjagin dual of Q^+. Now evidently Q^+ is a "restricted" direct product of countably many replicas of the group Z of all integers with one factor for each prime. Hence $\hat{Q}^+$ is the full direct product of countably many replicas of $\hat{Z} = T$ (the one dimensional torus). On this basis the multiplicative Fourier transform $\hat{\phi}$ of the problem function ϕ seems to be a function of infinitely many real variables which is periodic in each. However so far very little use has been made of anything but the restriction of $\hat{\phi}$ to a very special one dimensional subgroup of $\hat{Q}^+$. This is the subgroup of all characters of the form $r \to e^{-it \log r} = r^{-it}$ where t is a real number. The restriction of $\hat{\phi}$ to this subgroup is then $\sum_{n=1}^{\infty} \phi(n)n^{-it} = \sum_{n=1}^{\infty} \frac{\phi(n)}{n^{it}}$. Now just as with the additive Fourier transform this series seldom if ever converges for number theoretically interesting functions ϕ. However once again we are able to save the day by bringing in generalized characters. This means replacing it by an arbitrary complex number s and writing

$$\hat{\phi}(s) = \sum_{n=1}^{\infty} \frac{\phi(n)}{n^s}$$

For very many functions ϕ of number theoretical interest this series converges for all s with sufficiently large real part (that is, in a right half plane) and there defines an analytic function. It is this analytic function of one complex variable which we shall henceforth mean by the term multiplicative Fourier transform of ϕ and denote by the symbol $\hat{\phi}$. It seems to the author that it would be worthwhile to investigate the function of countably many complex variables to which the logic of our analysis has led but, until this is done, we may use our slightly inaccurate terminology without ambiguity.

Closely related to the fact, noted above, that $\widehat{\phi_1 * \phi_i} = \hat\phi_1\hat\phi_2$ is the fact that ϕ is multiplicative if and only if $\hat\phi$ has the following infinite product expansion

$$(6) \qquad \hat\phi(s) = \prod_p \sum_{n=0}^{\infty} \phi(p^n)p^{-ns}$$

where p varies over all primes. That this is so was first noted by Euler in 1737 in the special case in which $\phi(n) \equiv 1$ and s is real. Its truth in the general case is a more or less immediate consequence of the unique factorization theorem for positive integers. One usually refers to the right side of (6) as an "Euler product". The individual factors: $\sum_{n=0}^{\infty}\phi(p^n)p^{-ns} = \sum_{n=0}^{\infty}\phi(p)(p^{-s})^n$ in the Euler products are evidently power series in p^{-s}. Moreover, whenever $\phi(p^k)$ is zero for all but finitely many k then the power series $\sum_{n=0}^{\infty}\phi(p)(p^{-ns})$ is a *polynomial* in p^{-s}. It follows that whenever ϕ is a multiplicative function of finite order $m = 1, 2, \cdots$ then in the Euler product factorizaton of $\hat\phi$ all but finitely many factors are polynomials in p^{-s} of degree m. Similarly using the fact that $\hat\phi^{-1} = \dfrac{1}{\hat\phi}$ one sees that whenever ϕ is a multiplicative function of finite order $-m$ where $m = 1, 2, \cdots$ then in the Euler product factorization of $\hat\phi$ all but finitely many factors are the reciprocals of polynomials in p^{-s} of degree m. In particular the Euler product factorization of $\hat1$ is $\prod_p \dfrac{1}{(1-p^{-s})}$ and that of any completely multiplicative function ϕ is $\prod_p \dfrac{1}{1 - \frac{\phi(p)}{p^s}}$.

Euler used his factorization of $\hat1(s) = 1 + \dfrac{1}{2^s} + \dfrac{1}{3^s} + \dfrac{1}{4^s}$ to prove that $\sum_p \frac{1}{p} = \infty$ and thus obtain a refinement of the theorem that there are an infinite number of primes. Almost exactly one century later Dirichlet was able to adapt Euler's method to prove that there are an infinite number of primes in any "non degenerate" arithmetic progression. To do this he replaced $\sum \dfrac{1}{n^s}$ by $\sum \dfrac{\phi(n)}{n^s}$, where $\phi(n)$ is one for all n in the progression and zero for all n not in the progression. Since such a ϕ is not multiplicative in general this prevents writing $\sum \dfrac{\phi(n)}{n^s}$ as an Euler product. Dirichlet got around this difficulty by (in effect) using Fourier analysis on finite commutative groups to represent ϕ as a finite linear combination of completely multiplicative functions. In order to complete the argument it was necessary to study the behaviour as $s \to 1$ of $\sum \dfrac{\chi(n)}{n^s}$ where χ is any of the completely multiplicative functions that occur. There is a difficulty when χ is real valued. Dirichlet observed that all of these functions χ occur in formulas of the form $\phi_D = C_D(1 * \chi)$ from the theory of binary quadratic forms as developed earlier by Lagrange and Gauss. From this it is immediate that $\hat\phi_D = C_D\hat\chi\hat1$ so that $\hat\chi = \dfrac{\hat\phi_D}{C_D\hat1}$. Using the fact that $\hat\phi_D$ and $\hat1$ have positive coefficients and that both $\dfrac{\phi_D(1) + \phi_D(2) \cdots + \phi_D(n)}{n}$ and $\dfrac{1 + 1 + 1 \cdots + 1}{n}$ have positive limits as n tends to ∞ Dirichlet was able to carry his argument to completion. As a by-product he obtained an explicit formula for computing the "class number" for quadratic forms of discriminant

D. The details appear in a series of papers published by Dirichlet between 1837 and 1840. Infinite series of the form $\sum_{n=1}^{\infty} \frac{\phi(n)}{n^s}$ are now (most appropriately) called Dirichlet series.

Overlooking the work of Jacobi, Dirichlet is sometimes given credit for being the founder of "analytic number theory". This statement has considerable justification if one is thinking only of so-called "hard" analysis with its emphasis on limits and estimates. In any event it is interesting that these two men, who introduced analysis into number theory in such apparently different ways, were almost exact contemporaries and lifelong close friends.

One of the most important and useful properties of the multiplicative Fourier transform was not discovered until the twentieth century. Both for all ϕ_Q and all $\phi_{\mathcal{F},c}$ the multiplicative Fourier transform ϕ is not just analytic in the half plane of convergence of the series $\sum_{n=1}^{\infty} \frac{\phi(n)}{n^s}$. It has an analytic continuation which is meromorphic in the whole complex plane and is either entire or has a simple pole as its unique singularity. Moreover the proof of this fact is closely intertwined with the proof of a functional equation relating values in the left half plane to those in the right half plane. Such results were known for certain *explicit* functions ϕ before the twentieth century; the first being found by Riemann and published by him in 1859 in a short but extremely influential paper. Riemann considered only the (maximally explicit) function ϕ in which $\phi(n) \equiv 1$. He used the notation $\varsigma(s) = \sum_{n=1}^{\infty} \frac{1}{n^s}$ and ever since $\hat{1}$ has been referred to as the *Riemann zeta function*. He was continuing in the spirit of Euler and Dirichlet and sought to obtain even more information about the distribution of the primes by examining the behaviour of ς for *complex* values of s.

Riemann gave two proofs of the analytic continuation and functional equation one of which is capable of extensive generalization and leads among other things to the establishment of a close and important relationship between $\hat{\phi}$ and $\tilde{\phi}$ for quite general functions ϕ. We proceed now to explain this relationship. Let ϕ be any complex valued function on the positive integers which does not grow too rapidly at ∞. Then $\tilde{\phi}$ will be an analytic function in the upper half plane and as such is completely determined by its restriction to the positive imaginary axis $y \to \tilde{\phi}(iy)$. Since the positive imaginary axis is a locally compact commutative group under multiplication it makes sense to consider its Fourier transform and its extension into the complex domain. This is of course equal to

$$\int_0^{\infty} \tilde{\phi}(iy) y^s \frac{dy}{y}$$

where $y \to y^s$ is the general character of our group when $s = i\tau$ and dy/y is the relevant Haar measure. If we now replace $\tilde{\phi}(iy)$ by its infinite series definition and invert the order of summation and integration we get

GEORGE W. MACKEY   301

$$\sum_{n=1}^{\infty} \phi(n) \int_0^{\infty} e^{-2\pi n y} y^s \frac{dy}{y}.$$

Now for each n the integral

$$\int_0^{\infty} e^{-2\pi n y} y^s \frac{dy}{y} = \int_0^{\infty} e^{-2\pi n y} y^{s-1} dy$$

may be simplified by making the change of variable $u = 2\pi n y$. The result is

$$\int_0^{\infty} e^{-u} \left(\frac{u}{2\pi n}\right)^{s-1} \frac{du}{2\pi n} = \frac{1}{n^s} \left(\frac{1}{(2\pi)^s} \int_0^{\infty} e^{-u} u^{s-1} dn\right)$$

in which we recognize the integral form of the gamma function.

In short the Fourier transform of $y \to \tilde{\phi}(iy)$ is just

$$\frac{\Gamma(s)}{(2\pi)^s} \sum_{n=1}^{\infty} \frac{\phi(n)}{n^s} = \frac{\Gamma(s)}{(2\pi)^s} \hat{\phi}(s)$$

so that there is a simple and beautiful relationship between the additive Fourier transform $\tilde{\phi}$ and the multiplicative Fourier transform $\hat{\phi}$ of the same function ϕ. One multiplied by $\dfrac{\Gamma(s)}{(2\pi)^s}$ is just the analytically extended Fourier transform of the restriction of the other to the positive imaginary axis.

An easy consequence of this correspondence is that functional equations for the $\hat{\phi}$ can be deduced from those for the $\tilde{\phi}$. Of course, strictly speaking, a functional equation for $\hat{\phi}$ of the sort that actually holds does not make sense until one knows that $\hat{\phi}$ has an analytic continuation. On the other hand the proof of analytic continuation uses the functional equation. One escapes from this dilemma by proving both things at once using a simple trick based on writing the integral from 0 to ∞ as a sum of two integrals, one running from 0 to 1 and the other from 1 to ∞. Riemann used this argument in the very special case in which $\phi(n) = 1$ when n is a square and is zero otherwise. In that case $\tilde{\phi}(z) = \sum_{n=1}^{\infty} e^{2\pi i k^2 z}$ and $\hat{\phi}(z) = \sum_{n=1}^{\infty} \frac{1}{(n^2)^s} = \sum_{n=1}^{\infty} \frac{1}{n^{2s}} = \varsigma(2s)$. One gets a functional equation for $\hat{\phi}$ by applying the Poisson summation formula to e^{-az^2} and the positive integers and from this and the above described principle, a functional equation for $\varsigma(2s)$. The change of variable $s \to \dfrac{s}{2}$ converts thus into the well known functional equation for ς.

$$(7) \qquad \frac{\varsigma(1-s)\Gamma\left(\frac{1-s}{2}\right)}{\pi^{\frac{1-s}{2}}} = \frac{\varsigma(s)\Gamma\left(\frac{s}{2}\right)}{\pi^{s/2}}.$$

In 1882 Hurwitz modified Riemann's argument so that it gave the analytic continuation and a functional equation for $\hat{\phi}$ whenever ϕ is completely multiplicative and periodic. Of course if ϕ_1 and ϕ_2 are such that $\hat{\phi}_1$ and $\hat{\phi}_2$ have analytic continuations then $\widehat{\phi_1 * \phi_2} = \hat{\phi}_1 \hat{\phi}_2$ will also have an analytic continuation and if both have functional equations expressing $\phi_i(a - s)$ in terms of $\phi_i(s)$ for the *same* a then one can "read off" a functional equation for $\hat{\phi}_1\hat{\phi}_2$ from those for ϕ_1 and ϕ_2. In this way one can establish analytic continuations and functional equations for a number of other $\hat{\phi}$'s of number theoretical interest – but only after one has expressed ϕ as a convolution of explicit ϕ's. Thus it was a real breakthrough when Hecke succeeded, in 1917, in proving that $\hat{\phi}_{\mathcal{F}}$ has an analytic continuation *and* a functional equation for *all* algebraic number fields $\mathcal{F}$. Later the same year he extended his results to cover the $\hat{\phi}_{\mathcal{F},c}$ and in 1918 did likewise for some other $\hat{\phi}$ which we cannot stop to define. Just as above the proofs rest ultimately on the Poisson summation formula (via suitable generalizations of the canonical map of $\tilde{\phi}$ on $\hat{\phi}$) and harmonic analysis occurs in two different places.

The functional equation for $\hat{\phi}_{\mathcal{F}}$ is especially easy to state. It has the form

$$(8) \qquad A^{1-s}\Gamma(1 - s)^{r_2}\Gamma(\frac{1 - s}{2})^{r_1}\hat{\phi}_{\mathcal{F}}(1 - s) = A^{s}\Gamma(s)^{r_2}\Gamma(\frac{s}{2})^{r_1}\hat{\phi}_{\mathcal{F}}(s)$$

where r_1 and r_2 are positive integers and A is a positive real number.

Since Γ is a meromorphic function with known zeros and poles (8) can be used to obtain information about how the zeros of $\hat{\phi}_{\mathcal{F}}$ for $Re(s) < \frac{1}{2}$ relate to those for $Re(s) > \frac{1}{2}$. This information in turn almost determines (8). Indeed if $\hat{\phi}$ is suitably behaved at ∞ and its zeros for $Re(s) < \frac{1}{2}$ and $Re(s) > \frac{1}{2}$ are related in the same way as for $\hat{\phi}_{\mathcal{F}}$ then $\hat{\phi}$ satisfies (8) with A replaced by some other real number A'. Since the Euler product for $\hat{\phi}_{\mathcal{F}}$ converges for $Re(s) > 1$ one can show that $\hat{\phi}_{\mathcal{F}}$ has no zeros for $Re(s) > 1$ and hence explicitly determine all zeros and poles for $Re(s) < 0$. Only the zeros in the "critical strip" $0 < Re(s) < 1$ remain unknown.

We have gone in considerable detail into the analytic properties of the additive and multiplicative Fourier transforms of the functions ϕ and have emphasized the role played by harmonic analysis in deriving them. It is time to say a few words about how these properties can be used to obtain information about the functions ϕ themselves which is of direct number theoretical interest. The details are complicated but a few general principles can be briefly explained. Consider first of all the problem of showing that some problem function ϕ is a convolution of explicit functions or at least of functions of lower order $\phi = f_1 * f_2 \cdots * f_n$. This might be hard to establish even if $f_1 \cdots f_n$ were all given but $\phi = f_1 * f_2 \cdots * f_n$ if and only if $\hat{\phi} = \hat{f}_1\hat{f}_2 \cdots \hat{f}_n$ and verifying this second equation reduces primarily to checking that $\hat{\phi}$ and $\hat{f}_1\hat{f}_2 \cdots \hat{f}_n$ have the same zeros and poles – taking due account of multiplicities. Of course the zeros and poles of $\hat{f}_1\hat{f}_2 \cdots \hat{f}_n$ can be read off at once from those of $\hat{f}_1, \hat{f}_2 \cdots \hat{f}_n$. Furthermore to the extent that one knows the zeros and poles of $\hat{\phi}$ for a problem function ϕ and the zeros and poles of $\hat{f}$ for a variety of explicit functions f one can make shrewd guesses about which explicit functions $f_1 \cdots f_n$ one might factor ϕ into. In short the whole question of factoring a multiplicative problem function as a convolution of explicit functions or problem functions of lower (absolute) order is much more approachable if one works with the multiplicative Fourier transforms $\hat{\phi}, \hat{f}_j$. At the same time one sees the utility of functional equations in giving information about zeros and poles. We emphasize that the fact that $\hat{\phi}$ has an analytic continuation is essential in carrying out the above program because the significant zeros occur *outside* the region of convergence. Functions analytic only in a half plane are very far from being determined by their zeros and poles.

The additive Fourier transform $\tilde{\phi}$ is much more useful for the ϕ_Q than for problem functions like the $\phi_{\mathcal{F},C}$ whose multiplicative constituents have negative order greater than 2 except when they coincide with the ϕ_Q. This is because of the simple functional equation (3) satisfied by the f_Q which implies that they satisfy (5) and hence are what are known as "modular forms". The fact that the f_Q are modular forms is significant for at least two reasons. In the first place it compensates for the fact that f_Q is only analytic in a half plane in making f_Q essentially determined by its zeros and poles. Since the factor $(cz + d)^{2k}$ never vanishes (5) implies that the set of zeros and poles is invariant under the action of the discrete group Γ on the upper half plane and hence is completely determined by those zeros which lie in a "fundamental region" for Γ. Since this fundamental region can be compactified by adding points at a finite number of "cusps" one finds that modular forms satisfying suitable boundedness conditions (as the f_Q do) are in fact determined by the specification of a finite number of zeros. This last fact suggests that one can hope to use the methods of complex analysis to determine concretely all possible modular forms with a fixed weight k and subgroup Γ and use these concrete representations to obtain precise information about the ϕ_Q. This program was inaugurated by Hecke in the 1920's and has had many successes.

Because of the foregoing the systematic study of modular forms has come to be regarded as a branch of number theory – even when dealing with forms which are not known to be of the form $\tilde{\phi}_Q$. Indeed studies of modular forms for their own sake have more than once led to interesting theorems about the arithmetic of quadratic forms. A case in point is Hecke's study in the late 1930's of the finite dimensional vector space of all (appropriately bounded) modular forms with a given weight and subgroup Γ. His results are easiest to state in the special case in which $\Gamma = SL(2, Z)$. In that case any modular form f is necessarily periodic with period 1 and so has a Fourier series $f(z) = \sum_{n=-\infty}^{\infty} c_n e^{2\pi i n z}$. The boundedness restriction implies that $c_n = 0$ for $n < 0$ and the condition $c_0 = 0$ picks out a subspace of codimension one called the space of *cusp forms*. Each cusp form $\sum_{n=1}^{\infty} c_n e^{2\pi i n z}$ is just $\tilde{\phi}$ for the ϕ defined by the equation $\phi(n) = c_n$. Hecke's main result is the existence of a basis $f_1, f_2 \cdots f_m$ for the vector space of cusp forms so that the corresponding Fourier coefficient functions $\phi_1, \phi_2, \cdots \phi_m$ are all multiplicative of order -2. Of course this implies that the Dirichlet series $\hat{\phi}_1, \hat{\phi}_2 \cdots \hat{\phi}_m$ all have Euler products in which almost all of the factors are the reciprocals of second degree polynomials in p^{-s} and it was this that Hecke emphasized. He also obtained simple explicit expressions for the two end coefficients in these polynomials. This theorem and its analogues for modular forms associated with proper subgroups Γ of $SL(2, Z)$ imply the theorem cited in #4 about ϕ_Q's being finite linear combinations of multiplicative functions.

In dealing with the more difficult theory of modular forms with respect to Γ where Γ is a proper subgroup of finite index of $SL(2, Z)$, it turns out to be useful to apply the unitary representation theory of the finite quotient group $SL(2, Z)/\Gamma_0$ where Γ_0 is the largest normal subgroup of $SL(2, Z)$ which is contained in Γ. Let f be a modular form of weight k for Γ and let $L_{\begin{pmatrix} a & b \\ c & d \end{pmatrix}}$ denote the linear operator $f \to (cz + d)^{-2k} f(\frac{az + b}{cz + d})$. Then the defining equation of a modular form of weight k may be rewritten as $L_{\begin{pmatrix} a & b \\ c & d \end{pmatrix}}(f) = f$ for all $\begin{pmatrix} a & b \\ c & d \end{pmatrix}$ in Γ. Let V_f denote the vector space spanned by all $L_{\begin{pmatrix} a & b \\ c & d \end{pmatrix}}(f)$ for all $\begin{pmatrix} a & b \\ c & d \end{pmatrix}$ in $SL(2, Z)$. It is easy to see that each $L_{\begin{pmatrix} a & b \\ c & d \end{pmatrix}}$ maps V_f into itself and that $\begin{pmatrix} a & b \\ c & d \end{pmatrix} \to L_{\begin{pmatrix} a & b \\ c & d \end{pmatrix}}$ is a representation of $SL(2, Z)$ in V_f which reduces to the identity on Γ and hence on Γ_0. Since Γ is of finite index in $SL(2, Z)$ one proves easily that V_f is finite dimensional. Thus we obtain a finite dimensional representation of the finite group $SL(2, Z)/\Gamma_0$. Let $V_f = V_1 \oplus V_2 \oplus \cdots V_r$ define the unique decomposition of this representation into disjoint primary components; that is, into disjoint subrepresentations each of which is a multiple of a single irreducible. Let f_j be the component of f in V_j. It is clear that each f_j is a modular form of weight k for Γ but is special in being intrinsically associated with a particular irreducible representation of $SL(2, Z)/\Gamma_0$. In a paper published in 1928 Hecke suggested the strategy of looking at the irreducible representations of $SL(2, Z)/\Gamma_0$ one at a time and for each one making a separate study of the associated modular forms. A closely related and to some extent more elegant point of view is to replace the study of modular forms for proper subgroups Γ of $SL(2, Z)$ by the study of vector valued modular forms for $SL(2, Z)$. If L is any finite dimensional unitary representation of $SL(2, Z)$ one can define a vector valued modular form of type L and weight k to be an analytic function f the upper half plane to $\mathcal{H}(L)$ such that $f(\frac{az + b}{cz + d}) = (cz + d)^{2k} L_{\begin{pmatrix} a & b \\ c & d \end{pmatrix}}(f(z))$ for all z and all $\begin{pmatrix} a & b \\ c & d \end{pmatrix}$ in $SL(2, Z)$.

An even earlier application of the representation theory of finite groups to number theory was made by Artin in 1923. It may have been the first application of group representation theory outside of group theory itself. At any rate it was early enough for Artin to find it necessary to summarize the whole Frobenius theory in his introduction. Let $\mathcal{F}$ be an algebraic number field and let G be its group of automorphisms (the so-called Galois group of $\mathcal{F}$). For each pair consisting of a subgroup H of G and a finite dimensional representation L of H Artin introduced a multiplicative problem function ϕ_L^A. This problem function depends only on the equivalence class of L and Artin showed that the assignment $L \to \phi_L^A$ has the following three fundamental properties.

(a) If L^1 and L^2 are representations of the same subgroup H then $\phi_{L^1 \oplus L^2}^A = \phi_{L^1}^A * \phi_{L^2}^A$.

(b) If $H_1 \subset H_2$ and L is a representation of H_1 then $\phi_{U^L}^A = \phi_L^A$ where U^L is the representation of H_2 induced by L.

(c) If L is the one-dimensional identity representation of H then $\phi_L^A = \phi_F$ where F is the subfield of $\mathcal{F}$ consisting of all elements left fixed by the automorphisms of H.

REMARK: Artin's definition of ϕ_L^A was "local" in the sense that ϕ_L^A was assumed to be multiplicative and the definition of $\phi_L^A(p^k)$ involved the properties of the prime p. In particular there were "exceptional" primes for which the normal definition did not make sense. A satisfactory definition for all primes was only found some seven years later. When simply omitting the exceptional primes from consideration one interprets (a), (b) and (c) as being true "almost everywhere"; that is modulo what happens at a finite number of primes.

Using the fact that the regular representation of any group is the representation induced by the identity representation of the identity subgroup it follows at once from (a), (b) and (c) that the problem function $\phi_{\mathcal{F}}$ is a convolution product of the Artin problem functions ϕ_L^A where L varies over the irreducible constituents of the regular representation of G. Moreover that factor ϕ_L^A in which L is the one-dimensional identity is just ϕ_Q where Q is the rational field. This result is a prime example of an *explicit partial factorization* of a number theoretical problem function.

More generally using (a), (b) and (c) with $H^2 = G$ one deduces that $\phi_{\mathcal{F}}$ may also be written as a convolution product in which the factors correspond to the irreducible representations of H_1 and that the factor corresponding to the one-dimensional identity representation of H_1 is $\phi_{\mathcal{F}_{H_1}}$ where $\mathcal{F}_{H_1}$ is the fixed field of H_1. In the particular case in which H_1 is commutative and normal $\mathcal{F}_{H_1}$ is an algebraic number field in the sense in which we have been using the term, and $\mathcal{F}$ is an extension of $\mathcal{F}_1$ whose relative Galois groups is commutative. In this case a rather similar factorization of $\phi_{\mathcal{F}}$ was already known from class field theory. However the factors were parametrized by the irreducible characters of an apparently quite different commutative group. Artin conjectured that these two factorizations were in fact the same and that there was a canonical isomorphism between the two commutative groups. Four years later in 1927 he succeeded in proving this conjecture and thereby putting classical class field theory in a new light. It turned out that his isomorphism theorem was the key to the reciprocity laws of class field theory. One now speaks of the Artin reciprocity law.

Artin expressed himself in terms of the multiplicative Fourier transforms $\hat{\phi}_L^A$ of the problem function ϕ_L^A and used the notation $L(s, \chi) = \hat{\phi}_L^A(s)$ where χ is the character of L. These Dirichlet series are known as Artin L functions. When χ is the character of an irreducible representation L of dimension greater than one then ϕ_L^A is a problem function of a new sort and it is natural to ask whether its multiplicative

Fourier transform – the Artin L function $L(s, \chi)$ – has an analytic continuation and a functional equation. The answer is yes but the proof (for general G) depends upon a difficult theorem in the representation theory of finite groups. This theorem was first proved by Brauer, in a paper published in 1947, with the application to Artin L functions in mind.

Applications to number theory of harmonic analysis on groups which are neither commutative nor finite began with the discovery of a remarkable relationship between modular forms and the unitary representation theory of the group $SL(2, R)$. The irreducible unitary representations of this group were found by V. Bargmann in a fundamental paper published in 1946. (They were also found at about the same time by Gelfand and Naimark but only Bargmann gave full details). These representations fall into three families called respectively, the principal series, the discrete series, and the complementary series. For our immediate purpose we need only be concerned with the discrete series. One useful description of this series realizes half of them in a Hilbert space of analytic functions in the upper half plane and the operator $T_{\begin{pmatrix} a & b \\ c & d \end{pmatrix}}$ corresponding to $\begin{pmatrix} a & b \\ c & d \end{pmatrix}$ in $SL(2, R)$ has the form $f(z) \to f(\frac{az + b}{cz + d})(cz + d)^{-2k}$ where k is a positive integer. (The other half are the complex conjugates of those in the first half). To any one familiar with the theory of modular forms the formula for $T_{\begin{pmatrix} a & b \\ c & d \end{pmatrix}}$ suggests an intimate connection between this theory and the discrete series representations of $SL(2, R)$. This connection seems to have been first noted by Gelfand and Fomin who used it to obtain some results in ergodic theory in a paper published in 1951. What it amounts to may be stated as follows: Let Γ be a suitable discrete subgroup of $SL(2, Z)$ and let I_Γ be the one-dimensional identity representation of Γ. Then one can construct the induced representation U^{I_Γ} of $SL(2, R)$ and inquire into the nature of its decomposition into irreducible components – that is, into its harmonic analysis. It can be shown that the number of times that the discrete series member T^k defined by the formula

$$T^k_{\begin{pmatrix} a & b \\ c & d \end{pmatrix}}(f)(z) = f(\frac{az + b}{cz + d})(cz + d)^{-2k}$$

is equal to the dimension of the space of all cusp forms for Γ of weight k. This fact is moreover a corollary of a theorem setting up a canonical isomorphism between the vector space of all cusp forms for Γ of weight k and the vector space of all intertwining operators for T^k and U^{I_Γ}. In short, we may say that a cusp form of weight k for Γ *is* an intertwining operator for the representations T^k and U^{I_Γ}. If one prefers to work with vector valued modular forms for the full group $SL(2, Z)$ as described above one simply replaces I_Γ by the appropriate irreducible unitary representation L of $SL(2, Z)$. This identification of modular forms with intertwining operators for unitary representations of $SL(2, R)$ turns out to be quite useful in developing certain aspects of the theory of modular forms which in turn have important applications to the number theoretical problem functions ϕ_Q.

The most striking example of the utility of this identification is the reformulation of Hecke's theory of modular forms with multiplicative Fourier coefficients, found by R.P. Langlands in the mid 1960's. Langlands reformulated Hecke's theory in the language of group representations and recognized that so reformulated one could see one's way to a vast generalization in which $SL(2, R)$ was replaced by an arbitrary reductive Lie group. Then he indicated how one might apply the generalized theory to number theoretical problem functions other than the ϕ_Q; in particular to the Artin functions ϕ_L^A for higher dimensional irreducible L.

Langlands reformulation of Hecke theory depends upon considering the representation theory of so-called "adele groups". For each prime p let Q_p denote the locally compact field of all p-adic numbers. This contains the field Q of all rationals as a dense subfield and the closure in Q_p of the ring of integers in Q is a compact open subring called the ring K_p of p-adic integers. The subset of $SL(2, Q_p)$ consisting of all elements $\begin{pmatrix} a & b \\ c & d \end{pmatrix}$ in which a, b, c and d are in K_p is then a compact open subgroup of $SL(2, Q_p)$ which we denote by $SL(2, K_p)$. Now consider the infinite product group $\prod_p SL(2, Q_p)$. This is *not* a locally compact group although its subgroup $\prod_p SL(2, K_p)$ is compact. We consider instead the subgroup $\prod_p' SL(2, Q_p)$ consisting of all sequences $\{x_p\}$ with $x_p \in SL(2, Q_p)$ for which $x_p \in SL(2, K_p)$ for all but finitely many p. The restricted group so defined has $\prod_p SL(2, K_p)$ as a subgroup of countable index and is a locally compact topological group with respect to the unique topology in which $\prod_p SL(2, K_p)$ has the product topology and is open as a subgroup. The associated adele group A is then defined to be $\prod_p' SL(2, Q_p) \times SL(2, R)$. Each factor contains $SL(2, Q)$ as a dense subgroup and so one has a natural embedding of $SL(2, Q)$ as a subgroup of the adele group $\prod_p' SL(2, Q_p) \times SL(2, R) = A$. A somewhat surprising and highly important fact is that the image of $SL(2, Q)$ in the adele group so obtained is closed. The countable group $SL(2, Q)$ appears in a natural way as a discrete subgroup of the adele group.

Using an argument different (superficially at least) from that of Langlands one can connect representation theory with Hecke's theory as follows. Let $\Gamma = SL(2, Q)$ and form the unitary representation U^{I_Γ} of A induced by the identity representation I_Γ of $\Gamma = SL(2, Q)$. Then consider the restriction of U^{I_Γ} to the subgroup $e \times SL(2, R)$ of A. If one considers the decomposition of U^{I_Γ} into multiples of irreducibles and compares this with a structure theorem for $U^{I_\Gamma} \mid e \times SL(2, R)$ which follows from the general theory of induced representations one is led quite naturally to the Hecke decomposition of spaces of cusp forms – at least to something very close to it. To be somewhat more precise one uses a basic theorem on the decomposition of restrictions to subgroups of induced representations and proceeds in two stages.

In the first we restrict only to $\prod_p SL(2, K_p) \times SL(2, R)$ and then on down to $e \times SL(2, R)$. The final result is that the restriction to $e \times SL(2, R)$ defines a representation of $SL(2, R)$ which is a direct sum of an infinite number of induced representations of the form U^M where each M is an irreducible unitary representation of $SL(2, Z)$ and U^M occurs $dim(M)$ times. The irreducible representations M which occur include all of those which reduce to the identity on some congruence subgroup and hence occur in the theory of vector modular forms. Confronting the decomposition of U^{Ir} with the decomposition of its restriction into the U^M one finds decompositions of the U^M into subrepresentations parametrized by the irreducible unitary representations of A which occur in U^{Ir}. These lead directly to decompositions of spaces of modular forms parametrized in the same way and are essentially the Hecke decompositions.

In generalizing all this, Langlands focused on the fact that one has multiplicative functions on the integers and hence Dirichlet series with Euler products canonically attached to certain irreducible representation of the adele group A. It is obvious how to generalize the construction of A to other reductive Lie groups and Langlands began by seeking a canonical way of attaching Euler products to irreducible representations of these more general adele groups.

As far as applications are concerned Langland's idea was to generalize the theorem stating that the additive Fourier transforms of the multiplicative constituents of the ϕ_Q are modular forms by proving that other multiplicative problem functions ϕ have multiplicative Fourier transforms $\hat{\phi}$ which may be identified with these new Euler products. This would be especially desirable since the Euler products of Hecke's theory are always the Fourier transforms of multiplicative functions of order -2 while in the proposed generalizations all negative integers occur. In his 1967 Whittemore lectures at Yale, "Euler products" (published in 1971) Langlands indicated that the Artin reciprocity law may be regarded as such an application in that the classical L function to which the Artin L function is proved equal (in the case of one dimensional L) is in fact a Langlands Euler product in which the group $SL(2, Q)$ is replaced by the multiplicative group of the relevant algebraic number field. Further support for the validity of Langlands' ideas was provided in a paper published by A. Weil in 1967. This paper was written without knowledge of its relevance to the above and contains strong reasons for making the conjecture that certain multiplicative problem functions of order -2 have additive Fourier transforms which like the ϕ_Q are essentially modular forms. These are the multiplicative functions whose multiplicative Fourier transforms are the so-called "zeta functions" of elliptic curves.

The implementation of Langlands' program has attracted many younger mathematicians and work on it now constitutes a major part of modern number theory. While progress has been made, much remains to be done.

The p-adic numbers were introduced by Hensel in 1901 and the construction and number theoretical use of infinite product groups similar to $\prod_p' SL(2, Q_p) \times SL(2, R)$ began with work of Chevalley in 1936 and 1940. Chevalley's groups were commutative and the role of $SL(2, Q)$ was played by the multiplicative group of all non zero elements of an algebraic number field $\mathcal{F}$. Non commutative examples were first introduced in 1957 by Ono and Tamagawa. Several striking applications were found between the work of Chevalley and that of Langlands. These include independent work in 1950 by Tate and Iwasawa giving elegant unified derivations of Hecke's functional equations for zeta and L functions and Weil's adele theoretic treatment in 1964 and 1965 of the celebrated work of C.L. Siegel on numbers of representations of one quadratic form by another. It is significant that both the work of Iwasawa and Tate and that of Weil involve applying the Poisson summation formula to adele groups.

Acknowledgements – The author has treated the theme of this paper in earlier publications, notably in his 1966-67 Oxford lectures published in 1978 as "Unitary Group Representations in Physics, Probability and Number Theory", in the survey article "Harmonic analysis as the exploitation of symmetry. A historical survey, Bull. Am. Math. Soc. 1980, pp. 543-698 (first published in Rice Univ. Studies 64 (1978) #2, 73-228 and in his contribution to the 1985 Haar Memorial Conference in Budapest. "The significance of invariant measures for harmonic analysis", Colloquia Mathematica Societas János Bolyai.

Much of #2 appears for the first time in the Haar article and most of #4 and part of #5 were worked out at the Max Planck Institut für Mathematik in Bonn in 1985-86 while the author was supported by a prize from the Alexander von Humboldt foundation. The Haar article was also written and typed at the Max Planck Institut during this period.

Department of Mathemtics,
Harvard University,
Science Center 325,
One, Oxford Street,
Cambridge Mass 02138
USA

Proceedings of Symposia in Pure Mathematics
Volume **50** (1990)

Von Neumann and the Early Days
of Ergodic Theory

GEORGE W. MACKEY

1. Background. Ergodic theory, as a self-conscious new branch of mathematics, began in 1931 with the mean and pointwise ergodic theorems of John von Neumann and G. D. Birkhoff respectively. However the central notion of "ergodicity", or "metric transitivity" as it was originally called, was introduced a bit earlier. It was defined in a paper of Birkhoff and Paul Smith entitled *On the structure of surface transformations* published in 1928. All three of these papers appeared in the Proceedings of the National Academy of Sciences of the U.S.A.

Of course the circle of problems to which the notion of ergodicity and the two ergodic theorems made such important contributions has a much longer history. The problems in celestial mechanics which inspired the paper of Birkhoff and Smith go back at least as far as G. W. Hill. It was Hill who first fully realized the necessity of studying the qualitative properties of the solutions of differential equations, directly from the equations themselves, without seeking to find explicit integrals and published a paper on the three-body problem in 1878 which stimulated the extensive work of Poincaré on the same subject. Birkhoff took up where Poincaré left off and his paper with Smith was a direct outgrowth of this continuation of the work of Poincaré. One could go back even further and implicate works of Lagrange in the 1770s on "small divisors" and their effects on perturbation theory.

The idea of an ergodic theorem goes back in another direction and originated in the work of Boltzmann on the "kinetic theory of gases". Continuing earlier work of Maxwell, Boltzmann wished to explain the thermodynamical properties of gases by applying Newton's laws and probability considerations to the motions of the gas molecules. Let Ω denote the phase space of the dynamical system describing the motion of the collection of all of the molecules of a gas. For each real number t let T_t denote the one-to-one transformation

---

1980 *Mathematics Subject Classification* (1985 *Revision*). Primary 28D99, 01A70.
This paper is in final form and no version of it will be submitted for publication elsewhere.

of Ω onto itself which is such that if $\omega \in \Omega$ is a point in phase space then a state of the system described by ω at some time t_0 will be described by $T_t(\omega)$ at time $t_0 + t$. If f is a real valued function on Ω describing some "observable" of the system one can study the changes in f with time and (if it exists) can ask for $\lim_{a \to \infty} \frac{1}{a} \int_0^a f(T_t(\omega))\, dt$. In order to carry out his program Boltzmann wanted a proof that this "time average" of f can be obtained, without knowing the T_t, by taking the "space average" of f over the set Ω_E of all points $\omega' \in \Omega$ with $H(\omega') = H(\omega)$. Here H is the real valued function on Ω whose value at each $\omega' \in \Omega$ is the total energy of the system in that state. One calls the Ω_E the constant energy hypersurfaces. It follows, of course, from the law of conservation of energy that the Ω_E are invariant under the T_t so that one can restrict oneself to a particular Ω_E in considering both kinds of average. There is a "natural" integration process on the Ω_E (and on Ω) which is also T_t invariant so that the average of f over Ω_E makes good mathematical sense whenever Ω_E is of finite "measure" with respect to this process. Boltzmann wished to deduce the equality of the space and time averages from the hypothesis (which he called the ergodic hypothesis) that the orbit of any point of ω (that is, the set of all $T_t(\omega)$) fills up the whole hypersurface Ω_E on which ω lies. When it was pointed out to him that this hypothesis is impossible on topological groups whenever the Ω_E are more than one dimensional he modified it to the "quasi-ergodic hypothesis" which asserts that the $T_t(\omega)$ are dense in Ω_E. Boltzmann began his work in the late 1860s and formulated the ergodic hypothesis in 1887. He died in 1906, without having found an acceptable argument for the equality of space and time averages, and with his program as a whole incomplete and in trouble because of lack of agreement with experiment.

2. The years 1931 and 1932. The immediate stimulus for the work of Birkhoff and von Neumann in 1931 seems to have been a paper of B. O. Koopman published in the same year and stimulated in turn by M. H. Stone's paper of 1930 on the spectral theory of one-parameter groups of unitary operators. Koopman's paper was based on a simple observation which was much less obvious in 1931 than it would be today. Let Ω be the phase space of a dynamical system with T_t as above and form the Hilbert space $\mathscr{L}^2(\Omega, \zeta)$, where ζ is the measure corresponding to the natural integration process in Ω alluded to above. Then for each t, the operator $f \to g$, where $g(\omega) = f(T_t(\omega))$, is a unitary operator V_t and $t \to V_t$ is a one-parameter group of such. Stone's theorem may be applied to this one-parameter unitary group and allows one to classify dynamical systems using the invariants supplied by spectral theory. Applying operator theory to dynamical systems in this way was a startlingly new idea at the time. André Weil reports in his collected works that he had the same idea independently of Koopman. Both Koopman and Weil communicated their observations to von Neumann in conversation and von Neumann, one of the leading experts in operator theory, was inspired to try operator theoretic methods on the ergodic problem. In a very short time he had conceived and proved his celebrated mean ergodic theorem. In its modern formulation this is a theorem about arbitrary one-parameter unitary groups $t \to V_t$ and asserts that for each vector ϕ in the Hilbert space, $\lim_{T\to\infty} \frac{1}{T} \int_0^T V_t(\phi)\,dt$ exists in the sense of the metric defined by the Hilbert space norm and is a vector ψ which is such that $V_t(\psi) = \psi$ for all t. Now when V_t is defined by the Koopman construction and Ω_E is of finite measure the limit function ψ will have the property that $V_t(\psi) = \psi$ for all t if and only if $\psi(T_t(\omega)) = \psi(\omega)$ almost everywhere in ω for all t. If ψ is not almost everywhere constant then it follows from elementary measure theory that Ω_E can be written as a union of two disjoint measurable sets of positive measure each of which is invariant under the transformations T_t. The impossibility of such a decomposition of Ω_E is precisely the definition of metric transitivity. Thus if the action of the real line on Ω_E given by the T_t satisfies the hypothesis of metric transitivity one can conclude that von Neumann's limit function ψ is a constant almost everywhere and it is easy to deduce that the constant is equal to the space average $(\mu(\Omega_E))^{-1} \int \phi(\omega)\,d\zeta_E(\omega)$ of ϕ. Thus the mean ergodic theorem implies that the time average exists and is equal to the space average whenever the hypothesis of metric transitivity holds on the Ω_E. This is true, of course, only if one interprets the time average as being defined in the sense of "convergence in the mean". However in a second paper, published in the same volume of the National Academy Proceedings as the first, von Neumann argued in detail that convergence in this sense is all that is needed in order to draw the physical conclusions that Boltzmann wanted.

The difficult point that remains is that of establishing the metric transitivity of the action of the T_t on the Ω_E. This is still a problem in 1988 in spite of many attempts to settle it. However it was an important conceptual advance to have isolated the concept of metric transitivity as the concept that is needed. The original ergodic hypothesis of Boltmann certainly implies metric transitivity but is much too strong. It is known *not* to hold in any physically interesting example. The quasi-ergodic hypothesis is too weak. The "right" concept is metric transitivity and von Neumann proposed renaming this important concept "ergodicity". Von Neumann's terminology has now become standard.

Of course it is natural to wonder whether or not the time averages $\frac{1}{T} \int_0^T \phi(T_t(\omega))\,dt$ converge to a limit, not only in the $\mathscr{L}^2$ mean, but also for almost all ω. Even if not needed for Boltzmann's program it is an interesting question mathematically and turns out to have important applications in probability theory. Von Neumann mentioned this question to Birkhoff and Koopman while communicating his mean ergodic theorem to them. Birkhoff rose to the challenge and, using methods quite different from those of von Neumann, found a proof in a rather short time. The theorem which resulted is now known as the pointwise ergodic theorem or as the Birkhoff ergodic theorem. Its proof is much more difficult than that of the mean ergodic theorem and finding it is one of Birkhoff's most celebrated achievements.

In addition to making the indicated contributions to Boltzmann's program and stimulating Birkhoff to prove the pointwise ergodic theorem, von Neumann wrote a long and very influential paper entitled *Zur Operatorenmethode in der klassischen Mechanik*. This paper was published in 1932 in the Annals of Mathematics and marks the beginning of the systematic study of the concept of ergodicity (metric transitivity) for its own sake. The paper begins with a fully detailed proof of the mean ergodic theorem and then presents the statement and proof of an important (and still inadequately appreciated) decomposition theorem establishing the fact that the ergodic measure preserving actions of the real line are the fundamental building blocks from which all measure preserving actions can be constructed. The nature of this theorem can perhaps be best appreciated by looking at an example where its truth is more or less obvious. Moreover for greater simplicity we shall consider an example of an analogue of the theorem in which the additive group of the integers replaces the additive group of the real line. Consider the unit disk in the x, y plane, $x^2 + y^2 \leq 1$, and denote it by the symbol D. Let T_θ be the transformation $x, y \to x \cos \theta + y \sin \theta, -x \sin \theta + y \cos \theta$. Then T_θ is a one-to-one transformation of D onto itself which preserves Lebesgue measure μ_D in D. Hence the iterates T_θ^n ($n = 0, \pm 1, \pm 2, \dots$) define a measure preserving action of the group Z on D. This action of course is far from ergodic. Indeed given any two real numbers r_1 and r_2 with $0 < r_1 < r_2 < 1$ the annulus $r_1^2 \leq x^2 + y^2 \leq r_2^2$ is a Z-invariant measurable set of positive measure whose complement is also of positive measure. On the other hand consider the fibering of D by the concentric circles $S_r: x^2 + y^2 = r^2$ where $0 \leq r \leq 1$. These are all Z-invariant sets into which D can be considered as decomposed but are unfortunately of measure zero. To compensate for this we notice that there exists a unique measure ν in the interval $0 \leq r \leq 1$ such that if μ_r denotes the usual linear Lebesgue measure in S_r then for every measurable subset A of D, $\mu_D(A)$ may be computed from ν and the $\mu_r(A \cap D)$ by the formula

$$\mu_D(A) = \int_0^1 \mu_r(A \cap D) \, d\nu(r).$$

In other words, we may reconstruct the action of Z on D, μ_D from its actions on the S_r, μ_r by integrating the latter with respect to r as indicated. Moreover whenever θ is an irrational multiple of π the action of Z on S, μ_r defined by T_θ is just that given by Birkhoff and Smith as an example of a metrically transitive action and hence is ergodic. We have decomposed our action as a direct integral of ergodic actions. Theorem 2 of von Neumann's paper states that every measure preserving action of the real line satisfying a mild regularity hypothesis admits an analogous decomposition into ergodic constituents and moreover that modulo inessential changes on sets of measure zero, this decomposition is *unique*.

The truth of Theorem 2 makes it clear that the problem of classifying one-parameter groups of measure preserving transformations reduces essentially to that of classifying those that are ergodic. Theorem 5 is rather easy but quite striking contribution to the latter problem. Before describing it, it will be useful to say a few words about a division of ergodic actions into two categories which differ fundamentally from one another. Let S, μ be any measure space on which we are given a one-parameter group $t \to T_t$ of measure preserving transformations and suppose that the triple S, μ, T is subject to certain, not very restrictive, regularity hypotheses. Suppose that for some s_0, the set of all $T_t(s_0)$—the so-called *orbit* of s_0—has positive measure. Then, if the given action is ergodic, the complement of this orbit is necessarily of measure zero. In other words by disregarding an (inessential) invariant subset of measure zero we obtain an action which is *transitive* in the sense that one orbit fills up the whole of S. Naively one might have supposed that all ergodic actions are *essentially transitive* in this sense. However, as the example of Birkhoff and Smith shows, this naive conjecture is false. There exist ergodic actions in which every orbit is of measure zero. We shall call such actions *properly ergodic*. The fact that they exist is of the greatest importance. If they did not, the Birkhoff ergodic theorem would be an easy consequence of von Neumann's decomposition theorem and the subject of ergodic theory would be exhausted by two theorems.

While it is trivial to determine all possible transitive measure preserving actions of the real line, and the same is true for actions of the integers, the corresponding statement is very far from true for the properly ergodic actions of either group. To this day, almost sixty years after the beginning of ergodic theory described above, we are still very far from any sembalance of a complete description of all possible ergodic actions of the real line or of the integers. We have many examples and at least two classes of examples within which a complete classification is known but not much more.

Theorem 5 in von Neumann's 1932 Annals paper is concerned with one of these classes. Let $t \to T_t$ be an ergodic one-parameter group of measure preserving transformations in a measure space S, μ and let $t \to V_t$ be the one-parameter unitary group associated with $t \to T_t$ by the Koopman construction. It is natural to say that the ergodic action has *pure point spectrum* if $t \to V_t$ can be decomposed as a *discrete* direct sum of (necessarily one-dimensional) irreducible representations of the real line. Each of these irreducible constituents is, of course, of the form $t \to e^{it\lambda}I$, where λ is some real number and I is the identity operator. The numbers λ which occur constitute the spectrum. If $\mathscr{L}^2(S, \mu)$ is separable, this spectrum is a countable set. The content of Theorem 5 can then be summarized as follows.

Given that an ergodic action has a countable pure point spectrum $\lambda_1, \lambda_2, \ldots$:

(1) Every λ_j occurs with multiplicity one.

(2) $\lambda_i \pm \lambda_j$ is equal to some λ_k for every i and j and each choice of $+$ or $-$, so that the spectrum is a countable subgroup of the group of all real numbers.

(3) Every countably infinite subgroup occurs for some ergodic action with a pure point spectrum.

(4) If two such actions have the same spectrum and the measures μ_1 and μ_2 are normalized so that $\mu_1(S_1) = \mu_2(S_2) = 1$, then there is an isomorphism B between S_1, μ_1 and S_2, μ_2 as measure spaces such that $BT_t^1 B^{-1} = T_t^2$ for all t.

Since one sees easily that the above described action is properly ergodic if and only if the group of all λ_j is dense in the real line one has continuum many inequivalent examples of properly ergodic actions with pure point spectrum.

In Theorem 6 von Neumann introduced a method for constructing an ergodic action of the additive group of the line out of an ergodic action of the integers and a positive measurable function f on the space S of the action. The space of the action of the line is the set of all $s, y \in S \times P$ with $y \leq f(s)$. Here P is the positive real axis. One defines $T_t(s, y)$ to be $s, y + t$ as long as $t \leq f(s) - y$ and for $f(s) - y < t < f(s) - y - f(\alpha(s))$ one defines $T_t(s, y)$ to be $\alpha(s), t - f(s) + y$. Here α is the measure preserving transformation of S associated with 1 in the given integer action. The definition for other t values should now be clear. In pictorial terms each point s, y is pushed up the interval $s, [0, f(s)]$ until it reaches the "top" $s, f(s)$. Then it is identified with $\alpha(s), 0$ and pushed up the interval $\alpha(s), [0, f(d\alpha(s))]$, etc. One speaks of the flow *built over* α, under the function f. Using this construction von Neumann was able to produce examples of properly ergodic actions with no point spectrum at all.

3. Afterwards. The long-range consequences of the work described thus far go considerably beyond clarifying and somewhat advancing Boltzmann's program in statistical mechanics. We shall first list some of these and then add explanatory comments in each case.

(A) It provided a tool for analyzing dynamical systems that could be applied to other problems than those of Boltzmann—in particular the celebrated three-body problem of celestial mechanics.

(B) It provided a powerful new tool for use in probability theory.

(C) By way of more or less straightforward generalizations from the real line and the integers it provided a new tool for analyzing arbitrary groups of diffeomorphisms of manifolds.

(D) By raising the question of classifying ergodic actions, of deciding which measure preserving actions are ergodic, and of how to decompose nonergodic actions into ergodic parts it created an important new branch of pure mathematics; a branch which is still thriving half a century later.

(A) An immediate consequence of von Neumann's decomposition theory was that one had an illuminating new way of looking at the old problem of seeking a "complete" set of "integrals of the motion" for any classical Hamiltonian dynamical system. Let $F_1, F_2, \ldots, F_N$ be any finite set of such integrals and for each N-tuple $c_1, c_2, \ldots, c_N = c$ of real numbers let Ω_c denote the subset of phase space Ω on which $F_j(\omega) = c_j$ for all j. The Ω_c are then invariant subsets under the group $t \to T_t$ of measure preserving transformations describing the time evolution of the system. Moreover a slight adaptation of the proof of the von Neumann decomposition theorem shows that one can decompose the given system as an integral over the vectors $c_1, c_2, \ldots, c_N = c$ of the various Ω_c. As in von Neumann's theorem one will have an invariant measure in each Ω_c uniquely determined up to a multiplicative constant for almost all c. It is then easy to see that, either these ergodic actions or almost all ergodic, or else there exists at least one further integral F_{N+1} which is *not* almost everywhere equal to $G(F_1, F_2, \ldots, F_N)$ for any function G of N real variables. Thus the system $F_1, F_2, \ldots, F_N$ is a "maximal" system of integrals if and only if the decomposition defined by the Ω_c coincides almost everywhere with von Neumann's decomposition into ergodic pieces. Moreover if in the von Neumann decomposition one has a set of constituents of *positive* measure each of which is *properly ergodic* then it is impossible to find a set $F_1, \ldots, F_N$ of integrals so large that knowing the values of all these integrals determines a single orbit. The search for all possible integrals must be replaced by an attempt to find the nature of the decomposition of the group $t \to T_t$ into its ergodic constituents and the failure of sufficiently many integrals to exist must be traced to the occurrence of *properly* ergodic constituents in this decomposition.

The distinction between ergodic actions with pure point spectrum and more general actions is also important here. In the so-called completely integrable systems studied by Liouville in 1855 one does *not* have a sufficient number of *globally defined* integrals in the sense we have been using it. One has instead a situation in which almost all the components of the von Neumann decomposition are either essentially transitive or have a pure point spectrum. Classically the existence of Ω_c's with pure point spectrum was perceived as the existence of what are sometimes called quasiperiodic motions. As far as the existence of von Neumann constituents, which are properly ergodic and have continuous spectrum, are concerned it is interesting to read the early work of G. D. Birkhoff on qualitative dynamics and even earlier work of Hadamard on "geodesic flows" where phenomena present themselves suggesting the need for considering something more general than quasiperiodic motions. [See, for example, G. D. Birkhoff, *Quelques théorèmes sur le mouvement des systèmes dynamiques*, Bull. Soc. Math. France **40** (1912), 305–323, and the paper of Hadamard cited therein.]

For more on geodesic flows see the comments under (D) below. There has recently been a certain amount of discussion and comment about the occurrence of "chaotic" phenomena in presumably deterministic dynamical systems. This occurrence is closely connected with the occurrence of components with no point spectrum in the von Neumann decomposition. How this comes about should become clearer after reading (B) below.

(B) Ergodic theory appeared on the scene just in time to be integrated with the modern measure theoretic approach to probability theory. The latter began with a paper of E. Borel published in 1909 and reached maturity of formulation in 1933 with the publication of Kolmogoroff's book, *Grundbegriffe der Wahrsheinlichheitsrechnung*. Important steps along the way were taken in Wiener's papers on Brownian motion published between 1920 and 1923 and in a paper of Steinhaus published in 1923. It is interesting that Kolmogoroff's book makes no mention of the ergodic theorem. It was only in 1934 that Doob, E. Hopf, and Khinchine independently recognized that the pointwise ergodic theorem can be looked upon as completely equivalent to a significant generalization of one of the basic theorems of probability theory—the so-called strong law of large numbers.

In the measure-theoretic approach to probability as systematized in Kolmogoroff's book one starts with a suitably restricted measure space Ω, μ such that $\mu(\Omega) = 1$. One thinks of the points ω in Ω as "events" in the "universe" Ω of all possible events. If E is a measurable subset of Ω, then $\mu(E)$ is the probability that the ω that actually occurs is in the set E. The space Ω, μ is given and fixed once and for all, and probability theory concerns itself with measurable functions defined on Ω. These are given the suggestive name of "random variables" and are usually (but not necessarily) real valued. If g is a real-valued random variable, then the probability that g takes on a value in the Borel set F of real numbers is just $\mu(g^{-1}(F))$. Thus the behavior of the random variable g taken in isolation is completely determined by the measure α on the real line $F \to \mu(g^{-1}(F))$. This is a "probability measure" in the sense that $\alpha([-\infty, \infty]) = 1$ and is what is called the distribution of the random variable g. When it exists, $\int_\Omega g(x)\, d\mu(x) = \int_{-\infty}^{\infty} x\, d\alpha(x)$ is called the "expected value" or expectation of the random variable g and if this is denoted by $\bar{g}$ then $\int (g(x) - \bar{g})^2\, d\mu(x) = \int_{-\infty}^{\infty} (x - \bar{g})^2\, d\alpha(x)$ is called the "variance" of g. It is clear that g has an expectation and a finite variance if and only if g is in $\mathscr{L}^2(\Omega, \mu)$. Of course, as long as only one random variable is involved, there is little point in introducing the "universe" Ω; everything can be expressed in terms of the measure α. It is in dealing with relationships between different random variables that the usefulness of Ω becomes clear—especially when there are infinitely many. For example, let g_1 and g_2 be two real-valued random variables and let α_1 and α_2 be their distributions. Then $\omega \to g_1(\omega), g_2(\omega)$ is a measurable function ψ from Ω to $R \times R$ and setting $\beta(F) = \mu(\psi^{-1}(F))$ defines a probability measure β in $R \times R$ which is far from uniquely determined by α_1 and α_2. It is called the joint distribution of the random variables g_1 and g_2. If $g_2 = g_1^2$ then β is supported by the curve $y = x^2$. On the other hand if g_1 and g_2 are so-called "independent" random variables then β is the product measure $\alpha_1 \times \alpha_2$ in which $\alpha_1 \times \alpha_2(F_1 \times F_2) = \alpha_i(F_1)\alpha_2(F_2)$. In fact, this is the definition of independence, and this definition seems to capture quite completely the intuitive notion of what it means for random variables to be independent.

To see the connection of the measure theoretic picture of probability theory with ergodic theory choose a probability measure space Ω, μ and let $\ldots, f_{-2}, f_{-1}, f_0, f_1, \ldots$ be a doubly infinite sequence of random variables. In probabilistic language this sequence is a "discrete parameter stochastic process". For each ω the doubly infinite sequence of real numbers $\ldots, f_{-2}(\omega),$ $f_{-1}(\omega), f_0(\omega), f_1(\omega), \ldots$ is called a "sample sequence of the process". Thus the function of two variables $n, \omega \rightarrow f_n(\omega)$ may be looked upon either as a family of sample sequences one for each ω or a family of random variables one for each n. For each pair n, d of integers with $d > 0$ let $\psi_{n,d}$ denote the mapping $\omega \rightarrow f_n(\omega), f_{n+1}(\omega), \ldots, f_{n+d-1}(\omega)$ of Ω into R^d and let $\mu_{n,d}$ be the probability measure in R^d, $F \rightarrow \mu(\psi_{n,d}^{-1}(F))$. $\mu_{n,d}$ is then the joint distribution of $f_n, f_{n+1}, \ldots, f_{n+d-1}$. In the special case in which $\mu_{n,d}$ depends only on d and is independent of n one says that the stochastic process is "stationary". Now in considering a single stationary stochastic process there is no loss in generality in supposing that the mapping $\omega \rightarrow \{f_n(\omega)\}$ of Ω into sample sequences is one-to-one and then a simple argument shows that there is an essentially unique measure preserving map β of Ω onto Ω such that $f_n(\omega) = f_0(\beta^n(\omega))$ for all n, and almost all ω. Conversely if β is a measure preserving transformation of a probability measure space Ω, μ onto itself and f is any real valued measurable function on Ω, then setting $f_n(\omega) = f(\beta^n(\omega))$ for all integers n defines a *stationary* stochastic process. In short one has a natural one-to-one correspondence (modulo the obvious equivalences) between all possible stationary stochastic processes on the one hand and all pairs consisting of a measure preserving action of the integers and a measurable function on the other. Of course the measure must be 1 on the whole space.

In understanding the significance of this correspondence it is useful to examine the meaning of ergodicity for the associated stochastic process. If one applies the von Neumann decomposition theorem to the action of the β^n on the underlying universe Ω, one divides all sample functions of the process into disjoint sets each associated with a single ergodic constituent. However in applying the theory one is usually given a sample function (actually a large segment of one) and wants to understand it by guessing the process from which it comes. For this purpose only the ergodic component in which it lies can be of any relevance since what goes on elsewhere has no influence on the sample function. Moreover once this ergodic component is known one has all one needs for further analysis of the sample. For these reasons there is seldom any loss in generality in restricting attention to those stationary stochastic processes in which the associated measure preserving action is indeed ergodic. Assuming ergodicity and assuming that the f_j are all L^1 functions so that they have expected values (necessarily the same for all j) one verifies easily that the pointwise ergodic theorem may be restated as follows: For almost all $\omega \in \Omega$ the corresponding sample sequence $\{f_n(\omega)\}$ has the property that

$$\lim_{N \to \infty} \frac{f_n(\omega) + f_{n+1}(\omega) + \cdots + f_{n+N}(\omega)}{N + 1}$$

exists and equals the common expected value of the f_j. This is the so-called law of large numbers. Before 1934 it was known to be true only when the f_j are independent.

Let $\ldots, f_{-2}, f_{-1}, f_0, f_1, \ldots$ be a stationary stochastic process in which the action of the β^n on Ω is ergodic and in which the f_j are all in $\mathscr{L}^2(\Omega, \mu)$. Let $p(j) = \int f_j(\omega) f_0(\omega) \, d\mu(\omega)$. It is easy to prove that p is a so-called "positive definite function" on the integers whence it follows from a well-known theorem of Herglotz that there exists a unique measure ν on the unit circle such that $\int e^{ij\theta} \, d\nu(\theta) \equiv p(j)$. This measure is an important invariant of the process called its *spectrum*. In another important application of the ergodic theorem one can use it to prove that the positive definition function p can be computed from almost any sample function by the simple formula

$$p(j) = \lim_{N \to \infty} \frac{a_{j-1}a_{-1} + a_{j-2}a_{-2} + \cdots + a_{j-N}a_{-N}}{N}.$$

This shows that p coincides with the well-known auto correlation function introduced earlier by students of the statistics of time series. Using the ergodic theorem in a somewhat more elaborate way one can prove the possibility of reconstructing the entire stochastic process (not just its spectrum) from almost any sample sequence.

There is a simple and interesting connection between the spectrum of a stationary stochastic process, as just defined, and the spectrum, in the sense of the spectral theorem of operator theory, for the Koopman unitary operator U_β associated with the underlying measure preserving transformation β. By the spectral theorem for unitary operators there exists a projection valued measure $E \to P_E$ defined for all Borel subsets of the unit circle in the complex plane such that for every f in $\mathscr{L}^2(\Omega, \mu)$, $(U_\beta(f) \cdot f) = \int_0^{2\pi} e^{i\theta} \, d(P_\theta(f) \cdot f)$. From this it follows easily that

$$p(j) \equiv \int_0^{2\pi} e^{ij\theta} \, d(P_\theta(f) \cdot f) \quad \text{for all } j.$$

In other words the spectrum of the process defined by β and f is just the measure $E \to (P_E(f) \cdot f)$. While this measure depends on both f and β it is necessarily absolutely continuous with respect to those measures whose sets of measure zero are just those sets E for which $P_E = 0$. In particular if β has a pure point spectrum then the spectrum of *any* stationary process based on β is a countable set contained in this point spectrum. Since the degree to which the sample functions of a process are "truly random" or have futures determined by their pasts is strongly influenced by the nature of the spectrum of the process, it must also be strongly influenced by the nature of the underlying ergodic actions. It is for this reason that we may speak of some ergodic actions as being more "random" or "chaotic" than others.

We have confined our attention here to discrete stochastic processes because they are technically easier to study and are sufficient to make clear the nature of the connection with ergodic theory. One can also consider families of random variables parametrized in other ways. When the integers are replaced by the real numbers one speaks of continuous parameter stochastic processes and these are connected with ergodic actions of the additive group of the real line in much the same way that discrete parameter stochastic process are related to ergodic actions of the integers. Of course sample sequences must be replaced by sample functions and the measure μ in the universe Ω defines a measure in the space of sample functions. The celebrated Wiener measure which occurs in Wiener's Brownian motion theory (mentioned above) is of this character. Indeed, Wiener's work can be interpreted as the first rigorously defined example of an important type of continuous parameter stochastic process and the underlying ergodic action turns out to be the simplest example of an important type of ergodic action. Another of Wiener's important papers is illuminated by the development of ergodic theory and its connection with probability. This is *Generalized harmonic analysis* published in 1930 in Acta Mathematica, the year before the two ergodic theorems were proved. Here Wiener is concerned with a class of real or complex valued functions on the real line which are like almost periodic functions in that neither ordinary Fourier series expansions nor Fourier transforms are applicable to them. However they are much more general and as was realized a decade or so later are best understood as sample functions of continuous parameter stochastic processes. It is only when the underlying ergodic action has a pure point spectrum that these sample functions can be almost periodic.

The connection between ergodic theory and probability theory outlined in the preceding pages is in large part due to work of Khinchine and Wold published in 1934 and 1938 respectively. For lengthier and more thorough expositions in the spirit of the foregoing the reader may consult the following articles by the present author: *Ergodic theory and its significance for statistical mechanics and probability theory*, Adv. in Math. **12** (1974), 178–268 (especially §§2, 5, 6, 7 and 8) and §18 of *Harmonic analysis as the exploitation of symmetry–a historical survey*, Bull. Amer. Math. Soc. **3** (1980), 543–698.

(C) From the point of view of pure mathematics there is no reason to confine attention to the study of measure preserving actions of the real line and integer groups. The basic concepts make sense and are interesting for a more or less arbitrary group. Moreover one can formulate generalizations of the ergodic theorem whenever the group is such that one has a notion of "mean value" for suitable functions defined on the group. A somewhat less obvious generalization lies in relaxing the requirement that the measure μ be invariant. One says that a measure μ is "quasi-invariant" if μ and its translates μ_x by group elements x have the same sets of measure zero for all x and one can generalize the Koopman construction to associate a unitary group representation $x \to U_x$ by making use of the Radon-Nikodým derivatives of the μ_x with respect to μ to compensate for the noninvariance of μ. One defines ergodicity in the obvious way and can prove a generalization of the von Neumann decomposition theorem when G is separable and locally compact and S satisfies mild regularity conditions. Rather surprisingly one can even formulate and prove a generalization of the pointwise ergodic theorem when the invariant measure is replaced by one which is only quasi-invariant (see Hurewicz, Ann. of Math. (1944)).

As far as applications of these two generalizations of ergodic theory are concerned the first two known to the author were made by von Neumann himself and by von Neumann and his collaborator Murray. In the first of the celebrated von Neumann-Murray papers on operator algebras (published in the Annals of Mathematics in 1936) the authors proved the existence of Type II factors by an explicit construction starting with a properly ergodic action of an arbitrary countably infinite discrete group G. (See elsewhere in this volume for the definition of factors of Type I, Type II, and Type III.) While it is true that choosing G to be the integers would still have given a Type II factor much more general examples could be gotten by making G general. The first example of a Type III factor was not found until after the completion of the second von Neumann-Murray paper. The third paper of the series, written by von Neumann alone, was published in the Annals of Mathematics in 1940. Its main result is that one can obtain Type III factors by generalizing the Type II construction mentioned above by allowing measures which are only quasi-invariant.

In dealing with measures which are quasi-invariant with respect to a group action it is important to realize that any measure with the same null sets as the given one is also quasi-invariant and that for most purposes it does not matter very much which measure one chooses as long as they all have the same null sets. Indeed it is sometimes clarifying to introduce the concept of a "measure class" and realize that one can apply the concepts of ergodic theory whenever one has a group acting on a space which preserves some naturally occurring measure class. A potentially far reaching class of examples is suggested by the fact that every smooth manifold satisfying mild regularity conditions has a canonical measure class—the one defined locally by the class of Lebesgue measure. This measure class is invariant under every subgroup of the group of all diffeomorphisms. It follows that one can use the generalized von Neumann decomposition theorem and other concepts from ergodic theory to analyze more or less general diffeomorphism groups in much the same spirit as explained above for classical dynamical systems.

Another more or less unexploited potential application of ergodic theory involving quasi-invariant measures lies in extending the connection between ergodic theory and stationary stochastic processes to one between ergodic theory and nonstationary stochastic processes.

(D) Finally we conclude this essay with some remarks about the development of ergodic theory as a branch of pure mathematics. This is a huge subject and we have time for only a few indications. Some of the earliest examples of a properly ergodic action of the real line after those in von Neumann's 1932 Annals paper were given in papers by Hedlund published in 1934 and 1935. Let $\mathcal{M}$ be a complete Riemann manifold and let $\mathcal{M}_T$ be the manifold of all tangent vectors to $\mathcal{M}$. If C is any pointed geodesic on $\mathcal{M}$ then each point m of C has a tangent vector $\mathbf{v}$ and the pair $m, \mathbf{v}$ defines a point in $\mathcal{M}_T$. Thus C is canonically associated to a curve C^T in $\mathcal{M}_T$. It can be shown that the curves C^T in $\mathcal{M}_T$ for the various geodesics in $\mathcal{M}$ are the orbits of a unique action of the real line on $\mathcal{M}_T$. This action is such that on each orbit the length of the vector $\mathbf{v}$ is a constant so that the subset $\mathcal{M}_T^0$ of $\mathcal{M}_T$ consisting of all $m, \mathbf{v}$ for which $\mathbf{v}$ is of unit length is an invariant subset. The above described flow restricted to this invariant subset is called the "geodesic flow" on $\mathcal{M}_T^0$ or the "geodesic flow" for the Riemannian manifold $\mathcal{M}$. Hedlund found two types of two-dimensional Riemannian spaces for which he could prove the proper ergodicity of the geodesic flow. His proofs were long and difficult. Since a "free" particle in a Riemannian space is supposed to move along the geodesics one can think of geodesic flows as the flows of certain dynamical systems restricted to constant energy hypersurfaces. Hedlund's results were proved in a unified manner by a different and somewhat simpler argument by E. Hopf in 1936. The study of the ergodic character of geodesic flows has been an important topic in ergodic theory ever since.

Another topic of enduring interest has been the classification of ergodic actions of "shift" type. Let F be a finite set and let Ω be the set of all functions from the group Z of all integers under addition to F. Let F be made into a measure space by assigning a nonnegative real number $p(f)$ to each f in F in such a way that $\sum_{f \in F} p(f) = 1$ and defining $\mu(E) = \sum_{f \in E} p(f)$. Then regarding Ω as an infinite product of replicas of F and calling the product measure μ^α we get a measure in Ω such that $\mu^\alpha(\Omega) = 1$. Regarding Ω once again as the set of all functions a from Z to F we may define an action β of Z in Ω by setting $\beta(a)(n) = a(n+1)$. It is obvious that β is measure preserving and hence that its iterates define a measure preserving action of Z on Ω. One sees easily that these actions are properly ergodic and that they are precisely the actions associated with those stationary stochastic processes whose random variables are independent and take on only a finite number of values. For all choices of F and p these actions have the same Koopman spectrum. On the other hand it is indicative of the difficulty of the problem of classifying ergodic actions that for twenty-five years no one was able to decide whether two actions of shift type based on distinct finite measure spaces F_1, p_1 and F_2, p_2 were isomorphic as ergodic actions are not. The problem was finally solved by work of Kolmogoroff and Sinai in 1958 and 1959. They introduced a new invariant of ergodic actions of Z called the entropy and proved that the entropy of the shift defined by F and p is $-\sum_{f \in p} p(f) \log p(f)$. Since shifts with different entropy cannot be isomorphic it follows that one can have nonisomorphic shifts. More than a decade elapsed before Ornstein proved a converse: two shifts with the same entropy are isomorphic.

In these comments to (A), (B), (C), and (D) there has been no attempt at completeness. Many more interesting and important things could have been said. Our aim rather has been to give a general idea in a few pages.

Acknowledgment. The author is indebted to G. Birkhoff and A. Gleason for helpful comments. Gleason in particular read the whole typescript with care.

Harvard University

Final Remarks

GEORGE W. MACKEY

1. Introduction.   We conclude this collection with a short account of some developments of the past quarter century which have proceeded more or less independently of the theory of unitary group representations and its applications but which are related to it in many ways. These new developments began with four major discoveries made in 1967 and 1968 which seemed to have nothing much to do with one another. Indeed they occurred in quite distinct areas of mathematics and physics; namely the theory of Lie algebras, strong interaction physics, two-dimensional lattice models in statistical mechanics, and the theory of a nonlinear partial differential equation of importance in classical hydrodynamics. Each discovery had extensive ramifications and before long these ramifications were found to have interesting and illuminating interconnections. One hopes to see these interrelated ramifications unified into a coherent whole. Like the theory of unitary group representations this complex of ideas has applications to both number theory and physics, and the exploitation of symmetry plays a significant role. Perhaps one can hope further that our putative coherent whole and the theory of unitary group representations can be blended into an even larger coherent whole. At any rate, working in this direction has been the author's chief preoccupation for several years; for this reason the following exposition seems an appropriate conclusion to this book.

In brief the four discoveries were: (a) the Kac-Moody Lie algebras, (b) the Veneziano model in strong interaction physics, (c) the exact solvability of the Korteweg de Vries equation $\frac{\partial u}{\partial t} = 6u\frac{\partial u}{\partial x} - \frac{\partial^3 u}{\partial x^3}$ by Gardner, Greene, Kruskal, and Miura, and (d) the exact solvability of the two-dimensional "ice problem" in lattice statistical mechanics by Lieb.

Before describing these discoveries more fully and giving an account of their early ramifications, it will be useful to present three sections of background material: one each on roots and weights in the theory of Lie algebras, the theory of the S operator in the quantum theory of scattering, and the nature of lattice statistical mechanics. In spite of their physical motivation the second and third topics will be presented as intrinsically interesting topics in pure mathematics, intelligible to mathematicians with little background in physics. The section on the S operator will be relevant to both discoveries (b) and (c)—but in quite different ways.

2. Roots and Weights in the Theory of Lie Algebras. Let $\mathcal{L}$ be any Lie algebra over the complex field and let $\mathcal{A}$ be a commutative subalgebra. Let V be a representation of $\mathcal{L}$ by linear operators in a vector space $\mathcal{H}(V)$. If $\varphi \in \mathcal{H}(V)$ is an eigenvector of V_x for all x in $\mathcal{A}$ so that $V_x(\varphi) = w(x)\varphi$ then $w(\varphi)$ is clearly a linear function of φ; that is, w is a member of $\mathcal{A}^*$, the dual vector space of $\mathcal{A}$ considered as a vector space. Such members of $\mathcal{A}^*$ are called *weights* of V with respect to $\mathcal{A}$. For a given w the set of all weight vectors, that is, the set of all φ such that $V_x(\varphi) = w(x)\varphi$ for all x in $\mathcal{A}$, is called the *weight space* $\mathcal{H}_w$. The $\mathcal{H}_w$ for distinct weights are easily seen to be linearly independent and, whenever their linear span is the whole of $\mathcal{H}(V)$, one says that V has a *weight space decomposition* with respect to $\mathcal{A}$. It amounts to the same thing to say that the V_x for $x \in \mathcal{A}$ are simultaneously diagonalizable.

Now let us specialize these concepts to the case of the adjoint representation of $\mathcal{L}$. This is the representation A whose space is $\mathcal{L}$ and in which $A_x(y) = [x, y]$ for each $x, y \in \mathcal{L}$. The commutative subalgebra $\mathcal{A}$ is evidently a subspace of the weight space for the weight zero. The nonzero weights are called the *roots* of $\mathcal{L}$ with respect to $\mathcal{A}$. Whenever $\mathcal{A}$ is the entire 0 weight space and $\mathcal{L}$ is the direct sum of $\mathcal{A}$ and the other weight spaces (which are called root spaces), one says that $\mathcal{L}$ has a root space decomposition with respect to $\mathcal{A}$. The following simple theorem plays a very important role in the theory.

THEOREM. *Let x be a member of $\mathcal{L}$ that belongs to the root space of a root r of $\mathcal{A}^*$ for some commutative subalgebra $\mathcal{A}$ of $\mathcal{L}$. Let φ be a weight vector for a weight w in $\mathcal{A}^*$ for some representation V of $\mathcal{A}$. Then either $V_x(\varphi) = 0$ or $w + r$ is another weight of V and $V_x(\varphi)$ is in the weight space of $w + r$.*

PROOF. Suppose $V_x(\varphi) \neq 0$. Let $a \in \mathcal{A}$. Then $V_a(V_x(\varphi)) - V_x(V_a(\varphi)) = V_{[a,x]}(\varphi) = V_{r(a)x}(\varphi) = r(a)V_x(\varphi)$. Hence $V_a(V_x(\varphi)) = V_x(V_a(\varphi)) + r(a)V_x(\varphi) = V_x(w(a)\varphi) + r(a)V_x(\varphi) = (w(a) + r(a))V_x(\varphi)$. Thus $V_x(\varphi)$ is a nonzero weight vector for $w + r$.

When we take V to be the adjoint representation we obtain the

COROLLARY. *If x and y are root vectors for the roots α and β, then either $[x, y] = 0$, or $\alpha + \beta$ is again a root, and $[x, y]$ is a member of its root space.*

Whenever one can find an $\mathcal{A}$ for which the adjoint representation has a weight space decomposition, one can use the preceding theorem and its corollary to deduce a great deal about the structure of $\mathscr{L}$, as well as about the nature outside of $\mathscr{A}$ of those of its representations that have weight space decompositions.

An illuminating example of the foregoing is provided for $n = 2, 3, \ldots$ by the Lie algebra $\mathscr{L}$ of all $n \times n$ complex matrices with trace 0 and $[x, y]$ defined as $xy - yx$. Let $\mathscr{A}$ be the commutative subalgebra of all diagonal matrices. For each positive integer pair i, j with $i \neq j$ and $i \leq n, j \leq n$ let e_{ij} denote the matrix whose elements are all zero, except that the element in the ith row and jth column is 1. One computes at once that e_{ij} is a root vector whose root is the linear functional $\lambda_1, \lambda_2, \ldots, \lambda_n \to \lambda_i - \lambda_j$. It is immediate that $\mathscr{L}$ is a direct sum of $\mathscr{A}$ and the root spaces with respect to $\mathscr{A}$, and that all the root spaces are one dimensional. $\mathscr{A}$ is of course $n - 1$ dimensional and there are $n^2 - n$ distinct roots. We note that the negative of every root is also a root. Slightly less obvious is the important fact that one can choose one member of each pair $\alpha, -\alpha$, where α is a root, in such a manner that the sum of two chosen roots is either again a chosen root or not a root at all. For example, one may choose those roots in which $i < j$. Once this choice has been made, one can speak of positive and negative roots and the set of all roots acquires a partial ordering.

Let $\mathscr{G}_{\mathscr{L}}$ denote the group of all automorphisms of $\mathscr{L}$ (as a Lie algebra). The subgroup of $\mathscr{G}_{\mathscr{L}}$ consisting of all automorphisms that map $\mathscr{A}$ onto itself has a normal subgroup consisting of those that take each element of $\mathscr{A}$ into itself. The quotient group is a group of linear transformations of $\mathscr{A}$, which is called the Weyl group of $\mathscr{L}$ (with respect to $\mathscr{A}$). It is easy to see that each root of $\mathscr{L}$ is transformed into another root by each member of the Weyl group so that the Weyl group acts as a group of permutations of the roots. Since it is easy to check that there are $n - 1$ linearly independent roots in $\mathscr{A}^*$, one sees that a member of the Weyl group is uniquely determined by its action on the roots. It may thus be defined as a group of permutations of the roots and is accordingly finite. In the case at hand it is the full permutation group on n objects.

It can be shown that any commutative subalgebra $\mathscr{A}'$ of $\mathscr{L}$, for which a root space decomposition exists, must be conjugate to $\mathscr{A}$ in the sense that some member of $\mathscr{G}_{\mathscr{L}}$ transforms $\mathscr{A}'$ into $\mathscr{A}$. Thus the algebra $\mathscr{A}$ in terms of which W and the root spaces are defined is essentially unique. $\mathscr{A}$ and $\mathscr{A}'$ are called *Cartan subalgebras* of $\mathscr{L}$.

This example is quite typical of a whole class of Lie algebras—the finite-dimensional complex Lie algebras which are *semisimple* in the sense that they admit no nontrivial solvable ideals. There are general theorems about semisimple Lie algebras which imply all the above facts about our example except, of course, those about the dimension of $\mathscr{A}$ and the number of roots. In particular, $\mathscr{A}$ is unique up to automorphisms and is called a Cartan subalgebra. One can show, moreover, that every finite-dimensional complex representation of a complex semisimple Lie algebra has a weight space decomposition and is determined to within equivalence by its restriction to $\mathscr{A}$ —that is, by the weights that occur and the dimensions of the weight spaces. For each *irreducible* representation V there is a unique weight w_0 which has the property that $w_0 - w$ is a positive root for all other weights w of V. This unique *highest weight* w_0 has a one-dimensional weight space and moreover determines V to within equivalence. Indeed there is an explicit formula for computing the dimension of each weight space from w_0.

3. The Concept of an S Operator. Let U and V be two unitary representations of the additive group of the real line in the same Hilbert space $\mathscr{H}$. For each $\varphi \in \mathscr{H}$ we may ask whether there exists $\psi \in \mathscr{H}$ such that $\lim_{t \to -\infty} (U_t(\varphi) - V_t(\psi)) = 0$. It is easy to see that ψ is unique if it exists. Indeed it follows at once that $\lim_{t \to -\infty} V_t^{-1} U_t(\varphi) = \psi$. Let D^- denote the set of all φ in $\mathscr{H}$ such that this last limit exists. It is easy to prove that D^- is a closed linear subspace of $\mathscr{H}$. Let S^- be the linear operator defined in D^- by the formula $S^-(\varphi) = \lim_{t \to -\infty} V_t^{-1} U_t(\varphi)$. It is more or less obvious that S^- is an isometry with a closed range R^- and hence a unitary operator from D^- to R^-.

For each fixed real number a, consider $V_t^{-1} U_t U_a(\varphi)$ where φ is in D^-. This is equal to $V_t^{-1} U_{t+a}(\varphi)$ which is equal in turn to $V_a V_{t+a}^{-1} U_{t+a}(\varphi)$. Evidently the final expression has a limit as $t \to -\infty$ which is equal to $V_a(S^-(\varphi))$. It follows at once that D^- is invariant under all U_a, that R^- is invariant under all V_a, and that S^- implements a unitary equivalence between the restrictions of U to D^- and of V to R^-. Evidently D^- reduces to $\{0\}$ whenever U and V are *disjoint* in the sense that they admit no equivalent subrepresentations. Moreover $D^- = R^- = \mathscr{H}$ is impossible unless U and V are unitarily equivalent representations.

Precisely the same considerations may be applied to the limit of $V_t^{-1} U_t(\varphi)$ as t approaches ∞ leading to an isometry S^+ whose domain D^+ and range R^+ define subrepresentations of U and V respectively. As above, S^+ implements a unitary equivalence between these two subrepresentations.

We shall be mainly if not exclusively interested in the special case in which $D^+ = D^- \overset{\text{def}}{=} D$ and $R^+ = R^- = \mathscr{H}$. When these conditions hold, we shall say that U and V are *asymptotic at* $\pm\infty$. In that case both S^- and S^+ define unitary equivalences of V with the restriction U^D of U to D, and $S = S^+(S^-)^{-1}$ is a unitary operator in $\mathscr{H}$ which lies in the commuting algebra of V. It is called the S operator or the scattering operator of U with respect to V. The simplest and one of the best understood cases is that in which V and U are of the form $t \to e^{iH_0 t}$ and $t \to e^{iH t}$, respectively, where $\mathscr{H} = \mathscr{L}^2(-\infty, \infty)$ and H_0 and H are selfadjoint operators associated with the formal differential operators $-\frac{d^2}{dx^2}$ and $-\frac{d^2}{dx^2} + v$ where v is a real-valued function on the real line that goes to zero rapidly enough at ∞ so that

$$(1) \qquad \int_{-\infty}^{\infty} (1 + |x|)|v(x)|\,dx < \infty.$$

In this case one can prove that S exists. Independently of this, one can prove that whenever S exists it can be completely described by a function from the positive real axis to the group SU(2) of all 2×2 unitary matrices of determinant one. Since the most general member of SU(2) has the form $\begin{pmatrix} a & b \\ -\bar{b} & \bar{a} \end{pmatrix}$, where a and b are complex numbers such that $|a|^2 + |b|^2 = 1$, it follows that S can be completely described by two complex-valued functions on the positive real axis. As we shall explain below, the determination of these functions for a given v can be reduced to the problem of determining the solutions of the equations

$$(2) \qquad -\frac{d^2\psi}{dx^2} + v\psi = \lambda^2\psi$$

for all real positive λ.

The fact that S can be described by a function from positive real numbers to 2×2 unitary matrices follows already from the fact that S is unitary and commutes with H_0. If H_0 had a discrete spectrum this would imply that S is a direct sum over the eigenvalues of H of unitary operators in the eigenspaces. Since H_0 has a continuous spectrum of uniform multiplicity 2 which consists of the entire positive real axis, S is a direct integral over the positive real axis of unitary operators in two-dimensional spaces. Fortunately it is possible to describe S as being in this form without appealing to the spectral theorem or the theory of direct integrals. We do so by using the Fourier transform $f \to \tilde{f}$ where $\tilde{f}(y) = \int_{-\infty}^{\infty} f(x)e^{2\pi ixy}\,dx$ for all f in a dense subspace of $\mathscr{L}^2(-\infty, \infty)$ and is defined for all f in $\mathscr{L}^2(-\infty, \infty)$ by continuous extension. Then $\widetilde{H_0(f)} = y^2\tilde{f}(y)$. To describe the most general linear operator T in $\mathscr{H}$ that commutes with H_0 it is useful to think of each $\tilde{f}$ as a function from the positive real numbers to the complex two vectors $(\tilde{f}(-|y|), \tilde{f}(|y|))$ (i.e., to think of a function on the whole real line as two functions on the positive real line). If $r \to T(r)$ is any measurable function from the positive real numbers to the 2×2 complex matrices then $\tilde{f} \to g$, where $g(|y|)$ is the two vector obtained by applying $T(|y|)$ to $\tilde{f}(-|y|), \tilde{f}(|y|)$, is a linear operator from $\mathscr{L}^2(-\infty, \infty)$ to certain functions from the positive real numbers to the vector space of all complex-valued functions on $[-\infty, \infty]$. When $r \to T(r)$ is suitably restricted these functions will define members of $\mathscr{L}^2(-\infty, \infty)$ and the operator will be bounded. It is easy to see that all such bounded operators are in the commuting algebra of H_0 and, conversely, that every bounded linear operator in the commuting algebra of H has this form. In particular, if our operator is the S matrix and hence is unitary, the matrices $T(r) = \|s_{ij}(r)\|$ which describe it as above must be unitary.

We now describe (without proof) how the functions $r \to s_{ij}(r)$ may be determined from the solution of (2) when v satisfies (1). If v is zero outside of the finite interval $A < x < B$ it follows from the general theory of linear differential equations that for $\lambda > 0$ there exists a unique solution ψ of (2) which is $e^{-i\lambda x}$ for $x > B$ and that this solution is of the form $c_1 e^{-i\lambda x} + c_2 e^{i\lambda x}$ for $x < A$ where of course the complex constants c_1 and c_2 depend upon λ. Thus v, through (2), defines two functions $\lambda \to c_1(\lambda)$ and $\lambda \to c_2(\lambda)$. If v does not vanish outside of a finite interval but only satisfies (1) one can still show that there exists a unique solution ψ of (2) which is asymptotic to $e^{-i\lambda x}$ for $x \to \infty$ and that this solution is asymptotic to $c_1 e^{i\lambda x} + c_2 e^{-i\lambda y}$ for $x \to -\infty$. Thus the association of $c_1(\lambda)$ and $c_2(\lambda)$ to v is valid for all v satisfying (1). It can be shown that $s_{11}(r) = \frac{c_2(r)}{c_1(r)}$, $s_{12}(r) = \frac{1}{c_1(r)}$, $s_{21}(r) = -\frac{1}{c_1(r)}$, and $s_{22}(r) = \frac{c_2(r)}{c_1(r)}$ from which it follows that $\left\| \begin{matrix} s_{11}(r) & s_{12}(r) \\ s_{21}(r) & s_{22}(r) \end{matrix} \right\|$ is not only unitary for all r but also has determinant equal to one. In particular $|s_{11}(r)|^2 + |s_{12}(r)|^2 = 1$. In the quantum mechanical motion of a particle in a one-dimensional space, governed by the potential v (and when suitable units have been chosen), $|s_{12}(r)|^2$ represents the probability that a particle emerging from $-\infty$ with momentum r will be transmitted through the potential barrier and proceed to ∞ and $|s_{11}(r)|^2$ is the probability that it will be "scattered" by being reflected back to $-\infty$.

In the special case in which v is the real constant P in the interval $0 \le x \le L$ and zero outside this interval it is possible to determine $c_1(\lambda)$ and $c_2(\lambda)$ explicitly with the aid of a mildly tedious computation. The answer is as follows. Let $\sigma = \frac{\sqrt{|P-\lambda^2|}}{\lambda}$, let $\tau = \lambda L$, and let $\eta = i$ or 1 according as $P - \lambda^2 < 0$ or $P - \lambda^2 > 0$. Then

$$c_1(\lambda) = \frac{e^{-i\tau}}{4}(2(e^{\eta\sigma\tau} + e^{-\eta\sigma\tau}) + i(e^{\eta\sigma\tau} - e^{-\eta\sigma\tau})(\frac{1}{\eta\sigma} - \eta\sigma)),$$

$$c_2(\lambda) = \frac{e^{-k\tau}}{4}(e^{\eta\sigma\tau} - e^{-\eta\sigma\tau})(\frac{1}{\eta\sigma} - \eta\sigma).$$

Note that $\lim_{\lambda \to \infty} \sigma = 1$ and $n = i$ and when $\sigma = 1$, $n = i$. $c_1(\lambda) = \frac{e^{-i\tau}}{4}(4\cos\tau + 4i\sin\tau) = 1$, $c_2(\lambda) = 0$. Thus when λ is very large the probability of scattering is very small and transmission is almost certain.

Using essentially the same method one can find explicit formulae for $c_1(\lambda)$ and $c_2(\lambda)$ whenever v is a step function that is zero outside of a finite interval and from this get information about quite general cases by limiting processes. The fact that in the step function case the explicit formulae are built out of elementary functions which are analytic in the whole complex plane suggests an important property of $c_1(\lambda)$ and $c_2(\lambda)$, which extends to much more general S operators and will be discussed below.

It turns out that whenever v is such that the operator H has no discrete spectrum then v is uniquely determined by S and hence by the functions $c_1(\lambda)$ and $c_2(\lambda)$. That is also the case in which $\mathscr{D}$ is the whole Hilbert space so that S^+ and S^- set up unitary equivalences between U and V and hence between H and H_0. In the contrary case the orthogonal complement $\mathscr{D}^\perp$ of $\mathscr{D}$ is precisely the invariant subspace on which H has a pure point spectrum, and one has to know something about the square summable eigenvectors of H, as well as $c_1(\lambda)$ and $c_2(\lambda)$, in order that v be uniquely determined. Actually, what one has to know is closely analogous to knowing $c_1(\lambda)$ and $c_2(\lambda)$ except that it is defined where λ is such that $-\lambda^2$ is in the discrete spectrum rather than when λ^2 is in the continuous spectrum. Specifically let us consider the negative eigenvalues of our differential operator $-\frac{d^2}{dx^2} + v(x)$. They are the solutions of the differential equation $\frac{-d^2\psi}{dx^2} + v(x)\psi = -\lambda^2 \psi$ and are asymptotic at $\pm\infty$ to functions of the form $\gamma_1 e^{-\lambda x} + \gamma_2 e^{\lambda x}$. In particular, for each $\lambda > 0$ there is a unique solution which is asymptotic to $e^{-\lambda x}$ at $x = \infty$, and this solution will be asymptotic at $x = -\infty$ to $\gamma_1(\lambda)e^{-\lambda x} + \gamma_2(\lambda)e^{\lambda x}$ for some real numbers $\gamma_1(\lambda)$ and $\gamma_2(\lambda)$. It is clear that this eigenfunction will be in $\mathscr{L}^2(-\infty, \infty)$ if and only if $\gamma_1(\lambda) = 0$ and that the real numbers λ such that $\gamma_1(\lambda) = 0$ are precisely those real numbers such that $-\lambda^2$ is an eigenvalue of H regarded as a selfadjoint operator in $\mathscr{L}^2(-\infty, \infty)$. One can show that there are only a finite number $\lambda_1, \lambda_2, \ldots, \lambda_n$ of such real numbers. To know $\gamma_1(\lambda)$ and $\gamma_2(\lambda)$ when $\lambda = \lambda_1, \lambda_2, \ldots, \lambda_n$ is the same as to know $\gamma_2(\lambda_j)$ for $j = 1, 2, \ldots, n$ since $\gamma_1(\lambda_j) = 0$ for all $j = 1, 2, \ldots, n$. Moreover, the numbers $\gamma_2(\lambda_1), \ldots, \gamma_2(\lambda_n)$ are precisely what one must know in addition to the functions c_1 and c_2 in order that v be uniquely determined. Notice that the difference between the two functions $c_1(\lambda)$ and $c_2(\lambda)$ on the one hand and the n vector $\gamma_2(\lambda_1), \ldots, \gamma_2(\lambda_n)$ on the other comes from the fact that the positive part of the spectrum is continuous and covers the whole positive real axis while the negative part is discrete and lies in the set $-\gamma_1^2, -\gamma_2^2, \ldots, -\gamma_n^2$, where γ_1 is identically zero.

A more interesting example from several points of view is an analogue of the first example in which space is three dimensional instead of one dimensional. Thus our Hilbert space will now be $\mathscr{L}^2(E^3)$ and the two selfadjoint operators generating V and U, respectively, will be defined by the differential operators

$$f \longrightarrow -\left(\frac{\partial^2 f}{\partial x^2} + \frac{\partial^2 f}{\partial y^2} + \frac{\partial^2 f}{\partial z^2}\right) \quad \text{and}$$

$$f \longrightarrow -\left(\frac{\partial^2 f}{\partial x^2} + \frac{\partial^2 f}{\partial y^2} + \frac{\partial^2 f}{\partial z^2}\right) + vf.$$

Here v is a real-valued function on E^3 which we shall assume to be invariant under rotations about $0, 0, 0$ so that it may be written in the form $x, y, z \rightarrow w(\sqrt{x^2 + y^2 + z^2})$ where w is a suitably restricted function defined on the positive real axis.

We exploit the special form of v by introducing the unitary representation L of the group R_0 of all rotations about $0, 0, 0$, defined by letting L_α denote the unitary operator $f(x, y, z) \rightarrow f(\alpha^{-1}(x, y, z))$ for each rotation α in R_0. We see at once that $L_\alpha U_t = U_t L_\alpha$ and $L_\alpha V_t = V_t L_\alpha$ for all α and t and it follows at once that the S operator, when it exists, commutes with all L_α. Now R_0 is a compact group so that L is a direct sum of irreducible representations. Collecting together the equivalent irreducibles, one obtains a canonical direct sum decomposition of $\mathscr{H} = \mathscr{L}^2(E^3) = \mathscr{H}_0 \oplus \mathscr{H}_1 \oplus \mathscr{H}_2 \oplus \cdots$ where the index l varies over the nonnegative integers and L restricted to $\mathscr{H}_l$ is a direct sum of replicas of the unique irreducible $(2l + 1)$-dimensional unitary representation of R_0. This representation is often denoted by the symbol D_l. Because of the commutativity with the L_α, each subspace $\mathscr{H}_l$ is invariant under U and V. Let U^l and V^l, respectively, denote the restrictions of U and V to $\mathscr{H}_l$. Then we may consider the question of the existence of an S operator separately for each pair U^l and V^l. Indeed it is easy to see that S exists if and only if each pair U^l, V^l has an S operator S^l and that S itself, when it exists, is just the direct sum of the operators S^l. It is not hard to show that the restriction of $-\left(\frac{\partial^2}{\partial x^2} + \frac{\partial^2}{\partial y^2} + \frac{\partial^2}{\partial z^2}\right)$ to each $\mathscr{H}_l$ is a multiplicity free selfadjoint operator whose spectrum is the whole positive real axis (and whose associated measure class is that of Lebesgue measure). Thus, since S^l commutes with this restriction it is defined by a function u_l from the positive real axis to the one-by-one unitary matrices, i.e., to the complex numbers of modulus one. Accordingly S, when it exists, is defined by a sequence $u_0, u_1, u_2, \ldots$ of functions $r \rightarrow u_l(r)$ from the positive real axis to the complex numbers of modulus one.

Just as in our first example it is possible to reduce the problem of finding the functions $r \rightarrow u_l(r)$ and hence S to the study of eigenfunctions of second-order *ordinary* linear differential operators. The fact that the problem is three dimensional rather than one dimensional has to be paid for by the fact that we have to study countably many different second-order ordinary differential equations – one for each $l = 0, 1, 2, \ldots$. The differential operator associated with S^l is obtained by considering the restriction of H to $\mathscr{H}_l$ and the associated eigenfunctions (with positive eigenvalues) satisfy the differential equation

$$(*) \qquad \frac{-d^2 y}{dr^2} + (w(r) + \frac{l(l + 1)}{r^2})y = k^2 y$$

where k is a positive real number. When $l = 0$ this is identical (modulo notational changes) to that of the first example. Even here, however, there is an important difference in that we now study it only on the half line $0 <$

$r < \infty$ and asymptotic properties as x tends to $-\infty$ must be replaced by asymptotic properties as r tends to zero. At the same time we shall be mainly interested in the case in which $|w|$ tends to infinity as r tends to zero.

In describing each S^l in terms of the properties of eigenfunctions of $(*)$ we make use of a theorem suggested by the easily checked fact that the general solution of $\frac{-d^2y}{dr^2} + \frac{l(l+1)}{r^2}y = 0$ is $c_1 r^{l+1} + c_2 r^{-l}$, where c_1 and c_2 are complex constants. This theorem asserts that when $|w|$ goes to ∞ sufficiently slowly as r approaches zero so that $\int_0^c r|w(r)|dr < \infty$ for some $c > 0$ then every solution of $(*)$ is asymptotic at 0 to a unique function of the form $c_1 r^{l+1} + c_2 r^{-l}$.

Let w have this property and consider the unique solution of $(*)$ which is asymptotic to r^{l+1} as r approaches zero. If $|w|$ goes to zero sufficiently rapidly at ∞ so that $\int_c^\infty |w(r)|dr < \infty$ for some $c > 0$ then this solution will be asymptotic at ∞ to a unique function of the form $c^1 e^{-ikr} + c^2 e^{ikr}$ where c^1 and c^2 are complex constants. Of course these constants depend upon l and k. We shall denote them by $c_l^1(k)$ and $c_l^2(k)$. It is from these two functions that it is possible to construct the function u_l from which one can recover the partial S operator S^l as explained above. The details are as follows. Since the complex conjugate of every solution of $(*)$ is also a solution we deduce at once that $c_l^2(k) = \overline{c_l^1(k)}$ for all l and k so that $\frac{c_l^1(k)}{c_l^2(k)} = \frac{c_l^1(k)}{\overline{c_l^1(k)}}$ is of absolute value one for all l and k. One proves that for all l and k

$$u_l(k) = \frac{c_l^2(k)}{\overline{c_l^2(k)}}(-1)^{l+1}.$$

Just as in the first example, the S operator may be used to compute the probabilities of various eventualities when a particle obeying the laws of nonrelativistic quantum mechanics is "scattered" by a spherically symmetric potential after coming from far away with a definite momentum. However, when the particle moves in three-dimensional space rather than in one-dimensional space there are a whole continuum of possibilities for the scattering rather than the reflection-transmission dichotomy of the one-dimensional case. Computing these from the sequence of functions $u_0, u_1, u_2, \ldots$ is correspondingly less straightforward. By an argument which we shall not reproduce one can prove the following. For each $E > 0$ and each real number θ define the complex number $g(E, \theta)$ by the formula

$$(**) \qquad g(E, \theta) = \sum_{l=0}^{\infty} \frac{2l+1}{2i(\sqrt{E})}(u_l(\sqrt{E}) - 1)P_l(\cos\theta)$$

where P_l is the lth Legendre polynomial. Then, when units are properly chosen, $g(E, \theta)$ is the "probability density amplitude" that a particle coming from ∞ with energy E will be deflected through an angle θ. Of course the probability of being scattered (or deflected) through any particular angle is zero in general. One can expect a positive probability only for deflection through an angle lying in an interval $\theta_0 < \theta < \theta_0 + \epsilon$ for some ϵ and θ_0. One obtains this probability for fixed E by integrating the probability density $\theta \to |g(E, \theta)|^2$ over the interval in question. In quantum mechanics the term "probability amplitude" refers to a complex number whose absolute square is a probability. It is also used somewhat loosely to denote what more properly should be called a "probability density amplitude".

It is important to notice that the transition from the sequence $u_0, u_1, \ldots$ of functions of one variable to the single function g of two variables is not just a special formula that happens to arise in physics. For each fixed E, the function $\theta \to g(E, \theta)$ is simply the function whose expansion in suitably normalized Legendre polynomials has $(l + 1)$st coefficient equal to $\frac{1}{2i\sqrt{E}}(u_l(\sqrt{E}) - 1)$. It follows then, that just as g is uniquely determined by u_l, so the u_l can be recovered from g by the standard formula for computing the coefficients of expansions in Legendre polynomials. Indeed,

$$(***) \qquad \frac{u_l(\sqrt{E}) - 1}{i\sqrt{E}} = \int_{-1}^{1} g(E, \theta)P_l(\cos\theta)d\cos\theta.$$

Using the fact that $|u_l\sqrt{E})| = 1$ it is customary to define $\delta_l(\sqrt{E})$ so that $e^{2i\delta_l(\sqrt{E})} = u_l(\sqrt{E})$ and $\delta_l(\sqrt{E})$ is a real number defined modulo integer multiples of 2π.

The "angle-valued" functions $\delta_0, \delta_1, \delta_2, \ldots$ are called the *phase shifts* and of course determine the S operator just as $u_0, u_1, u_2, \ldots$ do. One verifies easily by complex trigonometry that $u_l(\sqrt{E}) - 1 = e^{2i\delta_l(\sqrt{E})} - 1 = 2ie^{i\delta_l(\sqrt{E})}\sin\delta_l(\sqrt{E})$. Thus in terms of phase shifts (**) takes the slightly simpler form

$$(**)' \qquad g(E, \theta) = \sum_{l=0}^{\infty}\left(\frac{e^{i\delta_l(\sqrt{E})}\sin\delta_l(\sqrt{E})}{\sqrt{E}}\right)(2l + 1)P_l(\cos\theta).$$

The E dependent factor $a_l(\sqrt{E}) = \frac{e^{i\delta_l(\sqrt{E})}\sin\delta_l(\sqrt{E})}{\sqrt{E}}$ is called a *partial wave amplitude.* In physically interesting cases it often turns out that (to a good approximation) one can neglect all but the first few terms in $(**)'$. Correspondingly one can present data on scattering by means of graphs of a few of the real-valued functions $\delta_0, \delta_1, \ldots$.

For each E the variable θ (more precisely $\cos\theta$) determines and is determined by another variable Δ^2 called the *square of the momentum transfer.* Let p_1, q_1, r_1 and p_2, q_2, r_2 be two real three vectors with the same length $\sqrt{E}$. Then their difference $p_1 - p_2, q_1 - q_2, r_1 - r_2$ has a length Δ whose square is readily computed to be $\Delta^2 = 2E(1 - \cos\theta)$ where θ is the angle between the two vectors. In a scattering problem where units are chosen so that E is $\frac{1}{2}$ the square of the momentum, this new variable has an obvious physical meaning which explains the terminology. Replacing $\cos\theta$ by $1 - \frac{\Delta^2}{2E}$ on the right-hand side of $(**)'$ one obtains a series defining a function f of E and Δ^2 which may be used to describe scattering just as g was except that it computes probabilities for momentum transfer rather than angles. It is also called the scattering amplitude. It is defined for all pairs E, Δ^2 with $0 \le E$, $0 \le \Delta^2 \le 4E$.

We conclude this section with an account of the rather surprising and important facts about the analytic continuability of $f(E, \Delta^2)$ into the product of the complex E plane with the complex Δ^2 plane. The hypothesis that the relativistic analogue of f has a continuation with parallel properties is absolutely essential for the considerations of §6 below on the Veneziano model. In the infinite series defining f, each individual term is a product of a polynomial in Δ^2/E with a function of E alone, whose analytic continuability can be established without great difficulty. It is thus natural to hope that the convergence of the series for complex values of E and Δ^2 is sufficiently extensive and well behaved to permit one to define the desired analytic continuation by this route. Unfortunately this hope is not justified and more elaborate procedures must be found.

It is the great merit of Tullio Regge to have discovered a procedure which not only accomplished the continuation—for suitably restricted functions w—but led to a new and useful way of describing the S operator as a function of two complex variables. These two variables are not E and Δ^2 but E and l where l for nonnegative integer values is the discrete index l in u_l and δ_l. Because of the connection between l and angular momentum one speaks of the complex angular momentum plane. Regge's procedure was an adaptation of a technique apparently first used by Watson and sometimes referred to as the Sommerfeld-Watson formula. Consider the mapping $\{a_l\} \longrightarrow \sum_{l=0}^{\infty} (2l + 1)a_l P_l(y)$ from sequences of complex numbers $a_0, a_1, \ldots$ to the functions on $-1 \le y \le 1$ defined by the Legendre expansion whenever it converges. The basic idea is that an "analytic continuation" of the sequence of Legendre coefficients $\{a_l\}$ can be used to construct an analytic continuation of the function defined by the Legendre series, provided that certain conditions are satisfied. Here by an "analytic continuation" of the series $\{a_l\}$ we mean an analytic function ϕ such that $\phi(l) = a_l$. We begin by describing the method of construction assuming for the moment that ϕ is defined in the whole l plane and has no singularities. Recall that the Legendre polynomials have a generalization in which the index l is replaced by an arbitrary complex number and the P_l are no longer polynomials but hypergeometric functions. Moreover for each y, $P_l(y)$ is analytic in l in the whole complex plane with certain exceptions which need not be discussed here. Thus the function $\frac{(2l+1)\phi(l)P_l(y)}{\sin \pi l}$ is analytic as a function of l except for poles at the integers. One computes easily that the residue of this function at the nonnegative integer l is $(-1)^l$ times a constant A times the term $(2l + 1)a_l P_l(y)$. Hence the sum of the Legendre series is equal to $1/A$ times the integral of $\frac{(2l+1)\phi(l)P_l(-y)}{\sin \pi l}$ around a contour which includes all the nonnegative integers. Using Cauchy's integral theorem, one deforms this contour into a line parallel to the imaginary axis plus a "circle at ∞". If ϕ is suitably well behaved at ∞ the contribution of the latter is zero and one has demonstrated the important fact that the sum of the Legendre series is equal to $\frac{1}{A}\int_C \frac{(2l+1)\phi(l)P_l(-y)}{\sin \pi l}dl$ where C is a line parallel to the imaginary axis. The point of this integral representation of the sum of the Legendre series is that in this form the analytic continuation to complex y is much easier to deduce.

Actually the assumption that ϕ is an entire function is much more than is needed or is realized in the examples in which we are interested. We made it to be able to describe the nature of the method with a minimum of complications. What one needs is analyticity in a right half plane including the nonnegative integers and one can permit a finite number of poles. Indeed these poles are very important. When they exist in such positions that deforming the contour to a vertical line passes through them, the deformed contour must include a small circle about each one. This implies that in our final formula for the sum of the Legendre series the integral over C must have added to it a finite number of further terms, one for each of these poles. These terms can be explicitly computed. The term for a pole at $l = \alpha$ is $\frac{P_\alpha(-y)}{\sin \pi \alpha}$ times the residue of ϕ at α. At least in our applications these terms arising from the poles of ϕ are the dominant ones at ∞ and play a key role in studying the properties of the analytically continued sum of the Legendre series.

To apply this method to the analytic continuation of the scattering amplitude $f(E, \Delta^2)$ one notices that for each fixed E, $f(E, \Delta^2)$ is the sum of the series $\sum_{l=0}^{\infty} a_l(\sqrt{E})(2l + 1)P_l(1 - \frac{\Delta^2}{2E})$ and hence a Legendre series in which y has been replaced by $(1 - \frac{\Delta^2}{2E})$. Thus to apply the Sommerfeld-Watson formula to study analyticity in Δ one needs only to find a suitable analytic continuation of $a_l(\sqrt{E})$ as a function of l. Regge found this by using the fact that $a_l(\sqrt{E})$ is an invariant of the differential equation $-\frac{d^2y}{dx^2} + w(r) + \frac{l(l+1)}{r^2}y = Ey$ and that this makes sense even when l is not an integer. Regge carried through this analysis in a now famous paper published in 1959 and entitled "Introduction to Complex Orbital Momenta" (*Il Nuovo Cimento*, vol. 14, pp. 951–996). Under the additional assumption on w that it is of the form $w(r) = \int_m^{\infty} e^{-\mu r}d\sigma(\mu)$ for some positive m and some finite measure σ; that is, that it is a so-called "superposition of Yukawa potentials", Regge was able to prove that the scattering amplitude $f(E, \Delta^2)$ has an analytic continuation which is a very special kind of function of two complex variables in the sense that it has a so-called "Mandelstam representation". This means, roughly speaking, that its singularities are all in the product of the two real axes and that the function itself can be written as a double integral over this product of functions of the form $\frac{A}{(E-E_0)(\Delta^2-\Delta_0^2)}$. More precisely it differs from such a double integral by a finite sum of terms each of which is a product of functions of one variable of a rather simple special type.

In addition to this, Regge obtained important information about the asymptotic behaviour of the analytic continuation of f by using the representation as a sum of an integral term and a finite number of pole terms implied by the Sommerfeld-Watson formula. Since the function $\phi(l)$ will now depend on the parameter E, the poles of ϕ will be functions of E. They can be shown to vary continuously with E. They are called *Regge poles* in the physics literature on elementary particles. The curve in the complex plane traced out by any particular Regge pole as E varies on the real axis is called a Regge pole trajectory.

4. Lattice Statistics and Onsager's Solution of the Ising Problem. Lattice statistics is a discrete version of classical statistical mechanics and we shall be interested in the finite (but large) dimensional eigenvalue problems to which it gives rise. Recall Section 13 of Article 1 of this collection where the fundamentals of classical thermodynamics and statistical mechanics are briefly explained. The key idea is to explain the laws of thermodynamics as resulting from applying probabilistic ideas to matter conceived of as a dynamical system composed of an enormous number of interacting atoms or molecules. It turns out that the kinetic energy of such a dynamical system plays a rather trivial role and that its influence can be easily computed and "factored out" leaving the real difficulties to be dealt with in a reduced system. In this reduced system there are no velocities or kinetic energy— only configurations of particles and a "potential energy", which is a given real-valued function on the space of all configurations. Thus one can study abstract statistical mechanics for any system C, α, H in which C, α is a measure space and H is a real-valued measurable function. For each $c \in C$ one thinks of $H(c)$ as the energy of the system when its configuration is c. A "state" of the system is a probability measure μ on C which is absolutely continuous with respect to α and $\int_C H(c)d\mu(c)$ (if it exists) is the expected value of the energy when the system is in the state μ. In this general context one can introduce analogues of "temperature", "entropy", and "free energy" and verify that they are related to one another just as are the classical notions of thermodynamics.

We shall now give details in the special case in which C is a finite set and α is the counting measure. Then we shall concentrate on the very special subcase that occurs in lattice statistics. Let C be any finite set and let

H be any real-valued function defined on C. One defines a one-parameter family of probability measures on C by setting $\mu_\beta(E) = \frac{1}{P(\beta)} \sum_{c \in E} e^{-\beta H(c)}$

where β (the parameter) is a positive real number, $P(\beta) = \sum_{c \in C} e^{-\beta H(c)}$ and

E is an arbitrary subset of C. The function $\beta \to P(\beta)$ is called the partition function of the system. One supposes that it is impossible to know which configuration one is dealing with and that the probability measure μ_β represents the extent of what we can know. The parameter β is proportional to the reciprocal of the "absolute" temperature of the system. ($\beta = \frac{1}{kT}$ where k is a universal constant of nature known as Boltzmann's constant.)

For each value of β one can compute the expected value of H with respect to μ_β and this is denoted by $\mathscr{E}(\beta)$,

$$\mathscr{E}(\beta) = \sum_{c \in C} H(c) \mu_\beta(\{c\}).$$

This is supposed to give the energy-temperature relationship as observed from macroscopic measurements. It is useful to notice that $\mathscr{E}(\beta)$ can be computed directly from the partition function without knowing the system C, H from which both are defined. Indeed one shows easily that $\mathscr{E}(\beta) = -\frac{d}{d\beta} \log P(\beta)$. The expected value of $-\log \mu_\beta(\{c\})$ may be taken as a measure of the amount of "randomness" in μ_β. One computes that it is equal to $-\beta \frac{d}{d\beta} \log P(\beta) + \log P(\beta)$ or more simply to $\beta^2 \frac{dF(\beta)}{d\beta}$ where $F(\beta)$ is defined to be $-\frac{1}{\beta} \log P(\beta)$. $\mathscr{E}(\beta)$ may also be computed directly from F. We have $\mathscr{E}(\beta) = \beta \frac{dF(\beta)}{d\beta} + F(\beta)$. The formulae for computing $\mathscr{E}(\beta)$ and the degree of randomness of μ_β from F are identical with those for computing the energy and entropy from the free energy in classical thermodynamics. Moreover, the relationship between F and P is identical with that between the partition function and the free energy in classical statistical mechanics. Thus we define $F(\beta)$ to be the *free energy* of our discrete system and $S(\beta) = \beta^2 \frac{dF(\beta)}{d\beta}$ to be its *entropy*.

In the special case known as *lattice statistics* the configuration space C is the set of all functions from some finite set $\mathscr{L}$ to a two element set D— often taken to be $\{0, 1\}$ or $\{-1, 1\}$. Moreover for each c, $H(c)$ is a sum over certain finite subsets of $\mathscr{L}$ of a number computed from the values of c on the members of that subset.

In the most well-known example, the so-called two-dimensional Ising model, $\mathscr{L}$ is a direct product of two finite cyclic groups with orders n and m respectively. Let h and v respectively denote generators of these groups. Then the points of $\mathscr{L}$ are the pairs h^{l_1}, v^{l_2}. One can visualize $\mathscr{L}$ geometrically as a set of points in the plane arranged in a rectangular array with $m + 1$ rows and $n + 1$ columns but with the first and $(m + 1)$ st rows being identified as well as the first and $(n + 1)$ st columns,

$$
\begin{matrix}
\bullet & \bullet & \bullet & \bullet & \bullet \\
\bullet & \bullet & \bullet & \bullet & \bullet \\
\bullet & \bullet & \bullet & \bullet & \bullet
\end{matrix}
$$

so that there are nm distinct points. One calls pairs of the form $h^{l_1}v^{l_2}$, $h^{l_1+1}v^{l_2}$ horizontal nearest neighbors and pairs of the form $h^{l_1}v^{l_2}$, $h^{l_1}v^{l_2+1}$, vertical nearest neighbors. The function H which completes the definition of the model depends upon two real parameters J and J'. For each configuration c, $H(c)$ is by definition the sum over all nearest neighbor pairs x, y of a function of the pairs which is J, $-J$, J', or $-J'$ according as x, y is a horizontal pair with $c(x) = c(y)$, a horizontal pair with $c(x) \neq c(y)$, a vertical pair with $c(x) = c(y)$, or a vertical pair with $c(x) \neq c(y)$. It is more or less obvious how one would alter this definition to define Ising models of dimension n where $n = 1, 3, 4, \ldots$. Other variations on the Ising theme, such as including next nearest neighbors, readily suggest themselves.

Models of the Ising type may be applied to several rather different physical problems. In the three-dimensional case one may approximate a gas with a so-called lattice gas whose configurations are determined by which points in a lattice are occupied by a gas molecule. In this case the two elements of D correspond to occupation or nonoccupation and may be conveniently labeled with 1 and 0. In a crystallized alloy obtained by mixing two different metals a configuration is determined by an assignment of an atom of one kind or the other to each "site" in the crystal. Here the two elements of D refer to the two kinds of atom. In the theory of magnetism each atom in a metallic crystal has a "spin" which is a vector of a fixed length which may point in any direction. The crystal is "magnetized" to the extent that these spin vectors are aligned. One usually approximates a real magnet by choosing a direction and supposing that the only possible spins are parallel to it. Thus there are just two possible spins at each site and the elements of D correspond to these. It is usual to label the elements of D with 1 and -1 in this case.

A major issue in all of these applications is the existence or nonexistence of a so-called "phase transition". In the case of a real gas one knows that there is a certain temperature at which the gas "condenses" and becomes a liquid or a solid. One expects this observed behavior to manifest itself in appropriate mathematical models through some sort of nonsmooth behavior in the free energy function at a definite value of β. Actually, in view of the statistical character of the theory this behavior need not be very clear cut except when the number of degrees of freedom is very large and is best sought for in the limit of an infinite number of degrees of freedom when this can be defined in a reasonable manner. In the Ising model one can indeed define a "free energy per lattice site" by showing that $F(\beta)$ divided by the number of lattice sites (nm in the two-dimensional case) does have a limit as n and m both approach infinity in suitably regular manner. In seeking for evidence of a phase transition one looks for abrupt changes in this free energy per lattice site as a function of β.

The Ising model was introduced in 1925 by E. Ising (Zeitschrift für Physik). He was interested in applying it to magnetism. In the one-dimensional case he found a simple argument leading to an explicit summation of the sum defining $F_n(\beta)$ and an exact formula for $\lim_{n\to\infty} F_n(\beta)$. This formula made it clear that no phase transition was predicted. Ising also treated the two-dimensional case but made a mistake which invalidated his work.

Actually the two-dimensional case turned out to be quite difficult and will be our only concern here. The first progress was made by Peierls in 1936 who gave an argument indicating that a phase transition must exist. This argument was only made rigorous almost thirty years later. In the meantime an important observation was made in three independent papers. This is to the effect that the formula for the partition function in the two-dimensional Ising problem may be rewritten as the trace of the mth power of a matrix T_n with 2^n rows and 2^n columns. These papers were all published in 1941 by Kramers and Wannier, Montroll, and Lassettre and Howe respectively. The matrix T_n is called the *transfer matrix* of the problem. Kramers and Wannier considered only the case in which $n = m$ so that the lattice is square and in which $J' = J$. By exploiting certain symmetries of T_n they were able to show that if there is a phase transition at all it will occur when β is the unique solution of $\sinh J\beta = 1$; that is, $\beta = .8814/J$. However, they were not able to give a rigorous proof of the existence of a phase transition.

The real breakthrough came three years later with the publication of a now celebrated paper by Lars Onsager in early 1944 (Physical Review, vol. 65, page 117). Onsager accomplished the startling feat of explicitly diagonalizing the transfer matrix T_n for all values of n.

To see the nature of the problem which Onsager solved let us write down the matrix T_n more or less in the operator form in which he wrote it down near the beginning of his paper. Let $\mathscr{H}$ be any two-dimensional Hilbert space and let $\mathscr{H}^n$ denote the tensor product of $\mathscr{H}$ with itself n times. $\mathscr{H}^n = \mathscr{H} \otimes \mathscr{H} \otimes \cdots \otimes \mathscr{H}$. For each linear operator W in $\mathscr{H}$ and each $j = 1, 2, \ldots, n$ let W_j denote the operator $I \times I \times \cdots \times W \times I \times \cdots \times I$ in $\mathscr{H}^n$ where W occurs in the jth factor. Let s and c denote any two selfadjoint operators in $\mathscr{H}$ that satisfy $s^2 = c^2 = I$ and $sc + cs = 0$. It is well known and easy to prove that these relations determine the pair s, c up to unitary equivalence. In particular, a basis for $\mathscr{H}$ may be chosen with respect to which s and c have the matrices $\begin{pmatrix} 1 & 0 \\ 0 & -1 \end{pmatrix}$ and $\begin{pmatrix} 0 & 1 \\ 1 & 0 \end{pmatrix}$ respectively.

From s and c define operators s_j and c_j in $\mathscr{H}^n$ by the above rules and introduce

$$A = \sum_{j=1}^{n} s_j s_{j+1} \quad \text{and} \quad B = \sum_{j=1}^{n} c_j,$$

where s_{n+1} and c_{n+1} are defined to be s_1 and c_1 respectively. Then $T_n = e^{\beta J' A} e^{H^* B}$ $(2\sinh 2\beta J^{n/2})$ where $H^* = \tanh^{-1}(e^{-2\beta J})$.

Notice that any decomposition of $\mathscr{H}^n$ as a direct sum of subspaces that are both A and B invariant will necessarily be a decomposition into subspaces that are invariant under T_n for all choices of J, J', and β. Thus the somewhat complicated form of $T_n(\beta)$ can be ignored and one can think of the problem as that of finding a simultaneous direct sum decomposition of the two rather simple operators A and B. Onsager took this approach. If only A and B happened to commute with one another one could pass easily from direct sum decompositions of A and B separately to one for the two simultaneously. However, A and B do not commute and the next best thing one could hope for is that A and B generate a Lie algebra $\mathscr{L}(A, B)$ whose dimension is small relative to $2^n \times 2^n$ and which has a simple structure. Onsager computed this Lie algebra and found a basis for it consisting of $3n - 1$ elements satisfying very simple commutation rules. Indeed labeling them as $X_0, X_1, X_2, \ldots, X_n, Y_1, Y_2, \ldots, Y_{n-1}, Z_1, Z_2, \ldots, Z_{n-1}$ he showed that $X_r Y_r - Y_r X_r = -2Z_r$, $Y_r Z_r - Z_r Y_r = -2rX_r$ and that all other commutators vanish. Here one makes the convention that $Y_0 = Y_n = Z_0 = Z_n = 0$. It follows that for each $r = 1, 2, \ldots, n-1$, X_r, Y_r, and Z_r span a three-dimensional Lie algebras and that these three-dimensional Lie algebras commute with one another and with X_0 and X_r. Moreover, X_0 and X_n commute with one another. The structure of $\mathscr{L}(A, B)$ is thus made clear. Onsager's calculations also showed that for all r, $X_r Y_r = -Y_r X_r$ and $X_r^2 = Y_r^2 = Z_r^2 \overset{\text{def}}{=} R_r = R_r^2$. Starting with these facts he was able finally to decompose $\mathscr{H}^n$ as a direct sum of 2^{n-1} two-dimensional subspaces each of which was invariant under A and B and hence under T_n for all β, J, and J'. In this way he was able to reduce diagonalizing T_n to diagonalizing 2^{n-1} 2×2 matrices and hence to solving 2^{n-1} quadratic equations.

Onsager's paper was entitled "Crystal statistics I: A two dimensional model with an order-disorder transition". Crystal statistics II and III appeared five years later and were subtitled "Partition Function Evaluated by Spinor Analysis" and "Short Range Order in a Binary Ising Lattice", respectively. They were published on consecutive pages of vol. 76 of the Physical Review. Crystal statistics III was a joint paper of Onsager and Bruria Kaufman. Crystal statistics II was signed by Kaufman alone, but acknowledges Onsager's "constant interest and advice during the course of this work." Kaufman found a somewhat less computational and more conceptual path to Onsager's diagonalization of T_n based on the following considerations. For each $r = 1, 2, \ldots, M$ let $\Gamma_{2r-1} = c_1 c_2 \cdots c_{r-1} s_r$ and let $\Gamma_{2r} = c_1 c_2 \cdots c_{r-1} i s_r c_r$. One sees easily that one can compute the c's and the s's from the Γ's by simple formulae and that the Γ's obey the following commutation rules: $\Gamma_k \Gamma_l = -\Gamma_l \Gamma_k$ for $k \neq l$ and $\Gamma_k^2 = I$ for all k. By a theorem of Jordan and Wigner published in 1928 these commutation rules have a unique irreducible solution (to within equivalence) and this representation is 2^n dimensional. This makes it possible to identify the Hilbert space $\mathscr{H}^n$ with the space of the so-called spin representation of the orthogonal group in $2n$ dimensions and the operator T_n with the image of a member of this group.

In appreciating the merit of Onsager's original diagonalization, it is well to keep in mind the historical facts about the physicists' involvement with group theory and the Lie algebra concept. Although E. Wigner and H. Weyl introduced the theory of group representations into physics in independent papers published in 1927, it took a very long time for theoretical physicists to follow their lead. In 1928 J. C. Slater published a paper showing how to obtain Wigner's results without explicitly introducing the group concept and in their monumental classic "The Theory of Atomic Spectra" published in 1935, Condon and Shortley develop Slater's methods and boast about their complete avoidance of group theory. It was only in 1949 that G. Racah changed things to some extent by simplifying and refining Slater's methods to a point at which he found himself forced to bring one of the exceptional Lie groups, G_2, into the picture. A decade later Murray Gell Mann, in developing his ideas about "current algebra" which led him to $SU(3)$ symmetry, quarks, etc., had to be told by a mathematician that he was "generating Lie algebras" and that there was a well-developed mathematical theory of these objects (see page 164 of "The Life it Brings", by J. Bernstein). Only after that did knowledge of Lie algebras and their relationship to Lie groups become standard equipment for theoretical physicists (see A. Pais, "Inward Bound", page 555).

5. Kac-Moody Lie Algebras. This important new class of Lie algebras was discovered by V. Kac in the Soviet Union and R.V. Moody in Canada and announced in separate papers published in 1967. They are infinite-dimensional complex Lie algebras which share many of the properties of the complex finite-dimensional semisimple Lie algebras. Notably each Kac-Moody Lie algebra admits a *finite*-dimensional commutative subalgebra $\mathscr{A}$ with respect to which the adjoint representation A admits a weight space decomposition for which $\mathscr{A}$ is the zero weight space. Thus one has roots and root spaces and a complete analogue of the theorem of §1. There are infinitely many roots but the root spaces are all finite dimensional.

There is an important subclass of the class of Kac-Moody Lie algebras whose members are called untwisted affine Lie algebras or affine Lie algebras of type I. We shall concentrate attention on this class because its members can be described quite concretely in terms of finite-dimensional simple Lie algebras and their properties suggest those of the general case rather well.

Let $\mathscr{L}$ be an arbitrary Lie algebra. Let $\mathscr{L}^\infty$ denote the vector space of all functions f from the additive group of all integers to $\mathscr{L}$ such that $f(n) = 0$ for all but a finite number of integers n. It is easy to see that $\mathscr{L}^\infty$

becomes a Lie algebra if one defines the Lie bracket by

$$[f, g](n) = \sum_{k \in Z} [f(n - k), g(k)].$$

The Lie algebra $\mathscr{L}^{\infty}$ so defined is called the *loop algebra* of $\mathscr{L}$. It is clearly always infinite dimensional. Now let D denote the linear operator on $\mathscr{L}^{\infty}$ that maps f into the function $D(f)$ where $D(f)(n) = nf(n)$. One checks easily that D is a derivation of $\mathscr{L}^{\infty}$ in the sense that $D([f, g]) = [D(f), g] + [f, D(g)]$. Let $\mathscr{L}^{\infty,e}$ denote the vector space of all pairs f, λ where $f \in \mathscr{L}^{\infty}$ and λ is a scalar. Define a bracket operation in $\mathscr{L}^{\infty,e}$ by the rule $[(f_1, \lambda_1), (f_2, \lambda_2)] = [f_1, f_2] + \lambda_1 D(f_2) - \lambda_2 D(f_1)$. It follows easily from the fact that D is a derivation that $\mathscr{L}^{\infty,e}$ is a Lie algebra having $\mathscr{L}^{\infty}$ as an invariant subalgebra. We shall call it the *extended loop algebra* of $\mathscr{L}$.

Before proceeding to the next and final construction we pause to state an important theorem about the construction $\mathscr{L} \to \mathscr{L}^{\infty,e}$.

THEOREM. *Let $\mathscr{L}$ be any Lie algebra having a commutative subalgebra $\mathscr{A}$ with respect to which it admits a root space decomposition. Let $\mathscr{A}^{\infty,e}$ be the set of all $f, \lambda \in \mathscr{L}^{\infty,e}$ for which $f(0) \in \mathscr{A}$ and $f(n) = 0$ for $n \neq 0$. Then $\mathscr{L}^{\infty,e}$ admits a root space decomposition with respect to $\mathscr{A}^{\infty,e}$. The roots fall into two classes. Those of the first class correspond one-to-one to the pairs α, n where α is a root of $\mathscr{L}$ with respect to $\mathscr{A}$ and n is an integer. The root corresponding to α, n maps f, λ into $\alpha(f(n)) + \lambda n$. Its root space is the set of all $f, 0$ such that $f(n)$ is in the root space of α and $f(k) = 0$ for $k \neq n$. Hence it is isomorphic to the root space of α. The roots of the second class correspond one to one to the nonzero integers. That class corresponding to n has its root space equal to the set of all $f, 0$ such that $f(k) = 0$ for $k \neq n$ and $f(n) \in \mathscr{A}$. As a vector space it is isomorphic to $\mathscr{A}$.*

The proof of this theorem is an easy and straightforward consequence of the definitions.

When applied to the special case in which $\mathscr{L}$ is a finite-dimensional complex simple Lie algebra this construction yields an example of an infinite-dimensional Lie algebra $\mathscr{L}^{\infty,e}$ having a finite-dimensional commutative subalgebra with respect to which $\mathscr{L}^{\infty,e}$ has a root space decomposition with all roots spaces finite dimensional. Of course there are an infinite number of such examples, one for each finite-dimensional complex simple algebra.

The class of untwisted affine Lie algebras is closely related to the class described in the preceding paragraph. Indeed each of the former has a one-dimensional center the quotient by which is one of the latter. To describe the class of untwisted affine Lie algebras explicitly it is only necessary to define a canonical central extension of each $\mathscr{L}^{\infty,e}$ in which $\mathscr{L}$ is simple, complex, and finite dimensional. In defining these it may be somewhat clarifying to begin by considering central extensions of $\mathscr{L}^{\infty}$ and then noticing that each such extension of L^{∞} is trivially convertible into a central extension of $\mathscr{L}^{\infty, e}$. Let B be a symmetric bilinear form on any Lie algebra $\mathscr{L}$. We then define an antisymmetric bilinear form B^{∞} on $\mathscr{L}^{\infty}$ by the formula

$$B^{\infty}(f, g) = \sum_{n=-\infty}^{\infty} nB(f(n), g(-n))$$ and a bracket operation in the vector

space $\mathscr{L}^{\infty, B}$ of all pairs μ, f where μ is a scalar and f is an $\mathscr{L}^{\infty}$ by the formula

$$[(\mu_1, f_1), (\mu_2, f_2)] = B(f_1, f_2)\mu_1\mu_2, [f_1, f_2].$$

A simple computation shows that this definition converts $\mathscr{L}^{\infty, B}$ into a Lie algebra if and only if B satisfies the identity $B([x, y], z) = B([z, x], y)$. Symmetric bilinear forms on a Lie algebra that satisfy this identity are known as *invariant bilinear forms*. Thus for any Lie algebra $\mathscr{L}$ and any invariant bilinear form B on $\mathscr{L}$, $\mathscr{L}^{\infty, B}$, as defined above, is again a Lie algebra and is evidently a central extension of $\mathscr{L}^{\infty}$.

As far as the existence of invariant bilinear forms is concerned it is evident that when $\mathscr{L}$ is commutative, *every* symmetric bilinear form is invariant. Less evident, and vital for our purposes, is the fact that when $\mathscr{L}$ is finite dimensional one can define a symmetric form B_0 by the formula $B_0(x, y) =$ Trace$(A_x A_y)$ where A is the adjoint representation and prove that B_0 is always invariant. In particular, when $\mathscr{L}$ is both finite dimensional and simple, B_0 is nondegenerate and up to a multiplicative constant is the *only* invariant symmetric bilinear form on $\mathscr{L}$.

Given a Lie algebra $\mathscr{L}$ and an invariant bilinear form B_0 on $\mathscr{L}$ the two constructions $\mathscr{L}^{\infty} \to \mathscr{L}^{\infty, e}$ and $\mathscr{L}^{\infty} \to \mathscr{L}^{\infty, B}$ can be blended leading to a Lie algebra whose vector space consists of all triples μ, f, λ where μ and λ are scalars and f is in $\mathscr{L}^{\infty}$ and where the bracket operator is defined by the formula

$$[(\mu, f_1, \lambda_1), (\mu_2, f_2, \lambda_2)] = \mu_1\mu_2 B(f_1, f_2), [f_1, f_2] + \lambda_1 D(f_2) - \lambda_2 D(f_1), 0.$$

We denote this Lie algebra by $\widehat{\mathscr{L}}^{B}$ and in the special case in which $B = B_0$ by $\widehat{\mathscr{L}}$. It is clear that $\widehat{\mathscr{L}}^{B}$ is both a central extension of $\mathscr{L}^{\infty, e}$ and the semi-direct product of $\mathscr{L}^{\infty, B}$ with a one-dimensional Lie algebra defined by an extension of the derivation D of $\mathscr{L}^{\infty}$ to $\mathscr{L}^{\infty, B}$.

6. The Veneziano Model in Strong Interaction Physics. The so-called strongly interacting particles are the neutrons and protons out of which all atomic nuclei are constructed and the short lived "heavy" particles which emerge when the particles enter into collisions with one another and with the light particles such as electrons and photons. The most extensively studied of the latter are the π mesons (or pions) which are related to the interaction binding neutrons and protons (known collectively as nucleons) into a nucleus in a manner analogous to the way in which photons are related to the Coulomb interactions binding (negatively charged) electrons to the positively charged nucleus of an atom. Just as there are two kinds of nucleon: the positively charged protons and the neutral neutrons, so there are three kinds of pions which are respectively positively charged, neutral, and negatively charged. Except for sign the charges of the electron, the proton, and the nonneutral pions are all the same. One looks upon the well-understood electromagnetic interaction between the charged particles as superimposed upon a separate strong interaction and in studying the latter one subtracts off the electromagnetic effects.

In applying the theory of §3 to study the scattering of the strongly interacting particles certain modifications have to be made. This is because the energies involved are so large that relativistic effects can no longer be neglected and physics still lacks a satisfactory integration of quantum mechanics with the special theory of relativity. In particular, one cannot describe particle interactions by anything so elementary as a potential function such as the v of §3. For thirty years or so the most popular way out was something known as "quantum field theory". In 1928, Dirac created a sensation by showing that the "particles" of light we call photons emerge automatically if one regards Maxwell's equations as the equations of motion of a continuum and applies quantum mechanics to the resulting dynamical system. The basic idea of quantum field theory is to assume that not only the photon but also the electron, the proton, the neutron, etc., can be obtained as the "quanta" of suitable generalizations of the electromagnetic field—and then shift attention from the particles to the fields. It is relatively easy to define suitable relativistic interactions between fields—at least on a formal level—and pass from these to particle interactions. This program had some early stunning successes but soon ran into difficulties, which eventually led some physicists to a "disbelief in field theory" and a readiness to seek other solutions to the problem.

Turning our attention back to the S operator we note that the chief difficulty is that the field theoretic substitute for the one-parameter group $t \to U_t$ is not known until one has a relativistic theory of interacting particles. Moreover, the answer given by field theory is insufficiently well defined in any rigorous sense to make it possible to prove that an S operator exists in the sense that this can be done in nonrelativistic potential scattering (when v satisfies suitable conditions). On the other hand, *assuming* that U is such that S does exist and assuming invariance under translation and rotations one can apply the arguments sketched or referred to in §3 to show (in the case of spinless particles) that S can be completely described by a scattering amplitude, which is a complex-valued function of the two real variables E and Δ^2 and also by a sequence $u_0, u_1, u_2, \ldots$ of functions of a positive real variable, the two descriptions being related just as in §3. However, one can no longer reduce the determination of the u_j to the study of the solutions of second-order differential equations.

It is an old idea going back to independent work of Wheeler in 1937 and Heisenberg in 1943 and 1944 that one might replace field theory as a starting point by the assumption that the S operator exists but is not necessarily defined by any pair of one-parameter groups. Hypotheses about the nature of the underlying fields would be replaced by hypotheses about the nature of the S operator. However, this idea only began to be taken seriously by a substantial segment of the physics community in the middle 1950s. In 1947 pions (whose existence was predicted in 1935 by Yukawa) were detected in cosmic rays at high altitudes and a year later were produced in the laboratory. Before long it began to be possible to study collisions between pions and nucleons and in the middle 1950s a number of papers by different authors were published which attempted to correlate experimental results on the scattering of pions by nucleons with properties of scattering amplitudes implied by their analytic continuability. Of course analytic continuability had to be postulated but such a postulate was suggested both by earlier results on potential scattering and by quantum field theory. The whole story is too complicated even to be outlined here but one constituent that seems worth mentioning is that quantum field theory combined with perturbation theory and "renormalization" leads to a formal infinite series expansion for the S operator, which does not converge, but whose individual terms have analytic continuations which suggest axioms to impose upon the scattering amplitude associated with the postulated S operator.

As a result of this work, evidence began to accumulate that one could get new insights into the scattering behavior of the particles considered by postulating and then exploiting the suggested analytic behavior. At first only analyticity in one variable was considered, but in 1958 an epoch making paper of Stanley Mandelstam demonstrated the extra power provided by going into the complex plane in both variables simultaneously. He was led by considerations from field theory to conjecture the rather strong statement about the analytic continuation of $f(E, \Delta^2)$ which we have already referred to in §3 as satisfying the Mandelstam representation. We recall from §3 that in 1959 Regge proved that the Mandelstam representation holds in nonrelativistic potential scattering whenever the potential is a superposition of Yukawa potentials. Mandelstam took it as an axiom whose consequences were to be explored. The fact that $f(E, \Delta^2)$ is a Legendre series for each E, whose coefficients are built up in a simple way from a sequence $u_0, u_1, u_2, \ldots$ of functions from the positive real numbers to the unit circle in the complex plane (see §3), is a further constraint. It is called unitarity, since it is equivalent to the unitarity of the S operator. Finally, Mandelstam added a third condition called the substitution law which is also suggested by field theory and which has no counterpart in nonrelativistic quantum mechanics. It depends upon the concept of the *anti-particle* of a particle and in particular asserts the existence of an exact but somewhat subtle relationship between the scattering amplitude for a pair of particles A and B and the "scattering" amplitude that tells one the relevant probabilities when A and its anti-particle meet and transform into B and its anti-particle.

The notion of anti-particle goes back to Dirac's paper of 1928 in which he presented his celebrated relativistically invariant wave equation for the electron. This equation had apparently superfluous solutions which suggested to Dirac a positively charged counterpart to the electron. First conjectured to be the proton, it was later identified with a new particle discovered by C. D. Anderson in 1932 and known as the positron. The positron has the same mass and spin as the electron but opposite charge. Moreover, it has a peculiar affinity to electrons which makes them not only attract one another but to have a strong tendency toward mutual annihilation; that is, to disappear and be replaced by two photons (gamma rays). One speaks of the positron as the *anti-particle* of the electron. In the course of time the development of field theory suggested that other particles such as protons, neutrons, pions, etc., should also have anti-particles and indeed by 1956 both anti-protons and anti-neutrons had been produced in the laboratory. Anti-pions turn out to be pions of the opposite charge, neutral pions being, like photons, their own anti-particles.

The mutual annihilation of an electron and a positron with the formation of two photons is an example of the celebrated mass-energy conversion of the special theory of relativity. The classical law of conservation of mass fails because photons are massless. However one finds that the energy of the emerging photons is much greater than the energy of motion (the kinetic energy) of the electron-positron pair. The difference is $2mc^2$ where c is the velocity of light and m is the mass of the electron and of the positron (strictly speaking the rest mass). The lost law of conservation of matter is replaced by a more comprehensive energy conservation law. In the case of nucleon anti-nucleon annihilation the pair is replaced by two pions and these particles are not massless. However, their masses are less than one-sixth the masses of the incident nucleons and here, too, there is a conversion of mass to energy on a large scale. One also has the converse process (pair production) in which matter is "created" out of energy. Indeed, the positrons observed by Anderson were so produced. More generally, one can have processes in which two particles A and B of rest masses m_1 and m_2 meet and turn into two other particles C and D of rest masses m_3 and m_4, where $m_1 + m_2$ need not equal $m_3 + m_4$. Any discrepancy between $m_1 + m_2$ and $m_3 + m_4$ is made up by appropriate changes in the energy of motion.

It is straightforward to modify the notion of scattering amplitude so that it applies to such encounters. The substitution law mentioned above, but not yet explained, applies to these as well. It is most easily formulated in terms of an invariant formalism, introduced by Mandelstam in the paper cited above, and generalized by Kibble in 1960 to the A, B, C, D situation just outlined. Let p_j, q_j, r_j, E_j denote the momentum components and the relativistic energy of the particle with mass m_j for $j = 1, 2, 3, 4$ and choose units so that c, the velocity of light, is one. Then $E_j = \sqrt{m_j^2 + p_j^2 + q_j^2 + r_j^2}$. Let V_j denote the four vector p_j, q_j, r_j, E_j. Then its Minkowski length squared is $E_j^2 - p_j^2 - q_j^2 - r_j^2 = m_j^2$. Moreover, by conservation of energy and momentum, $V_1 + V_2 = V_3 + V_4$ so $V_1 - V_3 = V_4 - V_2$ and $V_1 - V_4 = V_3 - V_2$. Let s, t, and u be defined by the equations

$$s = (V_1 + V_2)^2 = (V_3 + V_4)^2,$$
$$t = (V_1 - V_2)^2 = (V_4 - V_2)^2,$$
$$u = (V_1 - V_2)^2 = (V_3 - V_2)^2,$$

where $(\ )^2$ denotes square of Minkowski length. One checks that $s + t + u = m_1^2 + m_2^2 + m_3^2 + m_4^2$ so that any two of s, t, and u determines the other. Moreover, it is not hard to show that the scattering amplitude depends only on the four masses and the three invariants s, t, and u. In the special case in which $C = A$ and $D = B$ so that $m_1 = m_3$ and $m_2 = m_4$ and ordinary scattering is involved, one verifies that s is the square of the total energy, and t is $-\Delta^2$ where Δ^2 is the square of the momentum transfer. Thus s and t are the analogues of the E and Δ^2 used in the nonrelativistic theory.

Since $u = (m_1^2 + m_2^2 + m_3^2 + m_4^2 - (s + t))$ one does not need to introduce it as a third variable. However, the substitution law, which we are proposing to formulate in its generalized form, has symmetries which are made more manifest if all three variables are kept on the same footing. This law relates all processes that can be obtained from $A + B \to C + D$ by replacing A and B by any two elements of A, B, $\overline{C}$, and $\overline{D}$ and by replacing C and D by the anti-particles of the other two. There are six such processes in all including $A + B \to C + D$ itself. This set of six processes is invariant under the transformation of order two, which maps each particle into its anti-particle and reverses the direction of the process. Accordingly, the set of six falls into three families of two, each of which is intrinsically associated with one of the three possible ways of dividing the set of four particles into two pairs. These three families are called channels. Channel I contains $A + B \to C + D$ and $\overline{C} + \overline{D} \to \overline{A} + \overline{B}$. It is associated with the pairing AB, CD. Channel II contains $A + \overline{C} \to \overline{B} + D$ and $B + \overline{D} \to \overline{A} + C$ and is associated with the pairing AC, BD. Channel III contains $A + \overline{D} \to \overline{B} + C$ and $B + \overline{C} \to \overline{A} + D$ and is associated with the pairing AD, BC.

The substitution law says that the scattering amplitudes for all six processes are essentially the same. If one knows one, all of the others can be written down from very simple rules. However, there is a significant complication in that the scattering amplitude only makes physical sense when all of the variables s, t, u are real and the two that are taken as independent lie in a certain region of the plane bounded by line segments. Recall the conditions on E and Δ^2 in nonrelativistic scattering from §3. Now the region of physical meaningfulness is the same for two processes in the same channel, but in passing from one channel to another one finds that these regions are not only different but disjoint. At first the substitution law seems devoid of content!! The way out is through analytic continuation. One makes the assumption that there is a single function of three complex variables s, t, u, subject to the constraint $s + t + u = m_1^2 + m_2^2 + m_3^2 + m_4^2$ which, as a function of any two, is analytic in the product of the two complex planes except for a few singularities in the spirit of the Mandelstam representation.

The assumption that each relativistic scattering amplitude must be related by analytic continuation in this way to two others, and that these two others are subject to the unitary and other constraints already mentioned above, puts rather strong constraints on all of them. These constraints were strong enough to give Mandelstam hope that they would determine the analytically continued scattering amplitude up to a small number of constants that could then be evaluated by experiment. G. F. Chew became particularly enthusiastic about the possibility of such a program and became the leader of a group of physicists trying to implement it. During the summer of 1960 he gave a series of lectures about it in two different summer schools and these were published in both conference proceedings. A slightly revised version of these lectures appeared as the lecture part of a lecture and reprint volume entitled "S Matrix Theory of Strong Interactions" published in 1961 by W. A. Benjamin. The reprints in the volume include the papers of Mandelstam and Kibble mentioned above, a sequel to the cited paper of Regge and several others.

The year 1961 not only witnessed the publication of Chew's book but a new development which raised interest in the program to a higher level. This was the independent discovery by Froissart and Gribov that the ideas of Regge concerning complex angular momentum and Regge poles could be adapted to the relativistic situation, and that doing so gave new tools and new insights into the situation. The details are very interesting but complicated and time and space constraints force me to pass them by. Let it suffice to say that the incorporation of Regge poles into the Chew program led to a further rash of lecture series in 1962, most, if not all of which, were published. I particularly recommend two publications in the Benjamin series, *Frontiers in Physics*, "Regge Poles and S-Matrix Theory", by S. C. Frautschi and "Mandelstam Theory and Regge Poles", by Omnes and Froissart. Both were published in 1963. Frautschi began collaborating with Chew in 1961 and became one of the most active participants in the program along with Mandelstam.

As the program developed over the next few years, doubts began to accumulate about its self-consistency. Were there in fact *any* functions of two complex variables having all the properties that seemed to be demanded of scattering amplitudes—especially certain properties of Regge pole trajectories? Once again we pass over some interesting but complicated details and announce the discovery that this section was written to explain. In 1968 the physicist, G. Veneziano, caused considerable excitement and removed many of the doubts about self-consistency by exhibiting an example of a function of two complex variables having all of the sought-for properties. This example was very simple, being just a mild modification of the classical Euler beta function $B(s, t) = \frac{\Gamma(s)\Gamma(t)}{\Gamma(s+t)}$ where Γ is the well-known gamma function. Specifically, $A(s, t) = B(-\alpha(s), -\alpha(t))$, where α is usually taken to be the linear function $x \to a+bx$, where a and b are constants. To get an example in which all channels have similar singularity structure one takes

$$A(s, t) = C_1 B(-\alpha(s), -\alpha(t)) + C_2 B(-\alpha(s), -\alpha(u)) + C_3 B(-\alpha(t), -\alpha(u)),$$

where C_1, C_2, and C_3 are constants and $u = m_1^2 + m_2^2 + m_3^2 + m_4^2 - (s + t)$ as above.

Using the simple known facts about the poles and zeroes of the gamma function one can explore the properties of A quite easily. Just as with the gamma function the beta function can be expressed by a definite integral: $B(s, t) = \int_0^1 x^s (1 - x)^t dx$ and this can be used to study the asymptotic properties of A. These play an important role in the Regge pole part of the theory which we were not able to discuss.

Of course, scattering in which two particles come in from ∞ and two particles emerge is rather special. In the end one wants to know about scattering in which n particles come in and m particles emerge. This raised the question of seeking a generalization of the Euler beta function which would apply to $n+m$ particles rather than $2+2$. In the terminology of workers in the field the problem was to generalize Veneziano's four-point function to an N-point function where $N = m + n$. This problem was solved rather quickly. There were two independent discoveries of a 5-point function, one in 1968 and one in 1969 and then, also in 1969, (essentially the same) N-point function was discovered independently by four distinct two-man teams.

The N-point function can be given by an integral formula which generalizes a variant of the integral formula for the Euler beta function. We shall not write this formula down because our main interest in the subject lies in an ingenious alternative way of defining the N-point function by means of the so-called "operator formalism". The operator formalism was discovered independently twice in 1969 and once in 1970 and has the great advantage of making certain properties of the N-point function much easier to prove; notably its cyclic symmetry and the so-called factorization condition. The author confesses that at this point he does not understand the factorization condition beyond the fact that it seems to be a consistency condition for different ways of breaking N particles up into subsets which then scatter one another.

To describe the operator formalism some preliminaries are necessary. We begin by considering a pair of selfadjoint operators P and Q in a Hilbert space $\mathscr{H}$ which satisify the celebrated Heisenberg commutation relations

$PQ - QP = iI$. It is known that such operators are necessarily unbounded and hence only densely defined but let us ignore such considerations for the time being and proceed formally. If $A = \frac{P+iQ}{\sqrt{2}}$ then $A^* = \frac{P-iQ}{\sqrt{2}}$ and one verifies that $AA^* - A^*A = I$. Moreover, $\frac{A+A^*}{\sqrt{2}} = P$ and $\frac{A-A^*}{\sqrt{2}} = iQ$. Thus operators A such that $AA^* - A^*A = I$ correspond one-to-one to pairs of self-adjoint operators P, Q such that $PQ - QP = iI$. We shall follow the physicists in referring to such operators A as annihilation operators and to their adjoints as creation operators. Evidently creation operators are characterized by satisfying the commutation relation $A^*A - AA^* = I$. One defines a unit vector φ such that $A(\varphi) = 0$ to be a *vacuum vector* or *vacuum* for the annihilation operator A.

Given an annihilation operator A and a vacuum φ_0 for A, let us define a sequence $\varphi_1, \varphi_2, \ldots$ of vectors inductively by the formula $\varphi_{n+1} = A^*(\varphi_n)$. Using the identity $AA^* - A^*A = I$ one proves easily that $A(\varphi_n) = n\varphi_{n-1}$ for $n = 1, 2, \ldots$ and hence that the linear span D_{φ_0} of $\varphi_0, \varphi_1, \varphi_2, \ldots$ is invariant under both A and A^*. If $\varphi_n = 0$ for some $n = 1, 2, \ldots$, then D_{φ_0} is finite dimensional and it is well known that the Heisenberg commutation relations cannot hold in finite dimensions. Thus φ_n is never zero and D_{φ_0} is infinite dimensional. One also checks easily that $A^*A(\varphi_n) = n\varphi_n$ so that the φ_n are not only linearly independent but orthogonal (since A^*A is selfadjoint). Henceforth we shall assume that A is *irreducible* in the sense that no closed invariant subspaces are invariant under both A and A^*. This implies that D_{φ_0} is dense in $\mathscr{H}$.

In order to make sense of and manipulate operators of the form $e^{\lambda A^*}$ and $e^{\lambda A}$ in spite of the unboundedness of A and A^* it is convenient to introduce a new kind of basis for $\mathscr{H}$ called a *coherent state basis*. Its members are the eigenvectors of A in $\mathscr{H}$ but since A is not selfadjoint they need not be orthogonal. In fact, they are not and there are continuum many of them. Let λ be any complex number and let ψ_λ be defined by the convergent series

$$\psi_\lambda = \varphi_0 + \lambda\varphi_1 + \frac{\lambda^2}{2}\varphi_2 + \frac{\lambda^3}{3!}\varphi_3 + \cdots .$$

One checks at once that $A(\psi_\lambda) = \lambda\psi_\lambda$ and that to within a multiplicative constant ψ_λ is the only eigenvector of A with eigenvalue λ. It is easy to see that D'_{φ_0}, the linear space of all ψ_λ, is dense in $\mathscr{H}$ just as D_{φ_0} is, but is a different dense subspace with more convenient properties for our purposes. For example, if one differentiates the defining equation for ψ_λ one obtains the formula

$$\frac{d\psi_\lambda}{d\lambda} = \varphi_1 + \lambda\phi_2 + \frac{\lambda^2}{2}\varphi_3 + \cdots$$
$$= A^*(\varphi_0) + \lambda(A^*)^2(\varphi_0) + \cdots$$
$$= A^*(\psi_\lambda).$$

Thus as a complement to $A(\psi_\lambda) = \lambda\psi_\lambda$ one has $A^*(\psi_\lambda) = \frac{d\psi_\lambda}{d\lambda}$. The defining equation for ψ_λ can be interpreted as stating that $\psi_\lambda = e^{\lambda A^*}(\varphi_0)$, thus defining $e^{\lambda A^*}$ at φ_0. But then $e^{\mu A^*}(\psi_\lambda) = e^{\mu A^*}e^{\lambda A^*}(\varphi_0) = e^{(\mu+\lambda)A^*}(\varphi_0) = \psi_{\mu+\lambda}$. In other words, one may define $e^{\mu A^*}$ on all of D'_{φ_0} by the formula $e^{\mu A^*}(\psi_\lambda) = \psi_{\lambda+\mu}$. Using the formula $e^{\mu A}e^{\lambda A^*} = e^{\lambda\mu}e^{\lambda A^*}e^{\mu A}$ one shows that $e^{\mu A}$ may be defined on all of D'_{φ_0} by the formula $e^{\mu A}(\psi_\lambda) = e^{\lambda\mu}\psi_\lambda$.

In the operator formalism one begins with several infinite sequences of annihilation operators. Each of these sequences $A_1, A_2, \ldots$ is strongly commutative in the sense that A_i commutes with both A_j and A_j^* whenever $i \neq j$ and admits a unit vector φ_0 which is a vacuum for all A_j. To each such sequence one associates a two-parameter family of operators called *vertex operators* by the formula

$$\widetilde{V}(p, t) = e^{-\sqrt{2}p(\sum_{n=1}^{\infty} \frac{1}{\sqrt{n}} A_n^* z^n)} e^{\sqrt{2}p(\sum_{n=1}^{\infty} \frac{A_n}{\sqrt{n}} z^{-n})}.$$

In the final construction of the N-point Veneziano amplitude, which we shall not give in detail, one has as many sequences $\{A_j^v\}$ as there are dimensions in space, their operators A_n^v all commute with one another in the strong sense and the vacuum vector φ_0 is the same for all. Each has a vertex operator $\widetilde{V}^v(p, z)$ and one constructs a vector vertex operator by taking a product $\widetilde{V}^1(p_1, z)\widetilde{V}^2(p_2, z)\cdots$. The product of this, with a slightly altered factor for the time dimension, leads finally to a $(d + 1)$-parameter family $\widetilde{V}(p_1, p_2, \ldots, p_d, z)$ of vertex operators from which the N-point function is constructed in a manner which it would be profitless to specify.

The key point for our purposes is that the study of the N-variable generalization of the Euler beta function led physicists to introduce the operators A_n^v and the vertex operator concept. The operators A_m^v were already familiar from quantum field theory where each one is associated with a harmonic oscillator. However, their occurrence in this particular context inspired Nambu (1969), Susskind (1970), and Nielsen (1970) to independent proposals of the idea that the strongly interacting particles might be better understood if one thought of them not as almost structureless points (spin implies some degree of internal structure) but as one-dimensional objects behaving like vibrating strings. This was the origin of the now celebrated new initiative in physics known as "String theory".

7. The Exact Solvability of the Korteweg De Vries Equation. The KdV equation (as its name is usually abbreviated) is the nonlinear first-order partial differential equation

$$\frac{\partial u}{\partial t} + u\frac{\partial u}{\partial x} + \frac{\partial^3 u}{\partial x^3} = 0.$$

It was introduced in 1895 by D. J. Korteweg and G. De Vries in their study of long water waves in a channel of finite depth. Interest in it revived in the 1960s when applications of it, and variants derived by simple changes of variable, turned up in a number of branches of applied mathematics. Studies, including computer simulations, were carried out and the latter suggested the existence of solutions of "soliton" type. However, it came as a complete surprise when Gardner, Greene, Kruskal, and Miura published a short note in *Physical Review Letters* (vol. 19, 1967, pp. 1095–1097) announcing a method for obtaining an explicit general solution. Using this method the authors were able to study the soliton nature of the solutions in depth.

The key discovery that made this exact solution possible is a most remarkable relationship between a variant of the KdV equation

$$(1) \qquad \frac{\partial u}{\partial t} - 6u\frac{\partial u}{\partial x} + \frac{\partial^3 u}{\partial x^3} = 0$$

and linear differential equations of the form

$$(2) \qquad \frac{-d^2 \psi}{dx^2} + v\psi = \epsilon\lambda^2 \psi$$

where v is a real-valued function on the real line, λ is a positive real number and $\epsilon = \pm 1$. Recall that (2) plays a central role in the first example of §3 and that certain data that can be derived from the solutions of (2) determine v uniquely whenever v vanishes sufficiently rapidly at $\pm\infty$. This data consisted in a function $r \longrightarrow \begin{pmatrix} a(r) & b(r) \\ -\bar{b}(r) & \bar{a}(r) \end{pmatrix}$ from the positive real numbers to SU(2) and a function c from the negative eigenvalues of the selfadjoint operator defined by the left-hand side of (2) to the complex numbers. Given any solution u of (1), define $v_t(x) = u(x, t)$. Then the v_t define a one-parameter family of equations (2) and correspondingly the functions a and b and c become functions of two variables, one of which is t. The remarkable relation between (1) and (2) consists in the following. Let u satisfy (1) and let the v_t vanish at ∞ so that the considerations of §3 are valid. Then

(a) the eigenvalue set on which c is defined remains fixed.

(b) The functions a, b, and c satisfy very simple *linear* differential equations:

$$\frac{\partial a}{\partial t} = 0, \qquad \frac{\partial b}{\partial t} = -8ir^3 b, \qquad \frac{\partial c(j)}{\partial t} = -4\gamma_j^3 c(j)$$

whose general solution can be written down at once. Here $-\gamma_1, -\gamma_2, \ldots, -\gamma_n$ are the fixed negative eigenvalues of the left-hand side of (2).

The existence of this relationship converts the problem of solving (1) (when the functions u are suitably restricted) to the problem of inverting the map $v \to a, b, c$—a problem which was solved in the period 1952–1958 by the content of one-dimensional quantum mechanical scattering. One refers to the method and various generalizations to be mentioned below as the *inverse scattering method*.

The detailed development of this unexpected new method for dealing with the KdV equation was carried out in a series of six papers coauthored by various subsets of the four authors of the original note and in one case (paper V) including Zabusky. The first five were published in the *Journal of Mathematical Physics* in the years 1968 through 1971. The sixth appeared in 1974 in *Communications on Pure and Applied Mathematics*. The first two of these, in particular, cast light on the sequence of events that led the authors to study (1) by connecting it with (2).

The very next year, in 1968, Peter Lax made an important contribution to the subject by, in a sense, starting from the other end. More specifically, inspired by the fact that the spectrum of $\frac{-d^2}{dx^2} + v_t$ remains fixed when $u(x, t) = v_t(x)$ satisfies (1) Lax inaugurated a general study of what it means for a one-parameter family $t \to L_t$ of selfadjoint operators to be "isospectral" in the following strong sense: There exists for each t a unitary operator $U(t)$ such that $U(t)^{-1}L_tU(t) = L_0$ and $U(t)$ (which is clearly not unique) can be selected in such a way that $U(t)$ is differentiable as a function of t. There is no loss in generality in assuming that $U(0) = I$. Differentiating the identity $U(t)^{-1}L_tU(t) = L_0$ and performing some simple algebra one is led quickly to the identity

$$\frac{d}{dt}L_t = U(t)^{-1}\frac{d}{dt}U(t)L_t - L_tU(t)^{-1}\frac{d}{dt}U(t)$$

which may be written in the form

$$(*) \qquad\qquad \frac{d}{dt}L_t = B(t)L_t - L_tB(t)$$

if one defines $B(t) = U(t)^{-1}\frac{d}{dt}U(t)$.

Notice that the function $t \to U(t)$ can be recovered from the function $t \to B(t)$ by integrating an operator differential equation, and that $(*)$ is just the definition of isospectrality written in "infinitesimal form". By definition the operators $B(t)$ are skew adjoint. Pairs of functions $t \to L_t$, $t \to B(t)$ that satisfy $(*)$ and where the L_t and $B(t)$ are selfadjoint and skew adjoint respectively are called Lax pairs.

After introducing these general notions, Lax examined the special case in which L_t is the selfadjoint operator $\frac{-d^2}{dx^2} + v_t$ in $\mathscr{L}^2(-\infty, \infty)$, where $v_t(x) = u(x, t)$ for some real function u of two real variables. He then sought functions B such that L and B satisfy $(*)$, confining his search to the manageable class in which each B_t is a skew symmetric ordinary differential operator

$$b_0(x, t) + b_1(x, t)\frac{d}{dx} + b_2(x, t)\frac{d^2}{dx^2} + \cdots + b_n(x, t)\frac{d^n}{dx^n}.$$

Given $u, b_0, b_1, \ldots, b_n$, the condition that L and B should satisfy $(*)$ reduces after straightforward but tedious calculation to a system of partial differential equations involving these $n + 2$ functions. The lowest value of

n for which interesting results occur is $n = 3$. In that case antisymmetry demands that

$$b_0 = \frac{1}{2}\left(\frac{\partial b_1}{\partial x} - \frac{1}{2}\frac{\partial^3 b_3}{\partial x^3}\right), \quad b_2 = \frac{3}{2}\frac{\partial b_3}{\partial x}$$

so that only b_1 and b_3 can be independently chosen. The partial differential equations equivalent to (*) can then be reduced to the following:

(1) $\quad \frac{\partial u}{\partial t} = \frac{1}{2}\frac{\partial^3 b_1}{\partial x^3} + b_1\frac{\partial u}{\partial x} + b_3\frac{\partial^3 u}{\partial x^3}$,

(2) $\quad \frac{\partial b_1}{\partial x} = -3b_3\frac{\partial u}{\partial x}$,

(2) $\quad \frac{\partial b_3}{\partial x} = 0$.

From (3) one deduces that b_3 must be a function of t alone and using this, (2) is equivalent to $b_1 = 3b_3u + \gamma$ where γ is some other function of t alone.

Substituting this value of b_1 in (1), it takes the form

(1$'$) $$\frac{\partial u}{\partial t} = \frac{1}{2}b_3\frac{\partial^3 u}{\partial x^3} - 3b_3u\frac{\partial u}{\partial x} + \gamma\frac{\partial u}{\partial x}.$$

From this we deduce the following

THEOREM. *If $-\frac{d^2}{dx} + u$ is isospectral, with a B that is defined by differential operators of at most the third order, then b_3 depends only on t, and $b_1 - 3b_3u = \gamma$ also depends only on t. Moreover u satisfies the partial differential equation*

(**) $$\frac{\partial u}{\partial t} = -\frac{1}{2}b_3\frac{\partial^3 u}{\partial x^3} - 3b_3u\frac{\partial u}{\partial x} + \gamma\frac{\partial u}{\partial x}.$$

Conversely, given any functions b_3 and γ defined (and suitably regular) for all t and any u that satisfies (**), $-\frac{d^2}{dx^2} + u$ will be isospectral with a B defined from u, b_3, and γ as indicated above.

Notice that, if one chooses $\gamma \equiv 0$ and b_3 to be a constant to the equation, (**) reduces to

(**)$'$ $$\frac{\partial u}{\partial t} = -\frac{1}{2}b\frac{\partial^3 u}{\partial x^3} - 3bu\frac{\partial u}{\partial x},$$

which is a variant of the KdV equation for all nonzero values of b. Of course, by making changes of variable of the form $x' = bx$, $t' = \mu t$, one can alter the two constant coefficients in (**)$'$ at will.

The connection of the KdV equation with isospectrality is considerably clarified by Lax's theory. One is led to the KdV equation quite naturally as one of the simplest examples in a systematic study of isospectrality. Starting from the KdV equation it seems miraculous that one can solve it in the indicated manner. At the same time, Lax's theory suggests a method for finding a vast number of other nonlinear equations which can be solved in an analogous manner. To start with, one can remove the restriction that $n = 3$ and as Lax discovered, obtain a "hierarchy" of higher-order solvable generalizations of the KdV equations. Many other generalizations are possible and there is now a huge literature on the subject. We shall give some more detailed indications in §10.

8. The Exact Solvability of the Two-Dimensional "Ice Problem" by Lieb. The "ice problem" is a problem in lattice statistics suggested by the molecular structure of crystals of frozen H_2O. In a note published in vol. 18 of *Physical Review Letters* (1967), Elliott Lieb announced an exact value for the entropy of the system at absolute zero. His determination was based upon considering the diagonalization of a suitable "transfer matrix" just as with Onsager's treatment of the Ising problem twenty-three years earlier. However, the transfer matrix in this problem cannot be treated by any simple adaptation of the method used by Onsager and the diagonalization cannot be made so explicit. Lieb faced a different and more difficult problem. In R. J. Baxter's book, "Exactly Solved Models in Statistical Mechanics", Chapter 10 starts off with the statement "Lieb's solution of the ice type, or six vertex, models was the most significant new exact result since the work of Berlin and Kac (1952) on the spherical model, and the pioneering work of Onsager (1944) on the Ising model."

Lieb's success was based on his fundamental observation that the transfer matrix for the ice problem has the same eigenvectors (but not the same eigenvalues) as another $2^n \times 2^n$ matrix that entered physics in 1928, only three years after Ising introduced the Ising model. In that year Heisenberg introduced a quantum theory of magnetism whose Hamiltonian operator can be described in terms of the very same operators used in §4 to describe the transfer matrix for the Ising problem. Let $\mathcal{H}^n = \mathcal{H} \otimes \mathcal{H} \otimes \cdots \otimes \mathcal{H}$ as in §4 and for $j = 1, 2, \ldots, n$ let c_j and s_j be the operators in $\mathcal{H}^n$ defined as in §4. Change notation slightly by setting $\sigma_j^1 = c_j$, $\sigma_j^3 = s_j$ and define $\sigma_j^2 = ic_js_j$. Then the Hamiltonian introduced by Heisenberg is of the form $-\frac{1}{2}(X\Sigma^1 + Y\Sigma^2 + Z\Sigma^3)$ where X, Y, and Z are real numbers and $\Sigma^k = \sum_{j=1}^{n} \sigma_j^k \sigma_{j+1}^k$. As in §4, σ_{n+1}^k is defined to be σ_1^k. It will be convenient to denote this operator by $H(X, Y, Z)$ and refer to it as the XYZ Hamiltonian or as defining the Heisenberg XYZ model. Notice that Σ^3 is equal to the A of Onsager's transfer matrix for the Ising problem. For the purposes of Heisenberg's theory it was important to diagonalize $H(X, Y, Z)$ for various values of X, Y, and Z but this turned out to be difficult to do.

A first and rather important step was made by the physicist, H. Bethe, in 1931, who investigated the special case in which $X = Y = Z$ and found a powerful method. This method is now known as the Bethe Ansatz method. It depends upon guessing that the eigenvectors are of a certain form, writing down the most general vector of this form and solving the equations that result from assuming that the vector is indeed an eigenvector. Of course any finite matrix can be diagonalized in this way by solving a polynomial equation where the degree is the number of rows or columns of the matrix. The point of the Bethe Ansatz is that it is much more efficient.

Thirty years later, in 1961, Lieb, Mattis, and Schultz found that they could diagonalize $H(X, Y, 0)$ for all X and Y by rather simple devices and, more significantly, in 1966 C. P. and C. N. Yang were able to adapt the Bethe Ansatz method to deal with the diagonalization of $H(X, Y, Z)$ whenever $X = Y$.

Lieb's breakthrough was to connect the ice problem with the diagonalization of $H(X, Y, Z)$ by showing that the transfer matrix of the ice problem had the same eigenvectors as those found by Yang and Yang for $H(X, Y, Z)$ when $X = Y = -2Z$. A few months later Lieb found that he could deal with two other familiar problems in lattice statistics using the same connection with $H(X, Y, Z)$ but different values of the ratio X/Z. These results were announced in two further notes in *Physical Review Letters*, which appeared in June and July of 1967. Inspired by the first Lieb note but unaware of the two later ones, Sutherland discovered a generalization of Lieb's work on the ice problem which included the results of all three of Lieb's notes. It was submitted to *Physical Review Letters* just four days before Lieb's third note was submitted.

It turns out that the transfer matrix of the general problem considered by Sutherland may be described as follows. For each $j = 1, 2, \ldots, n$ and each quadruple w_1, w_2, w_3, w_4 of real numbers let

$$\mathscr{L}_j(w_1, w_2, w_3, w_4) = w_1 \begin{pmatrix} 0 & \sigma_j^1 \\ \sigma_j^1 & 0 \end{pmatrix} + w_2 \begin{pmatrix} 0 & -i\sigma_2^j \\ i\sigma_2^j & 0 \end{pmatrix}$$
$$+ w_3 \begin{pmatrix} \sigma_j^3 & 0 \\ 0 & -\sigma_j^3 \end{pmatrix} + w_4 \begin{pmatrix} \sigma^4 & 0 \\ 0 & \sigma^4 \end{pmatrix}$$

where σ^4 is the identity operator in $\mathscr{H}^n$.

Then $\mathscr{L}_n$ is a 2×2 matrix whose elements are operators in $\mathscr{H}^n$ that depend on the four real parameters w_1, w_2, w_3, w_4. Now take the product $\mathscr{L}_n \mathscr{L}_{n-1} \cdots \mathscr{L}_1$ regarding each factor as a two-by-two matrix of operators in $\mathscr{H}^n$ and define $T_N(w_1, w_2, w_3, w_4)$ to be the trace of the resulting 2×2 matrix of operators in $\mathscr{H}^n$. Then $T_N(w, w, w_3, w_4)$ is the transfer matrix of Sutherland's generalization of the ice problem. One now refers to this generalization as the six vertex problem. As Sutherland showed, the eigenvectors of $T_N(w, w, w_3, w_4)$ will be eigenvectors of $H(X, X, Z)$ whenever X/Z has a certain value which can be computed from w, w_3, and w_4.

9. Connections between §5, §6, §7, and §8. Preliminary Remarks. On the face of it the advances described in §5, §6, §7, and §8 have very little to do with one another. Indeed, as already noted in the introductory §1, one is in pure algebra and the others are in elementary particle physics, classical applied mathematics, and classical statistical physics, respectively. Moreover, the full extent of their actual interrelationship appeared only gradually as the original discoveries were developed and exploited.

Actually, one does not have to penetrate too far below the surface to see at least the possibility of significant connections. First of all, both §7 and §8 deal with exact solutions of apparently very different problems. However, in reflecting on this difference, one must not forget that the passage from classical to quantum mechanics transforms solution of differential equations into diagonalization of matrices. Secondly, in the past, the key to exact solutions has almost invariably been the existence of symmetry exploitable by group theory and group representations. So a development in group theory such as discovering a class of infinite-dimensional Lie algebras having a theory parallel to that of the well-understood finite-dimensional semisimple Lie algebras, has conceivable applications to any situation in which exact solutions are sought. Thirdly, as we have seen in §6, the analysis of the N-particle generalization of the Veneziano model had led by 1969 to the operator formalism in which everything is based on an infinite-dimensional Lie algebra—already familiar from quantum field theory. This infinite-dimensional Lie algebra is not a Kac-Moody Lie algebra but has closely related properties. We shall return to the topic of interconnections after we have discussed the later developments of the discoveries of §5, §6, §7, and §8.

10. The Immediate Aftermath. When I began to write this concluding essay I had meant to break this section into four separate sections and treat the developments of the next decade in the same degree of detail that I have used so far. Unfortunately, I vastly underestimated the time and number of pages that my plans would require—as far as time is concerned, by at least a factor of 20. By now both my editors and I feel that I have tried their patience long enough and, with considerable regret, I have decided to present a good deal of the remaining material in severely condensed form. Sections 3 and 6 were the most troublesome and I have already made serious compromises with my original plans for them. The path from the S operator concept to the Veneziano model is much more complex than I had appreciated.

A. We explained in §8 how Lieb's work on the ice problem led rather quickly (in the hands of Sutherland) to a generalization which can be reformulated as the so-called "six vertex problem". The very formulation of the six vertex problem suggests a generalization known as the eight vertex problem. Recall that the six vertex problem is that of studying the diagonalization of the transfer matrix $T_n(w_1, w_2, w_3, w_4)$ in the special case in which $w_1 = w_2$. The eight vertex problem is the same problem but with the restriction $w_1 = w_2$ removed.

While the eight vertex problem is easy to formulate it is considerably more difficult to solve than the six vertex problem and R. J. Baxter created something of a sensation when he announced a solution in 1971. Like the solution of the six vertex problem it rests on a relationship with the Heisenberg problem of diagonalizing $H(X, Y, Z)$. However, now one uses the general case $X \neq Y$ in compensation for $w_1 \neq w_2$. In fact, Baxter used the relationship between $T_n(w_1, w_2, w_3, w_4)$ and $H(X, Y, Z)$ to generalize the work of Yang and Yang on $H(X, X, Z)$ to general $H(X, Y, Z)$ at the same time. His work was announced in *two* notes in *Physical Review Letters* entitled "Eight-Vertex Model in Lattice Statistics" and "One-dimensional Anisotropic Heisenberg Chain" received by the editors on the same day (Feb. 25, 1971) and published on pages 832–833 and 834, respectively, of the April 5, 1971, issue of the journal cited.

We mention these dates in part because of some overlap with independent work of Sutherland. In a paper published in the *Journal of Mathematical Physics* in November of 1970 Sutherland investigated the commutator of $H(X, Y, Z)$ and $T_n(w_1, w_2, w_3, w_4)$. He was able to show that for each choice of w_1, w_2, w_3, w_4 there exists a triple X, Y, Z such that $H(X, Y, Z)$ and $T_n(w_1, w_2, w_3, w_4)$ commute. Moreover, the triple X, Y, Z is unique up to multiplying all three components by the same number. When $w_1 = w_2$, $X = Y$ and the result is known from the theory of the six vertex model. When $w_1 \neq w_2$, it is a definite step toward the general case. Sutherland remarks that it is only potentially useful because he knows no analogue of the results of Yang and Yang when $X \neq Y$. If he did he could have used Lieb's method.

In his two notes Baxter does not mention Sutherland's paper although it had appeared before the notes were submitted. Baxter explained later that he missed seeing it because of a long sea voyage to Australia in the crucial period. His approach was different and was based on a rather ingenious analysis of conditions on w_1, w_2, w_3, w_4 and w_1', w_2', w_3', w_4' that are equivalent to the commutativity of $T_n(w_1, w_2, w_3, w_4)$ with $T_n(w_1', w_2', w_3', w_4')$. The conditions are expressed in terms of a reparametrization involving the Jacobi elliptic functions cn, dn, and sn. These functions contain a parameter l called the modulus so that $cn(v, l)$ is the value at v of the function cn with modulus l. The reparametrization is defined by the equations

$$(1) \qquad w_1 = \frac{\rho cn(v, l)}{cn(\zeta, l)}, \qquad w_2 = \frac{\rho dn(v, l)}{dn(\zeta, l)}, \qquad w_3 = \rho, \qquad w_4 = \frac{\rho sn(v, l)}{sn(\zeta, l)}.$$

Expressed in terms of the new parameters ρ, ζ, v, l and ρ', ζ', v', l' Baxter's commutativity condition is that $\zeta = \zeta'$ and $l = l'$. In other words, for each fixed pair ζ_0, l_0, $\tilde{T}_n(\rho, \zeta_0, v, l_0)$ commutes with $\tilde{T}_n(\rho', \zeta_0, v', l_0)$ for all $\rho, \rho', v,$ and v'.

Here $\tilde{T}_n$ is just the composite of T_n with the functions of ρ, ζ, v, l defined by the reparametrization equations (1).

Of course multiplying all w_j by the same real number ρ simply multiplies T_n by ρ^n so one might as well fix ρ at 1. The substance of Baxter's result is that for each ζ_0, l_0 one obtains a one-parameter family of $\tilde{T}_n$'s that commute with one another nontrivially; i.e., are not constant multiples of one another. Evidently the joint eigenvectors of the one-parameter family associated with each ζ_0, l_0 are "more intrinsic" than the eigenvectors of a random member of the family—and hence, hopefully, easier to find. Baxter obtains Sutherland's connection with the X, Y, Z problem by showing that for each X, Y, Z there is a unique pair ζ_0, l_0 such that $H_n(X, Y, Z)$ is a limit of members of the associated commuting family of $\widetilde{T}_n$'s. Indeed, there is an explicit formula for H_n involving $\frac{\partial}{\partial v} \log \widetilde{T}_n$.

Actually finding the eigenvalues of a particular $T_n(w_1, w_2, w_3, w_4)$ requires further work involving a significant extension of the Bethe Ansatz method and is in part only implicitly done. The results are sufficient, however, to solve certain concrete problems exactly. Full details have been given by Baxter in several long papers published in the *Annals of Physics* in 1972 and 1973 as well as in his book (cited in §8).

B. An interesting new point of view toward the exact solvability of the KdV equation was presented in paper IV of the six detailed papers mentioned in §7. It was written by Gardner alone and appeared in 1971. In it Gardner shows how to consider the KdV equation as a (somewhat generalized) Hamiltonian system which is completely integrable in the sense of having a sufficiently large commutative symmetry group—or, in more classical language, a sufficient number of integrals in involution. Later in the same year Zakharov and Faddeev in the Soviet Union published a short paper independently announcing the same idea and developing it somewhat differently. Soon after this, other examples were found that (a) could be generated by Lax pairs, (b) could be solved by the inverse scattering method, and (c) could be considered as completely integrable Hamiltonian systems.

Before explaining further it will be useful to give a brief account of Hamiltonian systems in classical mechanics and the meaning of complete integrability. Recall that in classical particle mechanics a system of N particles (such as the planets) whose motion is determined by the particle masses m_1, m_2, ..., m_N and a potential energy function V defined on the set of all possible N tuples $x_1, y_1, z_1; x_2, y_2, z_2; \ldots x_N, y_N, z_N$ of particle coordinates, one introduces a more convenient notation by relabeling the $3N$-tuple $x_1, y_1, z_1, x_2 \ldots y_N, z_N$ as $q_1, q_2, \ldots, q_{3N}$ and defines $M_{3n+k} = m_{n+1}$ for $k = 1, 2, 3, \ldots, n = 0, 1, 2, \ldots, N-1$. Then one determines the motion by integrating the $3N$ second-order differential equations

$$(1) \qquad M_n \frac{d^2 q_n}{dt^2} = -\frac{\partial V}{\partial q_n}.$$

It is convenient to replace these by a system of $6N$ first-order equations by setting $p_n = M_n \frac{dq_n}{dt}$ so that the equations become

$$(2) \qquad \frac{dq_n}{dt} = \frac{p_n}{M_n}, \quad \frac{dp_n}{dt} = -\frac{\partial V}{\partial q_n}.$$

One then observes that if one defines

$$H(q_1, \ldots, q_{3N}, p_1, \ldots, p_{3N}) = \sum_{n=1}^{3N} \frac{p_n^2}{2M_n} + V(q_1, \ldots, q_{3n})$$

the equations (2) may be given the more symmetrical form

$$(3) \qquad \frac{dq_n}{dt} = \frac{\partial H}{\partial p_n} = \frac{dp_n}{dt} = -\frac{\partial H}{\partial q_n}.$$

The function H is called the Hamiltonian of the system and the equations (3) are called the equations of motion in Hamiltonian form.

Let us denote the set of all $6N$-tuples $q_1, q_2, \ldots, q_{3N}, p_1, p_2, \ldots, p_{3N}$ by Ω. This is the so-called *phase space* of the system. If G is any differentiable real-valued function on Ω and $t \to q_1(t), q_2(t), \ldots, p_{3N}(t)$ is any solution of (3) then one can compute the rate of change of G along the path of this solution of (3) by applying the usual formula and using (3). One gets

$$\sum_{n=1}^{3N} \left(\frac{\partial G}{\partial q_n} \frac{\partial H}{\partial p_n} - \frac{\partial G}{\partial p_n} \frac{\partial H}{\partial q_n} \right)$$

evaluated at the appropriate point of the trajectory. It follows at once that G is a *constant* on all trajectories; that is, it is a so-called *integral of the motion* if and only if the above sum is identically zero as a function on Ω. One makes the definition

$$(4) \qquad (G_1, G_2) = \sum_{n=1}^{3N} \left(\frac{\partial G_1}{\partial q_n} \frac{\partial G_2}{\partial p_n} - \frac{\partial G_1}{\partial p_n} \frac{\partial G_2}{\partial q_n} \right)$$

for any two differentiable functions G_1 and G_2 on Ω and calls (G_1, G_2) the *Poisson bracket* of G_1 and G_2. The above result can then be restated in the form: a differentiable function G on Ω is an integral of the motion if and only if $(G, H) = 0$. Since $(G, G) = 0$ for any G it follows at once that the Hamiltonian itself is always an integral of the motion. It is the celebrated energy integral. Its value at any point ω of Ω is called the *energy* of the system when its *state* is ω. Note that $(q_n, H) = \frac{\partial H}{\partial p_n}$ and $(p_n, H) = -\frac{\partial H}{\partial q_n}$ so that (3) may be rewritten as

$$\frac{dq_n}{dn} = (q_n, H), \qquad \frac{dp_n}{dt} = (p_n, H).$$

In other words, the equations of motion are just special cases of the general fact that $\frac{dG}{dt} = (G, H)$ for all differentiable functions G on Ω.

It is obvious that the Poisson bracket satisfies the identity $(G_1, G_2) = -(G_2, G_1)$ for any two differentiable functions G_1 and G_2 and less obvious but easily proved that it also satisfies the identity

$$(5) \qquad ((G_1, G_2), G_3) + ((G_3, G_1), G_2) + ((G_2, G_3), G_1) \equiv 0$$

for any three differentiable functions G_1, G_2, G_3 on Ω. This has the immediate corollary that if (G_1, G_3) and (G_2, G_3) are both zero, then

$((G_1, G_2), G_3)$ is also zero. In the special case in which $G_3 = H$ this says that the Poisson bracket of any two integrals of the motion is again an integral of the motion. The system defined by H is said to be *completely integrable* or to be a *completely integrable Hamiltonian system* if there exist $3n$ linearly independent integrals of the motion such that the Poisson bracket of any two of them is zero.

Observe now that, for any differentiable function G on Ω, we may define a system of $6N$ first-order differential equations for the q_j and p_j by replacing H by G in our original equations in Hamiltonian form. When we do so we find that almost everything we have said about our original system still applies and it is natural to call it a *generalized Hamiltonian system*. In particular, since $(G, H) = -(H, G)$ we conclude that G is an integral of the motion for the Hamiltonian system defined by H if and only if H is an integral of the motion for the generalized Hamiltonian system defined by G. In the not so special case in which the system defined by G can be globally integrated (not necessarily explicitly), in the sense that there exists a one-parameter family α_t^G of self diffeomorphisms of Ω whose orbits are solutions of the system, then to say that H is an integral of the system defined by G is the same as to say that the α_t^G define a one-parameter group of symmetries of the system defined by H. Moreover, to say that $(G_1, G_2) = 0$ is the same as to say that $\alpha_t^{G_1}$ and $\alpha_s^{G_2}$ commute for all t and s. Thus, at least when global integrabilty holds, a completely integrable Hamiltonian system is one with a $3N$-parameter *commutative* group of symmetries.

The above continues to hold when our system is only a generalized Hamiltonian system, so it is a significant fact that the Korteweg de Vries equation can be regarded as a generalized Hamiltonian system that is completely integrable.

In Gardner's approach he restricts attention to solutions of the KdV equation that are periodic in x with period 2π for some (and hence all) t. Now a function u on the real line that is periodic with period 2π can be written as a Fourier series $u(x) = \sum_{n=-\infty}^{\infty} u_n e^{inx}$ so that u is completely determined by the sequence $\ldots u_{-2}, u_{-1}, u_0, u_1$ of Fourier coefficients. In particular, the KdV equation may be written as an infinite system of ordinary first-order differential equations of the form $\frac{du_n}{dt} = A_n(\ldots u_{-2}, u_{-1}, u_0 \ldots)$. Let

$$\widetilde{H}(u) = H(\ldots u_{-2}, u_{-1}, u_0 \ldots) = \frac{i}{2\pi} \int_0^{2\pi} (\frac{1}{6} u^3 - \frac{1}{2} u_x^2).$$

One can then show that $A_n \equiv n \frac{\partial H}{\partial u_{-n}}$. If one defines $q_n = \frac{u_n}{n}$, $p_n = u_{-n}$ it follows that our system of differential equations may be rewritten in the form $\frac{dq_n}{dt} = \frac{\partial H}{\partial p_n}$, $\frac{dp_n}{dt} = -\frac{\partial H}{\partial q_n}$. This is in (generalized) Hamiltonian form with Hamiltonian H. Gardner goes on to find enough integrals of the motion whose Poisson brackets with one another are zero to justify the assertion that the system is completely integrable. Because of the infinite number of variables one has to deal with convergence questions but the formal set up is clear.

The approach of Zakharov and Faddeev is complementary to that of Gardner in that they deal not with periodic solutions but with solutions which, for each t, vanish at ∞ with sufficient rapidity that $\int_{-\infty}^{\infty} (1 + |x|)|u(x)|\, dx < \infty$. In this case it is not possible to reduce the KdV equation to an infinite system of ordinary equations and they are forced to define Hamiltonian system and complete integrability for continuum systems. We shall not give details.

Another important advance made in 1971 was the discovery by Zhakarov and Shabat that the so-called nonlinear Schrödinger equation $i\frac{\partial u}{\partial t} + \frac{\partial^2 u}{\partial x^2} + \chi|u|^2 u = 0$ could be successfully attacked by the inverse scattering method in the Lax formulation. This equation had turned up in several applied mathematics problems in recent years and the authors found a linear differential operator which was related to their nonlinear equation just as $-\frac{d^2}{dx^2} + v(x)$ is related to the KdV equation. This linear differential operator is $i\begin{bmatrix} 1+p & 0 \\ 0 & 1-p \end{bmatrix}\frac{\partial}{\partial x} + \begin{bmatrix} 0 & \bar{u} \\ u & 0 \end{bmatrix}$ where p is a constant such that $\chi = \frac{2}{1-p^2}$. It operates on pairs of functions and is an analogue of the Schrödinger operator for particles with spin $\frac{1}{2}$ in one dimension.

Two years later, in 1973, H. Segur, M. J. Ablowitz, A. C. Newell, and D. J. Kaup published a short note in "Physical Review Letters" showing that a slight variant of the method of Zakharov and Shabat could be used to treat the "sine Gordan equation" in the form $\frac{\partial^2 u}{\partial x \partial t} = \sin u$. This equation had also recently been of interest in mathematical physics. Only a month later the same authors published a further note synthesizing their work on the sine Gordan equation with the earlier work of Zakharov and Shabat by relating both to a more general Schrödinger operator for particles of spin $1/2$ in one dimension. Their more general treatment included several other nonlinear equations of known interest as well. In 1974 Segur, Aldowitz, Newell, and Kaup published a long paper entitled "The inverse scattering transform-Fourier analysis for non linear problems" in "Studies in Applied Mathematics" 53 (249–315) developing the ideas of their two notes and emphasizing the view that the inverse scattering transform method can be viewed as an extension into the nonlinear realm of the Fourier transform method.

The papers cited so far turned out to just mark the beginning of a rather extensive development which is still going on. The story is much too complicated to be adequately summarized here. Suffice it to say that the equations that can be integrated by the method usually turn out to be interpretable as completely integrable Hamiltonian systems. For further information the reader is referred to the book of Faddeev and Takhtajan.

A development in a somewhat different direction began in 1974 with the publication of a paper by H. Flaschka. In this paper Flaschka observed that the Lax pair idea could be used to demonstrate the complete integrability of a rather special Hamiltonian system consisting of a finite number of particles moving in one dimension. This system had a Hamiltonian of the form

$$H = \sum_{k=1}^{n} \frac{p_k^2}{2} + e^{-q_k + q_{k-1}}$$ where $q_{n+1} = q_n$ and was known as the Toda lattice.

Its study had been begun by Toda in 1970 and inspired by computer studies. Henon had shown it to be a completely integrable system by direct means. Flaschka's contribution was to note that it could be regarded as a discrete analogue of the KdV equation and that its complete integrability could be demonstrated, as stated, by Lax pairs.

In 1975 Moser found further examples involving systems of particles in one dimension and, in particular, proved a conjecture of Calogero. Two years later Kostant found a remarkable connection between the Toda lattice and the simple roots of the Lie algebra of $SL(n, R)$, which turned out to be generalizable and to lead to the association of completely integrable systems to all the semisimple Lie algebras and their representations. Once again an elaborate and beautiful theory developed which we cannot begin to summarize here. Perhaps at least we should call attention to the work of M. Adler which overlapped with that of Kostant.

C. In 1971, shortly after the original proposal of string theory in 1969 and 1970, Neveu and Schwarz suggested an important modification which they called the dual pion model. From the point of view of this essay its chief interest lies in an important generalization of the Lie algebra notion to which it gave rise. Stimulated by a paper published in the same year by P. Ramond, Neveu and Schwarz augmented the annihilation and creation operators of the operator formalism by other operators which also played a role in classical quantum field theory and were also called creation and annihilation operators. They differed however, in that the commutation relation $AA^* - A^*A = I$ is replaced by the "anti-commutation relation" $BB^* + B^*B = I$ supplemented by the extra condition $B^2 = 0$. When it is necessary to distinguish the two kinds, one refers to boson annihilation (creation) operators and fermion annihilation (creation) operators respectively.

Starting with a sequence of strongly *anti-commuting* fermion annihilation operators and working through analogues of certain constructions in the earlier operator formalism, Neveu and Schwarz arrived at a doubly infinite sequence $\ldots G_{-5/2}, G_{-3/2}, G_{-1/2}, G_{+1/2}, G_{+3/2} \ldots$ of operators parametrized by the half integers. Let $\mathcal{M}$ denote the vector space consisting of all finite linear contributions of the operators G_m and let $\mathcal{M}^2$ denote the vector space consisting of all finite linear combinations of anti-commutators $GG' + G'G$ of elements G and G' of $\mathcal{M}$. Neveu and Schwartz did not introduce $\mathcal{M}$ and $\mathcal{M}^2$ or look at things in quite this way but the following theorem is an immediate consequence of what they did prove.

Theorem. *If T_1 and T_2 are any two elements in $\mathcal{M}^2$ then $T_1 T_2 - T_2 T_1$ is again in $\mathcal{M}^2$ so that $\mathcal{M}^2$ is a Lie algebra of operators. Moreover, if G is any element of $\mathcal{M}$ and T is any element of $\mathcal{M}^2$ then $GT - TG$ is again in $\mathcal{M}$ and the action of $\mathcal{M}^2$ on $\mathcal{M}$ so defined is a Lie algebra representation. Finally $\mathcal{M}^2 \cap \mathcal{M} = 0$.*

Consider now the direct sum of $\mathcal{M}$ and $\mathcal{M}^2$. One can make this vector space into a nonassociative algebra by defining a multiplication $\circ$ such that $A \circ B = AB + BA$ when A and B are both in $\mathcal{M}$ and $A \circ B = AB - BA$ when one of A and B is in $\mathcal{M}^2$. Other possibilities are taken care of by linearity. The algebra so defined is an example of what is known as a *Lie superalgebra* and sometimes as a *graded Lie algebra*. The latter terminology can be misleading because a graded Lie algebra need *not* be a Lie algebra.

We shall not pause to give the definition of a Lie superalgebra in general. It is not very difficult to guess it. We shall content ourselves rather with a few remarks. First of all, a Lie superalgebra need not be infinite dimensional. Indeed, if one replaces $\mathcal{M}$ by the two-dimensional subspace spanned by $G_{-1/2}$ and $G_{1/2}$ one finds that $\mathcal{M}^2$ is three dimensional. This five-dimensional example is discussed from a different point of view in section VIII of the reprint in this collection entitled "Weyl's program and modern physics".

Secondly, it is a curious fact that the concept of a Lie superalgebra, as a new mathematical object worthy of study in its own right and implementing a new and useful way of describing symmetries, is due only indirectly to Neveu and Schwarz. Schwarz has acknowledged that he and Neveu "missed the boat". The Lie superalgebra concept was first introduced in 1973 by Volkov and Akulov in the Soviet Union and independently in 1974 by Wess and Zumino in the West. Wess and Zumino acknowledge the inspiration of the work of Neveu and Schwarz but apply the notion in a field theoretic context. Volkov and Akulov also worked in a field theoretical context. By the time supersymmetry had once again made contact with string theory (1976) the latter had been abandoned as a theory of strong interactions but reborn as a possible "theory of everything" unifying gravity with the other forces of nature.

Thirdly, the Lie superalgebra notion turned out to be a rather natural generalization of the Lie algebra notion. In particular, one can still speak of roots and weights and develop a representation theory based on them.
D. The further development of the theory of Kac-Moody Lie algebras did not begin in a serious way until 1974. This is not surprising when one realizes that the motivation for further development was the comparison of two papers published in 1971 and 1972 by I. N. Bernstein, I. M. Gelfand, and S. I. Gelfand on the one hand and I. G. Macdonald on the other. Both of these papers were concerned with the fact, mentioned at the end of §2, that an irreducible representation of a finite-dimensional semisimple complex Lie algebra is uniquely determined by its highest weight. More specifically

they were concerned with the fact that there is an explicit formula for computing the "character" of the representation from its highest weight. This formula, found by Weyl in 1928 and known as the Weyl character formula, was proved by him using transcendental methods. Purely algebraic proofs were first found over two decades later by Freudenthal (1954) and Kostant (1959).

The 1971 paper of Bernstein, Gelfand, and Gelfand completes a program begun by D. Verma in his 1966 Yale thesis and announced in the 1968 volume of the Bulletin of the American Mathematical Society. As anticipated by Verma the completed program provides still another purely algebraic proof of the Weyl character formula. The 1972 paper of Macdonald is about a generalization of an identity that occurs in the analysis of the Weyl character formula. This formula takes the form of a quotient whose denominator is a product over all the positive roots α of $(e^{\alpha/2} - e^{-\alpha/2})$ and the identity gives the coefficients in the expansion of this product. Recall that a root system for a finite-dimensional semisimple Lie algebra is a finite subset of the dual of a vector space. Those finite subsets that arise as root systems have many special properties and may be characterized axiomatically in an elegant way. The axioms suggest a generalization which Macdonald calls an "affine root system". The second paragraph of the introduction to Macdonald's paper (*Inventiones Math.* 15, 19–143) begins with the sentence, "The main purpose of this paper is to establish an analogue of (0.1) for an 'affine root system' S ". Here (0.1) is the Weyl denominator identity cited above. The analogue which he establishes is close to Weyl's in form but differs from it in two important respects. First of all, since affine root systems have *infinitely* many elements the sum and product on the two sides of the identity are infinite sums and infinite products. Secondly, there is a rather mysterious "extra factor" on the right-hand side. The proof is by no means simple and occupies about a dozen pages. The theorem itself might be of limited interest, except to specialists in root systems, if it were not for the fact that Macdonald found a very striking application. Let $\eta(X) = X^{1/24} \prod_{n=1}^{\infty} (1 - X^n)$ be the classical Dedekind η function of importance in number theory and the theory of modular forms. Then for certain positive integers d there exists an affine root system whose associated generalized Weyl formula may be reduced to a formula that evaluates the coefficients in the expansion of $\eta(X)^d$ as an infinite series of powers of $X^{1/24}$. For each classical (finite) root system R there is a canonically defined affine root system $S(R)$. The affine root systems that occur are just the $S(R)$ and the power of η that occurs in the Weyl identity, for $S(R)$ is just the dimension of the Lie algebra whose root system is R. As far as published work is concerned, these formulae for η^d were all new except when $d = 3, 8$, and 10 and the case $d = 10$ was only a few years old. On the other hand, it seems that Dyson, in recent unpublished work, had found the Macdonald examples for all cases in which R is the root system of a nonexceptional complex simple Lie algebra. However, he used different methods and puzzled over the powers of d that occurred without noticing their connection with Lie algebra dimensions.

So much for the background. In 1974 V. Kac published a note in the Soviet journal, "Functional Analysis and its Application", the English translation of whose title was "Infinite-dimensional Lie algebras and Dedekind's η function". In the first sentence he asserts that it resulted from a comparison of the papers of Bernstein, Gelfand, and Gelfand and Macdonald alluded to above. The point is that the program of Verma as completed by Bernstein, Gelfand, and Gelfand is just what is needed for extending the Weyl character formula to Kac-Moody Lie algebras and that one can obtain the Macdonald identities and many others by using the root systems of Kac-Moody Lie algebras instead of the abstract root systems of Macdonald. Besides its greater generality, this path to the identities has the great advantage that the mysterious "extra factor" of Macdonald appears naturally and automatically. This is because there are essentially two different kinds of roots of a Kac-Moody Lie algebra and only one kind has a direct analogue among the roots of a classical finite-dimensional semisimple Lie algebra. This second kind of root occurs, in particular, in the special case of untwisted affine Lie algebras and for that case is described in detail in §5.

It is interesting to note that, in his paper, Macdonald makes the remark that "the list of Dynkin diagrams of reduced irreducible affine root systems which we obtain in §5 also occurs in the work of Moody on Euclidean Lie algebras." However, he does not push the matter any further. A long (five columns) and very informative review of Macdonald's paper by Verma (MR 50, 9996) mentions that Moody has noted the connection with Kac-Moody Lie algebras and interpreted the mysterious extra factor more or less as Kac did. Moody published a short paper on the subject in 1975. Apparently neither Verma nor Moody knew of the 1974 paper of Kac at the time.

On the other hand, Kac went much further than Moody in going on to adapt the methods of Verma and Gelfand, Gelfand, and Bernstein to outline a representation theory for a class of infinite-dimensional representations of Kac-Moody Lie algebras and derive an analogue of the Weyl character formula for these representations. A detailed development of this outline appeared rather later in Chapters 9 to 14 of Kac's book, "Infinite Dimensional Lie Algebras" (Birkhauser, 1983).

11. Further Connections Between §5, §6, §7, and §8. In §9 we mentioned potential connections between §5 on the one hand and §7 and §8 on the other, arising from the general link between groups, symmetry principles, and exact solutions and these, as far as §7 is concerned, became an actuality in later work on inverse scattering beginning with a short announcement published in 1979 in the Soviet Union by A. G. Reiman, M. A. Seminov-Tjan-Shanski and I. B. Frenkel. In English translation their note is entitled, "Graded Lie algebras and completely integrable dynamical systems."

A year earlier a quite surprising direct connection between §7 and §8 was found and developed by L. Faddeev and his collaborators. According to Faddeev the discovery was motivated by his own studies of a 1976 paper of Luther "in which Baxter's results and [22] are applied to the investigation of the Sine-Gordan quantum model"; he "noticed that a number of Baxter's formulae resembled the formulae, already familiar to us from the inverse problem method...." It could have been motivated by the following considerations. Many equations, which can be solved by the inverse scattering method, are completely integrable Hamiltonian systems that have quantum mechanical refinements. Because of the relationship between quantum and classical mechanics it is not unreasonable to hope that many completely integrable classical systems have completely integrable quantum refinements and that when this complete integrability comes via inverse scattering that this is reflected in the quantum mechanics. In short, it is natural to seek a quantum analogue of the method of inverse scattering. Inspired by the parallelism between Baxter's formulae and inverse scattering mentioned above, Faddeev and collaborators sought and found such a quantum analogue. A first announcement was published in 1978 in a joint paper with E. K. Skylanin entitled (in English translation) "Quantum mechanical approach to completely integrable models of field theory" (Sov. Phys. Dokl. 23 (1978), 902–904), and an extensive development followed quickly. Of particular relevance to this essay is a lengthy paper by Faddeev and Takhtajan published in 1979 and entitled "The Quantum Method of the Inverse Problem and the Heisenberg XYZ Model" (Russian Math. Surveys 34 (1979), 11–68). In it, they show in detail that the quantum version of the inverse scattering method can be directly used to give solutions to the Heisenberg XYZ problem and the Baxter eight vertex problem. They also work out in detail the applications of their new method to the easier six vertex problem of Lieb and Sutherland. In terms of unifying §7 and §8 one could hardly ask for more. The quote from Faddeev given above can be found in the introduction to this paper.

Independently of this work of the Russian school and a few months earlier, a closely related connection of §7 with §8 was described in a paper published in the *Physical Review* by H. B. Thacker in the February 17 issue. His emphasis was on the connection of the Bethe Ansatz with inverse scattering and later developments led to a long article in *Reviews of Modern Physics*. Unfortunately, the author has not yet found the time to give Thacker's work the attention it deserves and must refrain from further comment.

Significant connections between §5 and §6 were implicit from the time of the formulation of the operator formalism because of the role that was played in the emerging "string theory" by the infinite-dimensional Lie algebra generated by the A_n^v and their adjoints and that generated by the infinite binary products of the former known as Virasoro operators. Neither of these Lie algebras is a Kac-Moody Lie algebra. However, they both share many of the properties of Kac-Moody Lie algebras. In particular, they have weight space decompositions with analogous properties and, accordingly have analogous representation theories. In fact it is not difficult to formulate a generalization of the notion of Kac-Moody Lie algebra for which much of the theory still holds and which includes the Virasoro algebra and the algebras generated by annihilation operators and their adjoints as special cases. This being the case it is not surprising that the further development of §5 has had serious applications to the further development of §6. One interesting connection along these lines is that a close analogue of the vertex operator construction of §6 turned up in the work of Lepowsky and Wilson on the representation theory of the simplest untwisted affine Lie algebra.

The Lie algebras generated by strongly commuting families of annihilation operation and their adjoints have long been familiar objects in quantum field theory but had not traditionally been looked at as Lie algebras—especially from the point of view of examining their root systems and studying their representations by the weight root method. If one does so, one is led to the usual Fock space from a somewhat novel point of view. The author found it quite helpful in understanding the Verma-GGB approach to the representation theory of semisimple Lie algebras mentioned in §10 D to apply it to this infinite-dimensional Lie algebra in the way Kac applied it to the Kac-Moody Lie algebras.

12. Conclusions. The story we set out to tell is by no means finished. There were a very great many interesting developments after 1978. We mention especially the evolution of the quantum version of the inverse scattering method into the theory of quantum groups in the hands of Drinfeld, Manin, and others (see Drinfeld's Berkeley Congress lecture of 1986 for an early account), and the connections with braid and knot invariants discovered and developed by Vaughan Jones from about 1983. We also mention the somewhat different approach to the same complex of ideas which was inaugurated by Jimbo, Miwa, and Sato in a series of sixteen notes published in the *Proceedings of the Japan Academy* in 1977, 1978, and 1979 with the running title, "Studies in holonomic quantum fields" and continued with the collaboration of Date and Kashiwara. However there is much more. We had originally hoped to outline the whole fascinating story but as time wore on it because clearer and clearer that, given the time and space constraints we were under, we had "bitten off much more than we could chew." Perhaps at least we have succeeded in convincing the reader that four apparently quite distinct discoveries are well on their way toward a beautiful unity.

ACKNOWLEDGMENT

The author's interest in the complex of ideas discussed in this essay was first aroused by his attendance at the Summer Seminar on Applications of

Group Theory in Physics and Mathematical Physics held at the University of Chicago in 1982. He was especially interested by the lectures of L. Dolan, I. Frenkel, and J. Lepowsky on the theory and applications of Kac-Moody Lie algebras. In the academic year 1983-84 he was stimulated to pursue this interest further by participation in the MSRI program on Kac-Moody Lie algebras organized by Lepowsky and Kac. He is grateful to MSRI and the Mathematics Department of the University of California at Berkeley for making this participation possible.